THE THEORY OF MEASUREMENTS

THE THEORY OF MEASUREMENTS

BY

LUCIUS TUTTLE

B.A. (YALE); M.D. (JOHNS HOPKINS)

FORMERLY ASSOCIATE IN PHYSICS, JEFFERSON MEDICAL COLLEGE, PHILADELPHIA

AND

JOHN SATTERLY

M.A. (CAMBRIDGE); D.SC. (LONDON); A.R.C.SC. (LONDON); F.P.S.L., F.R.S.C.

PROFESSOR OF PHYSICS, UNIVERSITY OF TORONTO

WITH FOLDING-OUT PLAN AND DIAGRAMS IN THE TEXT

UNIVERSITY OF TORONTO PRESS

TORONTO, CANADA

REPRINTED, 1952, 1960
Reprinted in 2018
ISBN 978-1-4875-7295-2 (paper)

Lithoprinted in U.S.A.
EDWARDS BROTHERS, INC.
ANN ARBOR, MICHIGAN

PREFACE

For the student of mathematics this book is intended to furnish an introduction to some of the applications of the exact sciences and their relation to the "practical" sciences and useful arts, and is primarily intended to give him a knowledge of facts and methods, but without neglecting the accurate exercise of his reasoning powers.

For the student of physical science it is intended especially to emphasize general considerations of measurement, theory of errors, general methods of procedure, quantitative accuracy, adjustment of observations, etc.—topics that are often merely mentioned in the introduction or appendix of a laboratory manual, but that need laboratory work and drill quite as much as the measurements of the individual quantities that the student will take up in his later work. Where it is impossible to find time for a complete course of the kind described here it may be helpful to use selected chapters of the book as occasion arises, or the student may be directed to use it as a reference book, or even to read it through without performing any of the experimental work.

A book that demands more or less vigorous mental exercise from a class of students who take a special interest in the subject will naturally need more elementary exposition—more detailed statement and less exercitational questioning—if it is to be used in larger classes where there is a greater likelihood of finding that some of the students are lacking in interest or ability or elementary mathematical training.* Accordingly, explanations and directions have been given with considerable detail, partly in order to avoid the necessity for continuous oral assistance on the part of the instructor, and partly to help the student to learn with a minimum of deliberate memorizing. The course is progressively graded in difficulty, with the object of developing the student's ability as he proceeds from the easier exercises to those that require more independent thought.

The book may be used in connection with courses of mathematics as well as for courses in physics, and for this reason the requirements of the mathematician have been especially kept in mind during the preparation of the present book. No knowledge of trigonometry, however, is presupposed, and none is imposed upon the reader of the book, the terms "function," "tangent," "cosine," etc., that will

* Such detailed directions as the instructions in regard to round numbers, p. 15, may seem superfluous, but they indicate faults that have been found in the work of many students.

occasionally be found being used merely as convenient abbreviations for ideas that would otherwise need a more cumbersome description.

In the introductory chapter the commonest mathematical deficiencies of the student are reviewed and an opportunity is given him to test his weak points. A lesson on logarithms is included, which can be omitted, if preferred, by a class that is familiar with the subject; but often there are members of such a class who cannot make practical use of logarithmic tables readily, or even accurately, without additional practice, and to any one who does not need the practice it will not be at all irksome. Care has been taken to make the tables in the appendix both accurate and convenient.

We have replaced the perpetually misleading name for the common representative value of a set of residuals by one which does not have this objectionable quality and at the same time suggests the nature of the quantity in question. A few other innovations will be found here and there in the text, but for the most part the book follows fairly well-beaten lines.

The matter in this book has been used for some years in a class on elementary practical mathematics taken by the students in the mathematical and physical classes, and in the honour classes in science in the University of Toronto. It has been found that science students need almost a perpetual drill in calculations and errors to keep them in the right path. The course taken up in the book is given in the Physics Department and in conjunction with the regular classes in Physics, but the attempt has been made to give all science students a training in elementary mathematics, apart from mathematical courses as such, that will be of use to them in their later and more specialized work. The student is trained how to arrange his experimental work in order to deduce the best result possible from that work, together with a measure of its accuracy, also how to represent the results of his work by a graph and to deduce the equation of the graph.

Especially has an attempt been made to open up for biological (and medical) students the elements of bio-mathematics. A fairly full but elementary treatment is given of the normal frequency curve, the representative values of the deviations, the principle of least squares, the subject of correlation and the calculation of the coefficient of correlation, graphical and algebraical work going hand in hand.

A large number of exercises are given. Some are worked out in the text, but the great majority are given at the ends of the chapters in the hopes that students will practise on these until they are quite familiar with their solution and be able to tackle their own problems as they arise. Many of the exercises are original, having arisen in the class and laboratory experiences of the authors, others have been taken from papers published in the scientific and technical periodicals.

By a wide choice of problems it is hoped to enlist the interests of all types of science students and to afford them the means of dealing with their own problems, especially when in course of time they take up their own research work and publish papers.

CONTENTS

CHAPTER I

PAGE

INTRODUCTORY 1

Object. Purpose. Continuity. Results. Forethought. Mental Attitude. Notes. Material Equipment. Mental Equipment. Variation. Physical Arithmetic. Abridged Division. Abridged Multiplication. Gradient. Exercises I.

CHAPTER II

WEIGHTS AND MEASURES 12

C.G.S. System. Units of Length. Units of Area and Volume. Units of Mass and Density. Unit of Time. Practice in Using the C.G.S. System. Rule for Rounding Off a Half. The Hand as a Measure. Measurement of Area. Measurement of Volume. Measurement of Mass. Measurement of Density. Exercises II.

CHAPTER III

ANGLES AND CIRCULAR FUNCTIONS 21

Unit of Angle. Circular Measure. Numerical Measure of an Angle. The Angle π and the Unit Angle. The Protractor. The Diagonal Scale. Measures of Inclination. Use of a Table of Tangents. Experimental Determination of Sines. Definition of Function. The Cosine of an Angle. Circular Functions. Generalized Idea of Angle. Exercises III.

CHAPTER IV

SIGNIFICANT FIGURES 31

Estimation of Tenths. Practice. Mistakes. Value of π. Physical Measurement. Ideal Accuracy. Decimal Accuracy. Significant Figures. Relative Accuracy. Calculation of Relative Errors. Rule for the Relative Difference of Two Measurements. Accuracy of a Calculated Result. Accuracy of the Abridged Methods. Standard Form. Exercises IV.

CHAPTER V

PAGE

LOGARITHMS 44

Definitions. Fundamental Properties. Common Logarithms. Use of Tables. Natural Logarithms. The Probability Function. Exercises V.

CHAPTER VI

SMALL MAGNITUDES 50

Negligible Magnitudes. Formula for Powers. Properties of Deltas. Transformation of Operands. Recapitulation. Rate of Increase. Exercises VI.

CHAPTER VII

THE SLIDE RULE 60

Addition with Two Scales. Multiplication with Logarithmic Scales. The Slide Rule. Reading a Logarithmic Scale. Multiplication. Division. Ratio and Proportion. Equivalent Measures. Reciprocals. C and D Scales. Squares and Square Roots. Compound Operations. Determination of Circular Functions. Determination of Logarithms and Antilogarithms. Cube and Higher Roots. Other Exercises. Exercises VII.

CHAPTER VIII

GRAPHIC REPRESENTATION 70

A. Graphic Diagrams: Indication of a Point by Two Numbers. Representation of Two Variables by a Line. Graphic Diagrams. Practice in Plotting Points. Choice of Scales. General Principles of Plotting. Representation of Tabular Values. Smoothing of a Graphic Curve. *B. Curves and Equations:* Graphic Representation of a Natural Law. Graph of an Equation. General Procedure. The Straight Line. The Parabola. The Hyperbola. The Exponential Curve. The Logarithmic Curve. The Probability Curve. Equation of a Graph. Change of Scales. Definitions of Circular Functions. Rate of Increase and Rate of Growth. Space-time Curves. Exercises VIII.

CHAPTER IX

GRAPHIC ANALYSIS 96

Interpretation of Equations. The Graph of $y=a+bx$. The Straight Line Law. The "Black Thread" Method. Intercept

PAGE

Form of a Linear Equation. The Graph of $y = a + bx + cx^2$. Law of Density-Variation of Water. Typical Curves. Linear Relationship by Change of Variables. Logarithmic Paper. Semi-logarithmic Paper. Summary. Exercises IX.

CHAPTER X

INTERPOLATION AND EXTRAPOLATION 118

Definitions. The Principle of Proportionate Changes. Examples of Linear Interpolation. Graphic Interpolation Interpolation along a Curve. Insufficiency of Data. Extrapolation. Interpolation Formula. Exercises X.

CHAPTER XI

CO-ORDINATES IN THREE DIMENSIONS 134

Co-ordinates of a Point in Space. Convention in Regard to Signs. Loci of Simple Three-Dimensional Equations. Contour Lines. Use of Contour Maps. Exercises XI.

CHAPTER XII

ACCURACY 142

Significant Figures. Infinite Accuracy. Relative Errors. Uncertain Figures. Superfluous Accuracy. Finer Degrees of Accuracy. Possible Error of a Measurement. Possible Error after a Calculation. "Probable" Error. Illustrative Examples. A Common Experiment. Exercises XII.

CHAPTER XIII

THE PRINCIPLE OF COINCIDENCE 153

Effect of Magnitude upon Accuracy. Measurement by Estimation. Measurement by Coincidence. The Vernier. Use of the Vernier Caliper. Slide-Rule Ratios. Averaging in Experimental Work. Generalizations. Exercises XIII.

CHAPTER XIV

MEASUREMENTS AND ERRORS—STATISTICAL METHODS . . . 164

Direct and Indirect Measurements. Independent, Dependent, and Conditioned Measurements. Harmony and Disagreement of Repeated Measurements. Errors of Measurements. Classification of Errors. Accidental and Constant Errors. Errors and Variations. Measurement of Variates. Frequency Distributions. Class Interval. Types of Frequency Distribution.

PAGE

The Probability Curve. Representative Magnitudes. The Average. The Median. The Mode. Choice of Means. Deviations. Average by Symmetry. Average by Partition. Average from the Diagram. Quartiles. Median and Quartiles from the Diagram. Semi-Interquartile Range. Exercises XIV.

CHAPTER XV

DEVIATION AND DISPERSION 184

Characteristic Deviations. Total Range. Average Deviation. Standard Deviation. Dispersion. Significance of the Dispersion. Advantage of the Dispersion. Calculation of the Dispersion. Rule for the Accuracy of the Average. Use of the Table of Dispersions. Sigma Notation. Dispersion of an Average. The Statement of a Measurement. Characteristic Values. Relative Dispersion. Comparison of Characteristic Deviations. Chances of Particular Deviations. Probable Error. Exercises XV.

CHAPTER XVI

THE WEIGHTING OF OBSERVATIONS 199

Necessity of Weights for Observations. Weights for Repeated Values. The Weighted Average. Arbitrarily Assigned Weights. Weight and Dispersion. Limitations of the Last Formula. Exception to the Rule. Exercises XVI.

CHAPTER XVII

REJECTION OF DOUBTFUL OBSERVATIONS 205

Observational Integrity. Importance of Criteria. Chauvenet's Criterion. Probability Integral Table. Graphic Approximation to Chauvenet's Criterion. Irregularities of Small Groups. Justification of the Criterion. Wright's Criterion. Exercises XVII.

CHAPTER XVIII

INDIRECT MEASUREMENTS 215

Importance of Indirect Measurements. Probable Error of a Sum. Probable Error of a Difference. Probable Error of a Multiple. Associative Law. Probable Error of a Product. Probable Error of a Power. Distributive Law. Recapitulation. Graphs of Propagated Errors. Relative Importance of Compounded Errors. General formula for Probable Error in an Indirect Determination. Worked Examples. Exercises XVIII.

CHAPTER XIX

PAGE

LEAST SQUARES 226

The Average as a Least-Square Magnitude. Least Squares for Conditioned Measurements. Best Relation between x and y. Least Squares and Proportionality. Least Squares for a Theoretically Constant Value. Consecutive Equal Intervals. Alternative Method for Equal Intervals. Comparison of Least Squares and Other Methods. Simultaneous Indirect Measurements. Least Squares when both x and y err. Angles of a Triangle. Solutions by Graphs. Exercises XIX.

CHAPTER XX

SYSTEMATIC AND CONSTANT ERRORS 247

Definitions. Test for Systematic Errors. Example of a Systematic Error. Example of a Periodic Error. Example of a Progressive Error. Constant Errors. Exercises XX.

CHAPTER XXI

APPLICATIONS TO BIOLOGY 254

A. Frequency Problems: Biological Examples. Variability. Equation of Normal Curve. Symmetry and Asymmetry. Correlation. Galton's Method. Pearson's Method. Coefficient of Correlation. Probability Paper. Probable Error in the Expectancy of an Event. *B. Biological Results represented by Equations:* Mathematical Theory of Wounds. Law of Fatigue. Racing and other Sporting Records. Growth and Temperature. Sensations, Weber and Fechner's Law. Magnitude and Similitude. Representation of Statistics. Exercises XXI.

CHAPTER XXII

AREAS 289

Formulæ for Areas. Simpson's and other Rules. Counting Squares. The Hatchet Planimeter. Exercises XXII.

MISCELLANEOUS EXERCISES FOR PRACTICE IN THE USE OF LOGARITHMS AND THE SLIDE RULE 298

ANSWERS TO THE EXERCISES 303

APPENDIX: TABLES 311

BIBLIOGRAPHY 327

INDEX 329

FOLDING-OUT PLAN

Nomogram showing duration of sunlight for every day in the year . . . . *facing page* 326

CHAPTER I

INTRODUCTORY

1. Object.

The object of a course in the theory of measurements is not only to give a certain knowledge of the scientific facts that are studied, but also to develop the thinking and reasoning powers and to furnish the special kind of mental training that results, in the first place, from practice in making various kinds of measurements with particular care for their accuracy, and, in the second place, from the consideration of accuracy in its quantitative aspects,—from realizing that accuracy itself can be made a subject of measurement, that there are relative degrees of accuracy, that accuracy is important in one place and means only a waste of effort in another, that absolute accuracy is an impossibility, that a measurement by itself is of much less value than when accompanied by a statement of its precision.

When a course of physical measurements is used as a preliminary to or conjointly with laboratory work or practical work in some subject such as physics, chemistry, or biology, it is further of value in giving the student a certain familiarity with apparatus and a facility in handling it in such a manner that he acquires the habit of utilizing it to the best advantage and of keeping it in such condition that it is most fully utilizable when needed.

2. Purpose.

It is advisable to point out to the student at the beginning of the course that his conscious purpose, throughout all of his work, should be to *learn*, rather than to accomplish, the assigned exercise. In order to help him in his education and training he is permitted to do certain laboratory work which will make his learning easier and its effects more lasting and more useful to him. He should tell himself that the work is allowed, rather than that it is required, in case he has any tendency to look upon his course of practical work as tedious and irksome.

3. Continuity.

In any graded course of study, where each exercise is an advance beyond the previous ones and requires a knowledge of the earlier work, it is a decided handicap for the student to miss any of the work. If an absence cannot be avoided he should take particular pains, for his own sake, to make up the work by outside study. This is especially important for any study that is at all mathematical in character. See that each topic is thoroughly comprehended before going on to the next. Usually the work will be easier if each lesson is read over before coming into class, so that it is not necessary to begin the class-room work as a new and unfamiliar subject.

4. Results.

In order to obtain good results it is necessary for the student to preserve an attitude of alert attention towards his own work, and especially not to omit any part of it or postpone it. Be thorough without being in haste ; better to have half of the day's work done and done well than to try to take in all of it without having any of it more than half assimilated.

5. Forethought.

Whenever any piece of apparatus is used it must be kept in mind that it may be needed again later, and it is as important to keep it in good order as it is to use it efficiently. It does not take long for the consequences to show in the student's work if he is accustomed to pick up an instrument where he happens to find it, and drop it as soon as its immediate need has passed. A certain orderliness in the handling of apparatus is a habit that is well worth cultivating.

6. Mental Attitude.

In addition to keeping in mind the fact that it is much more important to learn principles than to " work through " each day's lesson, the student should adopt the motto that in all kinds of studying it is better to *think* than to memorize. For some students it seems only too easy to get into the habit of concentrating upon individual items and memorizing isolated statements of fact without ever understanding their bearings or realizing their interrelationships or acquiring a larger comprehension of the body of scientific knowledge which is built up on them. The best help to a broader vision lies in *thinking over* the facts that one comes across. Just as important as the question " What is true ? " is the further question " Why is it true ? " Better than a brain packed full of facts is a mind that can reason out what the facts must necessarily be in particular cases. Memory with little reasoning power is useless for any highly organized living being ; but reasoning power with little memory would be perfectly practicable, as long as such things as paper and pencil can be had. Furthermore, the ability to think for oneself is of the greatest utility in enabling the student to rely upon his own observational powers. The untrained student is prone to ask "Is my result right?" in circumstances where the student who has learned to stand on his own feet knows that no one has a better knowledge of what the "right" result is than himself.

7. Notes.

For any practical work, or laboratory work, it is important that the student's notes shall be *accurate* and that they shall be *complete*. Neatness is usually worth while, but it is distinctly secondary in importance to thoroughness and accuracy. Time spent on beautifying the notebook by means of elaborately shaded drawings or painstaking arrangement of matter is usually time wasted. All notes should be clear and intelligible, and in such shape that they can be readily understood by any one who has an ordinary knowledge of the subject that they deal with. Each day's work should be dated, and it is advisable to write a heading in such a way that it will catch the eye at once, and show at a glance where one day's work ends and the next begins, as well as indicating the nature of the following matter after the manner of a title. A certain amount of "display," by underlining or otherwise should also be given to two other things, the statement of *original measurements*, especially when further calculations depend upon them, and the *final results* of such calculations.

The matter of making one's notes *thorough* needs little explanation; they should be made a digest of all that the student learns and does in the laboratory course, but without copying or duplicating matter in his textbook that he can easily turn to when it is wanted. Any questions that are asked in the text should be answered in the notebook.

The matter of accuracy is one that requires some care and alertness. It is necessary to make it a rule that all work done with a pen or pencil *must* be done in the notebook, and everything in the notebook must be put down consecutively, in its natural order. If the student uses the last pages of the notebook for miscellaneous calculations it is usually impossible to find a particular piece of work when it happens to be wanted at some later time. Under no circumstances is any work with pen or pencil to be done on scraps of paper, or in a "temporary notebook" or anywhere else except in its proper place and order in the permanent notebook. The slight gain in neatness of the student's notes, which is usually the object of such procedures, is not nearly important enough to counterbalance the possibility of errors in copying data and the probability of being later obliged to hunt in vain for statements that are not in their proper place. The notebook of the scientist, like that of the accountant, should be a book of "original entry," and for reasons that are as important for a scientific investigation of natural phenomena as they are for a legal investigation of indebtedness. Furthermore, no measurement that has been written down in the notebook should ever be rubbed out with an eraser; * if there is a reason for doubting its value, or even if it is obviously wrong it may be cancelled by drawing a line through it, but this should be done in such a way as not to obscure what is written down but to permit it to be utilized later if it is found desirable to do so.

8. Material Equipment.

The student's *notebook* should be of such size and character as will be best adapted to his work. If it is not furnished by the Department directions will be given in regard to the kind of notebook that should be used. *Pen and ink* will be needed, for notes that are taken with a pencil are almost always unsatisfactory. A fountain pen is advisable, although not necessary. A piece of *blotting paper* should be obtained which is long enough to reach across the page of the notebook. A *hard pencil* with the point kept well sharpened will also be needed.

9. Mental Equipment.

The student of the theory of measurement should have had a good course in algebra as far as the solution of equations of the first degree; also a sufficient knowledge of plane geometry to include the properties of perpendiculars, equal triangles, isosceles and similar triangles, the theorem of Pythagoras, and the properties of similar figures. Certain arithmetical processes, such as the method of extracting square root or cube root, are not needed for purposes of physical measurement, but a good grasp of certain others, such as proportion and variation, is almost a necessity. An intelligent comprehension of principles is as important as a memory of rules and formulæ.

10. Variation.

If the price of a commodity remains constant the value is said to vary in accordance with the weight, or, shortly, to vary *as* the weight, or, more explicitly, to vary *directly as* the weight. Here the weight is considered to be a *variable quantity*, that is, we may consider any weight we please, the weight of the substance may assume any numerical value for the purposes of the discussion. Under these circumstances the cost will also vary. Doubling the weight will double the cost; cutting the weight in half will reduce the cost by 50 per cent.; etc. When any change in one quantity that can vary is always accompanied by an equal relative change in a second quantity the variables are said to be proportional, or to be *directly proportional*, and each is said to vary *directly as* the other one.

* It is a pity that pencils are sold with indiarubber fastened to the end. The student who has the power to rub out a mistake tends to become careless.

If a given mass of a gas is subjected to compression it will be found that doubling the pressure exerted upon it will cause its volume to decrease to only one half of its former amount : * multiplying the pressure by five will reduce the volume to one-fifth, etc. This kind of variation is called *inverse*, and the pressure and volume are said to be *inversely proportional*; the volume is said to vary *inversely as* the pressure. Suppose that the volume is 6 cubic ft. when the pressure is one atmosphere ; then if the pressure is raised to 3 atmospheres the volume will be reduced to 2 cubic ft., but if it is diminished to 1/2 atmosphere the gas will expand enough to occupy a space of 12 cubic ft. If we write an equation

$$v=k\frac{1}{p}$$

it will be evident that any increase in the size of the denominator, p, will cause a relatively equal decrease in the size of the term, v; and we have already seen that this equation means the same as

$$v_1 : 1/p_1 :: v_2 : 1/p_2 :: v_3 : 1/p_3 \ldots$$

Clearing the equation of fractions gives

$$pv=k$$

and the ordinary way of expressing the fact that two variables, such as p and v, are inversely proportional is to write down an equation in which their product is stated to be equal to a constant ; just as direct proportionality is expressed by making their quotient equal to a constant.

It is perhaps worth noticing here that there may easily be other forms of variation, in which there is no proportionality at all. The distance travelled by a train is not usually strictly proportional to the number of hours that elapse during the process, nor is an individual's wealth proportional to his age. The matter of irregular variation will be taken up later.

11. Physical Arithmetic.

Before performing any written calculation with numbers that have been obtained from physical measurements it is advisable to make a rough *mental* calculation of the approximate value of the final result. For example, at £3·31 per lb. what will 5½ lbs. cost ? One of these factors is a little more than three ; the other is somewhat less than six. Their product accordingly, must be in the neighbourhood of £18.

A train travels 155·8 miles in 2¾ hours ; what is its rate in miles per hour ? Here the time is less than 3 hours, so the speed must be greater than 155·8÷3 ; and still greater than 150÷3. Ans. : Somewhat faster than 50 miles per hour.

* Temperature remaining constant. This is Boyle's Law.

If a still closer approximation should be desired it could be obtained by noticing that the actual distance is about 4 per cent. greater than 150, and the actual time is one-twelfth (say 8 per cent.) less than the assumed time. Increasing 50 by 12 per cent. gives 56 miles per hour for a closer value of the speed. The arithmetically accurate value is 56·65454545.

To multiply by 25 multiply by 100 and divide by 4; to multiply by 125 multiply by 1,000 and divide by 8; to multiply by some number near 100 or 1,000 multiply by 100 or 1,000 and add or subtract a few per cent.

Questions

(1) Is
$$\frac{4{\cdot}71 \times 13{\cdot}8}{9{\cdot}06}$$
equal to about 7, or about 70, or about 700?

(2) What is the approximate value of $7{\cdot}26 \times 0{\cdot}0328$?

(3) Point off 00000928800000 so as to make it equal to the product of 0·0216 and 0·0043.

(4) Point off the right-hand side of the equation $2{\cdot}/{\cdot}3 = 00666$.

(5) Reduce the ratios in the following expressions to approximate percentages, performing the calculation mentally: "8 lbs. in every 23 lbs. of sea water is solid salt." (Ans.: 8 in 24 would be $33\frac{1}{3}$ per cent.; 8/23 is a little greater and must be 34 or 35 per cent.)

"Seven inhabitants out of every 38 are voters." (Ans.: 7 out of 35 would be 20 per cent., $7/42 = 16\frac{2}{3}$ per cent., 7/38 must have some intermediate value, say 18 per cent.)

"Fourteen-carat gold is 14/24 pure." (Ans.: 14/24=56/96; this fraction has its numerator about half as large as its denominator and so will not be much changed by adding 4 to the latter if 2 is added to the former, 56/96=58/100=58 per cent.)

"Boiling water will dissolve 0·000022 of its weight of silver chloride." (Ans.: 0·000022=0·0022 per cent.)

"A saturated salt solution has a strength of 5 : 13." (Ans.: a little less than $5 : 12\frac{1}{2}$, or 40 : 100, say 38 per cent.)

"A steep railway grade may have a rise of as much as 180 feet per mile." (Ans.: 180 ft. per 5,280 ft. is less than 180 per 5,000 or 360 per 10,000 or 36/1000 or 3·6 per cent.)

Notice the different expressions that are in common use to denote the comparison between a definite fractional part and the total. The same meaning is expressed by each of the following phrases as by any of the other ones: 22 per million, 22 out of a million, 22 in 1000000, 22/1000000, 0·000022, 22 : 1000000, and 0·0022 *per centum* or 0·0022 per cent.

Observe also that the fractional notation is more convenient than the word *per* in naming compound units of measurement, and has the same significance. Thus, a speed of 40 miles per hour is customarily written 40 $\frac{\text{mi}}{\text{hr}}$, or 40 mi/hr, meaning 40 times 1 mile per hour or 40 $\frac{\text{1 hour}}{\text{1 mile}}$.

12. Abridged Division.

A number that is obtained as the result of a physical measurement is frequently needed for some kind of a calculation. When this is the case it is a fact (as will be shown later) that the final result never needs to be expressed with a greater number of figures than the original data contained. Thus, it may be possible to measure the width of a table so carefully as to make sure that the measurement is 62 centimetres+3 millimetres+8 tenths of a millimetre. Such a quantity is preferably written as a number of centimetres, and in this case is 62·38, a number consisting of four figures. Suppose it is necessary to find out what one-third of the width will amount to. One-third of 62·38 is 20·793333 . . . , and, as stated above, four figures of this result, namely 20·79, are all that are necessary. As a matter of fact, to keep more than four figures would be decidedly objectionable. If the original measurement gave the correct number of tenths and hundredths of a centimetre without pretending to state any knowledge of the correct number of thousandths how could any calculation assume to give correct figures in thousandth's and tens-of-thousandth's places? Similarly, in § 11, the "arithmetically accurate" value would be wrong if the given distance were even a thousandth of an inch longer or if the time varied from an exact $2\frac{3}{4}$ hours by as much as a millionth of a second. Here one of the numbers (155·8 miles) has four figures while the other can hardly be considered to have more than three (2·75 hours.) In such cases it is a fact that the *final result will have only as many trustworthy figures as there are in the shortest number from which it is derived*. When a number having four figures is divided by a number of three figures there should be only three figures kept in the quotient.

The principle just stated makes it possible to employ the "abridged" processes of multiplication and division, which will automatically give just the right number of figures in the answer, and will also save considerable labour on the part of the computer.

The first example shown in the margin has been worked out by ordinary "long" division; in the second one the abridged method has been used. The latter process differs from the former in only one respect: Whenever the process of "bringing down" a zero would be employed the last figure of the divisor is cancelled instead. In order that the temporary dividend shall be larger than the divisor one method stretches out the dividend by affixing a cipher; the other shortens the divisor by trimming off its last figure. A comparison of the two examples will show that the same result is achieved in each case. The beginner should work out the quotient of the two numbers given above, cancelling the last figure of the divisor whenever he would otherwise "bring down" a zero, but not referring to the illustration until he was finished. After the first subtraction, when the divisor, 275, is not contained in the remainder the divisor is shortened by crossing off the final 5. Then, 27 is contained in 183 six times. The next step is to multiply 27 by 6. Before saying $6 \times 7 = 42$ notice that if the 5 had not been cancelled there would have been 3 to carry, resulting in 45 instead of 42. Accordingly 5 is written down instead of 2 and the rest of the multiplication proceeds as usual. After the next subtraction gives a remainder of 18 the divisor is shortened to 2 instead of having the new dividend lengthened to 180. Then, 2 would be contained in 18 just 9 times, but considering the figure that was last cancelled it is plain that 2·7 will not be contained much more than 6 times. Six times the cancelled 7 would be 42 and would give 4 to carry, so 6 times the 2 (plus 4) is written 16. The next subtraction and cancellation puts an end to the work. In the specimen given above it has been noticed that the third figure of the quotient is to be the last one, and it has been written down as a 7 instead of a 6 because the next product, 19 (7 times 2 plus 5), comes nearer in value to the required 18 than would the number 16. In other words, the quotient is nearer 567 than 566 and so the larger number represents it more accurately than the smaller.

```
275)1558(566545
    1375
     1830
     1650
      1800
      1650
       1500
       1375
        1250
        1100
         1500
         1375
          125

275)1558(567
    1375
     183
     165
      18
      19
```

Abridged Division.—After each subtraction the next step is to shorten the divisor instead of to lengthen the dividend.

The quotient is to be pointed off by making a preliminary mental calculation, as explained in § 11. For example, if the original numbers had been 2·75 and 155·8 the answer would have been 56·7.

13. Abridged Multiplication.

The trustworthiness of a product of two or more numbers follows the same rule as that of a quotient: no more figures of the product are "significant" than the number of them which the shortest factor contains. Thus if the diameter of a circle is found to be 8 centimetres + 0 millimetres + 0 tenths of a millimetre but nothing is stated about hundredths of a millimetre, so that only three figures, 8·00, of the diameter are known, it will not be possible to obtain more than three figures of the circumference, even if the other factor, 3·14159265359, contains a dozen figures. Here too the ordinary arithmetical process can be abridged so as to save time and work, and, what is more important, to avoid being misled by figures that have been kept when they should have been discarded.

The illustration (*a*) shows the ordinary process. It is just as easy, although not customary, to use the figures of the multiplier

(*a*)	(*b*)	(*c*)
65·97	65·97	65·97
24·13	24·13	24·13
19791	13194	13194
6597	26388	2639
26388	6597	66
13194	19791	20
1591·8561	1591·8561	1591·9

GENESIS OF THE ABRIDGED METHOD.—(*a*) Ordinary long multiplication. (*b*) The same with the figures of the multiplier used in the reverse order. (*c*) The same as (*b*), but the partial products are kept from stringing out to the right by progressively shortening the multiplicand.

in the reverse order, multiplying first by 2, then by 4, 1, and 3, and "stepping" the partial products successively one more place to the right instead of to the left. This has been done in illustration (*b*). Examine it closely, and see that you understand just how the process (*b*) is carried out and why it must necessarily give the same result as (*a*).

The abridged method is shown at (*c*). The first multiplication is by the left-hand figure 2 as in (*b*). Then the last figure of the multiplicand, 7, is cancelled, and the next multiplication (by 4) is begun directly under the first. The process of cancelling and multiplying is continued in the same way until either the multiplicand is entirely cancelled or the multiplier has been entirely utilized, and the result is pointed off in accordance with the directions in § 11. The student should work out the same example independently, remembering to investigate how much there is "to carry" from the figure last cancelled. In the last partial product of (*c*) the $3 \times 6 = 18$ is increased by two units because 3×5 gives just 1·5 to carry, but it is evident that 3×59 must give a number nearer to 2 than to 1.

14. Gradient.

If a road that goes uphill rises 2 feet for every 5 feet of horizontal distance it is said to have a slope of 2 : 5, or 2/5, or 0·4, or 2 in 5. That is, the ratio of any vertical rise to the corresponding horizontal distance is taken as a numerical measure of its steepness. Of course the slope could also be measured in degrees; in the case just mentioned the "*grade angle*," or angle which the slant line makes with a horizontal line, would be twenty-two degrees—too steep to be satisfactory for a roadway,—but the usual custom is to state the measure of a slope in terms of vertical rise per horizontal distance. Another way of looking at the same thing is to consider it as the amount of rise per *unit* of horizontal distance; thus, a road that rises two feet in every five will of course have a rise of 2/5 of a foot for a single foot of horizontal distance.

A level road is one that has no slope. That is, its slope, when measured as rise per level distance, amounts to zero. If a line is made to slant more and more steeply the ratio that represents its slope will obviously become greater and greater without limit; thus, a slope of 1000 would be hardly distinguishable from a true vertical, and yet between these two there must be, for example, a slope of 1000000000. The diagonal of a square is inclined to any of the sides at an angle of 45°, and its slope is of course unity. These facts may be abbreviated as follows: slope of $0°=0$; slope of $45°=1$; slope of $90°=\infty$.

It is sometimes convenient to consider the steepest possible "slope" (a vertical line) as having 100 per cent. of steepness. This will be the case if we measure the amount of slant not by the ratio of rise to level distance but by the ratio of rise to slant distance. A road which has a grade angle of 22° will rise nearly 3 feet for every 8 feet of distance along its slanting surface, and this measure of steepness may be called *per cent. slope* to distinguish it from the slope, or gradient, as previously defined.

On good turnpikes the grades are almost always kept below 3 per cent. Two per cent is decidedly steep for a railway grade, and in modern good railway construction 1 per cent is about the maximum. For slight inclinations, such as railway grades, the difference between rise per horizontal distance and rise per slant distance is negligibly small. Thus, for 1° they are respectively 0·017455 and 0·017452, or 92·16 feet per mile and 92·15 feet per mile.

If one end of a long bar has been heated for some time in a furnace a temperature gradient will be established along the bar. This gradient is usually measured in degrees per inch or per cm. Its value is large near the hot end of the bar and decreases to nearly zero at the cold end. Compare this with the gradient of a river which is steep near the source and flat near the sea.

Exercises I

1. Divide 180·000 by 3·1416.

2. Divide 6764309 by 236453.

3. Multiply 28·60741 by 236453.

4. The value of $180/\pi$ is 57·29578. Multiply this by 3·141593 and see if you obtain 180·0000 correct to seven figures.

5. Multiply any number that has five figures by some number having only

two figures. Repeat the multiplication, using the shorter number for multiplicand and the longer one for multiplier. Which method is preferable, in view of the statement at the beginning of § 13 ?

6. Draw roughly an equilateral triangle that has one side horizontal. Draw a vertical from its apex to the middle of its base, and prove that the gradient of a sixty-degree slope is equal to $\sqrt{3}$; also that if the angle is half as large as this the slope will be only one-third as much.

7. Prove that the "per cent. slope" of 45° is $\frac{1}{2}\sqrt{2}$; of 0° is zero ; of 90° is 1 ; of 30° is $\frac{1}{2}$ (use the same bisected equilateral triangle).

8. The sum of two numbers is constant. When is their product greatest ? Illustrate graphically.

9. The product of two numbers is constant. When is their sum least ? Illustrate graphically.

10. Prove that of all plane figures having the same length of boundary the circle has the greatest area. Test it with a loop of thread.

11. When is the sum of a number and its reciprocal least ?

CHAPTER II

WEIGHTS AND MEASURES

15. C.G.S. System.

The older units of measurement, such as the length of a barley-corn, the width of a man's palm, or the length of a foot or a pace, were objectionable chiefly on account of their lack of uniformity. Not only did different countries use different units for measuring quantities of the same kind, but even when a unit of the same name was used in different localities its value was not the same. In many countries these older units have been entirely superseded by a new system of weights and measures, and in all countries this system has come into general use for every kind of scientific work. It is usually called the C.G.S. System, from the initial letters of the units of length (the centimetre), of mass (the gram), and of time (the second). These three units are called *fundamental*, because they have been arbitrarily fixed in size, while all the other units of the system have been so chosen as to make them depend upon these three in as simple a manner as possible. For example, the *derived* unit of velocity is such as will denote movement through a single centimetre of distance in a single second of time, thus making the measure of a velocity numerically equal to the quotient of space traversed divided by time elapsed during the process. Similarly, the density of an object is defined as its mass in grams divided by its volume in cubic centimetres, so that, although no name has been given to the unit, the density which is numerically equal to unity must be the density of such a substance as will weigh one gram for each cubic centimetre of its volume ; the unit of force is the force that must act for one second of time in order to produce a change of one unit (one centimetre per second) in the velocity of a unit mass.

16. Unit of Length.

The scientific unit of length is the *centimetre*. It is equal to about half a finger-breadth and is often found on tape-measures, rulers, etc. These are simply copies of accurate standards belonging to the manufacturer, which in turn owe their accuracy to a careful comparison with the standards of the government. In the case of the governments that subscribed to the Metric Convention, the standards, which are called *national prototype metres*, are lengths of one metre (*i.e.*, 100 centimetres) carefully laid off between lines near the ends of certain bars of artificially aged platinum-iridium alloy which are 102 centimetres long and have a cross-section that somewhat resembles a letter X (Fig. 1). The greater length is used instead of a single centimetre because it can be measured more accurately, and the particular cross-section is for the purpose of giving rigidity, and in order to allow the scale to be marked on a surface that would be neither stretched nor compressed if the bar should be slightly bent. The standards were constructed at Paris and distributed by lot among the signatories to the Metric Convention about 1889, after being carefully tested and compared with one another so that their relative errors and equations were accurately known.* One of these, which is kept at the International Bureau of Weights and Measures, near Paris, is known as the *international prototype metre* and corresponds in length with the original flat platinum bar (100 cm. × 0·4 cm. × 2·5 cm.) constructed for the French Government by Borda and called the *mètre des archives*. The Borda standard was intended to equal one ten-millionth of the length of the meridian quadrant passing through Paris from the north pole to the equator, the earth itself thus furnishing the original standard. The metal bar, however, is now taken as the fundamental standard, not only because a microscopic measurement of it can be made more easily and more accurately than a geodetic survey, but also because the actual length of the earth's quadrant is not constant. Its average length, according to the best estimations, is about 10,002,100 metres.

FIG. 1.—TRESCA CROSS-SECTION.

A bar of this shape has great rigidity for a given weight of metal; and a slight bending, such as would stretch the lower part and compress the upper part, has no effect on the "neutral web," N, where the scale is engraved. The diagram shows the exact size of the cross-section of the prototype metres.

The multiples and subdivisions of the metre that are in actual use are the kilometre (1 km. is 1000 metres, or about 5/8 of a mile; closer approximations are given in the appendix), the centimetre (1 cm.), the millimetre (0·1 cm.), and the micron (0·0001 cm.). These are more convenient than a single unit in some cases, but in scientific work it is desirable to express all lengths in terms of the accepted unit, 1 cm., in order to avoid possible confusion or

* Modern processes of measurement are so accurate that a difference can generally be found between two standards that were intended to be equal, no matter how carefully they were constructed. The "equation" of a prototype metre expresses the way in which its length varies when its temperature is changed. Thus at any ordinary temperature, t, the length in centimetres of prototype metre No. 18 is $99{\cdot}9999 + 0{\cdot}0008642t + 0{\cdot}000000100t^2$.

serious error in case the denomination of a quantity should be accidentally omitted. Remember 1 inch=2·540 centimetres to an accuracy of one in a million.

17. Units of Area and Volume.

The scientific unit of area is the square centimetre (1 cm.^{2}), the area of a square each of whose sides is 1 cm. in length. A square foot is about 1000 cm.^{2} The unit of volume is the cubic centimetre (1 cm.^{3}), the volume of a cube that measures 1 cm. on each edge. Another unit of volume is the litre (see below).

18. Units of Mass and Density.

The scientific unit of mass, or for practical purposes the unit of weight *in vacuo* at sea-level, latitude 45°, is the *gram* (1 gm.), which is divided into 1000 milligrams (mgm.) just as the metre is divided into 1000 millimetres. It is derived from kilogram prototype standards (1 kgm.=1000 gm.=2·2 lbs. approx.) established at the same time as the standards of length and was originally intended to be equal to the mass of one cubic centimetre of water under standard conditions. More careful measurements, however, on water that has been freed from dissolved air have shown that even at the temperature of its greatest density (3·98° C.) a gram of water occupies a trifle more space than one cubic centimetre, although the excess is only one-sixtieth as great as it is at ordinary room temperature. In cases where the slight change of volume that is produced by heating or cooling can be neglected it may be considered that water has a density of one gram per cubic centimetre.

The volume of a kilogram of water at its greatest density is exactly 1 litre. 1 litre=1000·027 c.c.; this is so nearly 1000 that in most work the litre and the 1000 c.c. are taken as synonymous. The litre is about 1¾ pints.

19. Unit of Time.

The scientific unit of time is the *second*, which is the 1/86400 part of the length of an average day from noon to noon. As the length of the solar day varies at different seasons of the year the second is determined in practice as 1/86164·1 of the time of a complete rotation of the earth with respect to the fixed stars. This unit of time was in use before the adoption of the C.G.S. System and is familiar to every one.

It is worth noticing that fairly accurate seconds can be counted off by repeating, at ordinary conversational speed, "one thousand and one, one thousand and two, one thousand and three," etc. Test this with a watch.

20. Practice in Using the C.G.S. System.

The scientific system of units is so largely used that the student should not be satisfied with the mere ability to translate measurements from one system to the other. He should practise first estimating (guessing) and then measuring the dimensions of various objects that he comes across, until he has acquired a certain ability to "think" in centimetres, cubic centimetres, grams, etc., as well as in inches, quarts, and pounds.

Across the top of one page of the notebook draw a horizontal line just ten centimetres long and rule two short perpendicular lines across its ends in order to indicate the exact length clearly. In the same way, along the right-hand edge of the page, draw a line twenty-five centimetres in length. Under the first line draw five others of various lengths without measuring them. After they have been drawn measure each one with a scale of centimetres and millimetres, and record its length to the *nearest* millimetre. For example, if the length appears to be about 152¼ mm. it should be recorded as 15·2 cm.; if about 152¾ mm. it should be called 15·3 cm. Write each length as a number of *centimetres*, not 153 mm. nor 15 cm. 3 mm., and make it a rule to see that the denomination of a measurement is never omitted.

21. Rule for "Rounding Off" One Half.

It may happen that the measured length is so near 152½ mm. that it is impossible to decide between 152 and 153. A fraction perceptibly less than a half should be discarded and more than a half should always be considered as one more unit, but when it is uncertain which figure is the nearer one the universally adopted rule is to *record the nearest even number* rather than the odd number that is equally near. The reason for this procedure is that in a series of several measurements of the same quantity it will be as apt to make a record too large as it will to make one too small, and so in the average of several such values will cause but a slight error, if any. If the rule were that the half should be always increased to the next larger unit the errors would not balance one another and the average would tend to be brought up to a larger value than it should have. The same advantage would of course be obtained if the nearest odd number were always used, but the even number has one slight additional merit, namely, that in case it should have to be divided by two a recurrence of the same situation would be avoided.

By comparison with the lines already drawn make a mental estimate of the length and width of the notebook; then verify the estimate by measuring with a scale. Remember to record clearly all the experimental work that is done; thus, the completed notes should show at a glance which number is the actual width and which is the rough estimate.

22. The Hand as a Measure.

Lay your hand across a ruler or a metre stick and either spread the fingers slightly or crowd them closer together, as may be necessary, so as to make a whole number of finger-breadths occupy the same amount of space as a whole number of centimetres. Then hold the hand in a similar manner while using it for practising approximate measurements of various objects. Record also the exact measurements of the same objects as they are obtained later with the graduated scale.

Separate the thumb and little finger as far as can be conveniently done without special effort and measure your span in centimetres and millimetres. Repeat this measurement five times, being careful not to let the sight of the scale under your hand influence the extent to which the fingers are spread, and decide which is the most satisfactory value. Measure the length and the breadth of the table by means of successive spans, using hand-breadths or finger-breadths for the final fraction of a span, and compare the result with the actual length and breadth.

23. Measurement of Area.

Ask the instructor to draw an irregular outline in your notebook (Fig. 2). Count the number of the small squares of the ruled paper

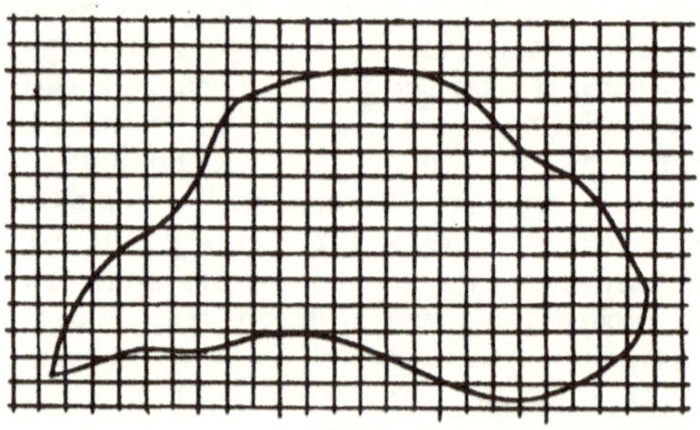

FIG. 2.—IRREGULAR AREA.—Area by Squares.

which are entirely included within it, but do not outline them on the diagram. To their total add half the number of the squares that are cut by the boundary of the figure. The result will be the area of the irregular outline, not in square centimetres, of course, but in terms of the small ruled squares, and its denomination may be written "□'s." Try to obtain a more accurate value for the total area by estimating as closely as possible how many tenths of each cut square is included within the boundary and adding these actual fractions. The first result should agree quite closely with this one because any one of the fractions of a square is as likely to be less than one-half as it is to be more than one-half, and the most probable value for the average of the fractions is just one-half.

As an alternative method, block off an equal area on the same irregular figure by drawing several rectangles and triangles to cover it (Fig. 3). If one corner of a triangle projects considerably beyond the irregular line see that one of its sides is drawn so as to include less than the requisite amount and try to make the two opposite errors balance as nearly as possible. Then find the total area of the geometrical figures by mensuration, measuring their dimensions not by means of the centimetre scale but with a scale copied from the ruling of the notebook, or by transferring each length to the ruled page with a pair of dividers, so as to obtain the area in the same units as before.

Chapter XXII gives a summary of rules for finding areas.

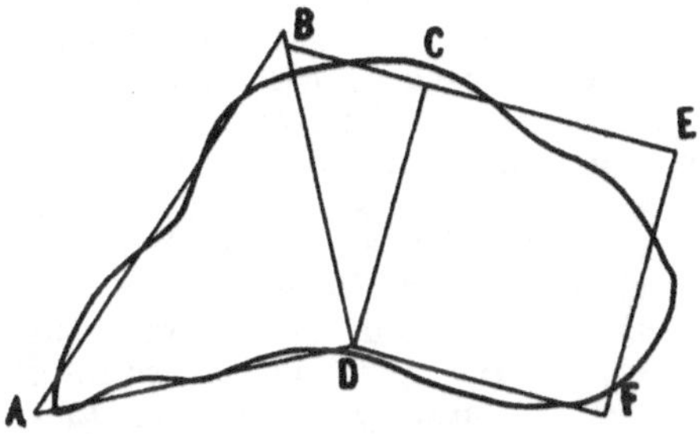

FIG. 3.—AREA BY MENSURATION.

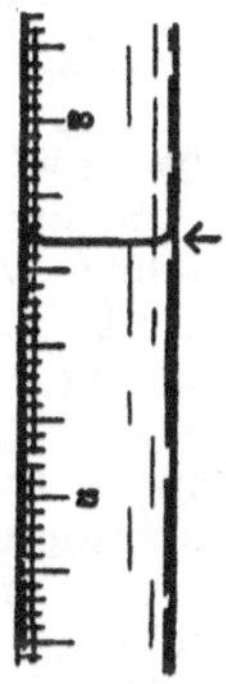

FIG. 4.—GRADUATED CONTAINER.

24. Measurement of Volume.

Examine a graduated cylinder and a graduated pipette and notice the volume occupied by one cubic centimetre in each. Do they seem larger or smaller than the space that would be enclosed if an imaginary square centimetre were drawn beside one linear centimetre of the ruler and an imaginary cube were then built up on the square? Pour into a test-tube an amount of water which you think will be 10 cm.^{3}; then measure it carefully.*

Put an irregular block of metal (*not* one of a set of standard weights) into a graduated cylinder partly filled with water and determine its volume by the change in the water-level. In reading a cylinder or a pipette the result is to be obtained by noting the height of the lowest part of the meniscus, or curved surface, of the liquid (Fig. 4).

* See that a towel is at hand when water is used in any experiment, and wipe up immediately any that is spilled on the table.

25. Measurement of Mass.

Examine the brass weights in a set extending from one gram to 500 grams, and observe especially the size of the 10-gram weight. What would its volume be if its density were ten gm. per cm.^{3}? If the density of brass is only 8·5 gm. per cm.^{3} (*i.e.*, if it is less compact) will its volume be greater or less than this?

Examine a platform balance * and notice that there are two wooden wedges that hold the pans away from the beam and the beam away from its support, so as to remove their weight from the accurately ground bearings when the apparatus is not in use. With a fine analytical balance this is such an important matter that a mechanism is provided by means of which the user keeps the scale pans and the beam "supported" while arranging the weights and the object to be weighed, and only lowers them upon their bearings for a few moments at the time of actual weighing. Such care is not necessary with an ordinary platform balance, but it is always advisable to support the scale pans after one has finished using the apparatus. Remove the wedges carefully and notice where the moving pointer comes to rest on the arbitrary scale. This need not be in the centre but may be at any point on the scale, and in weighing if the weights are so added as to return the pointer to this same position the result will be the same as if the pointer were made to take a central position in the first place. Notice the counterpoise, which can be screwed toward one side or the other in order to adjust the position of equilibrium, but do not attempt to move it unless you are sure that the apparatus is on a level part of the table and the scale pans are free from dust or other adherent matter. Notice whether the balance has a sliding weight that is used for weighing fractions of a gram.

In recording fractions of a gram write them as decimals of a gram, not as milligrams, thus 7·34 gms., not 7 gms. 340 milligrams.

Find the mass, in grams, of a four-ounce avoirdupois weight, weighing it first on the left pan of the balance and then on the right. Notice that when it is on the right-hand pan the reading of the sliding weight must be subtracted from the sum of the other weights instead of being added to them.

Draw up a few cubic centimetres of water in a graduated pipette, retaining it by pressing the *dry* finger tip on the *upper* end of the tube. Practise letting it run out slowly until you have no trouble in delivering any exact amount. Then fill it again, weigh some kind of container, run just 7 cm.^{3} of water from the pipette into the container and weigh the latter again.

* Sometimes called a druggist's balance.

26. Measurement of Density.

Calculate the density of water from the data obtained in the last experiment.

Find the volume of a 200-gram weight by measuring its height and diameter as carefully as possible, *but do not immerse it in water.* Suppose that the handle of the weight were soft, like wax, and could be flattened out and spread uniformly over the top of the cylindrical part, and estimate as well as you can how much this would add to the height of the cylinder. Then calculate the density of the brass weight, remembering that multiplication and division are always to be done by the abridged methods. The result may turn out to be less than the usual density of brass (about 8·5 gm. per $cm.^3$) if the handle is a separate piece screwed into the body of the weight so that an air space is left between the two parts.

Find also the mass of an irregular solid whose volume has been determined by immersion and calculate its density.

27. Equivalents.

In most English-speaking countries the C.G.S. System is very little used except for academic scientific purposes, the older system, known as the Foot, Pound, Second System, still holding its own in spite of such obvious disadvantages as possessing an ounce (avoirdupois) that weighs about 28 gm. and another ounce (Troy) that is equal to a little more than 31 gm.; furthermore, the U.S. fluid ounce of water weighs more than an avoirdupois ounce and less than a Troy ounce, and is 4 per cent. larger than the imperial fluid ounce of England. Accordingly, a knowledge of the approximate relationship between the units of the old system and the new is almost a necessity for the scientific student of to-day.

Both the F.P.S. and the C.G.S. systems have their advantages, and there is little sign yet that the people generally and the engineers and manufacturers of the British Empire and the United States of America are willing to discard the older system.*

Turn to the tables of equivalent weights and measures in the appendix of this book and use the *approximate* equivalents for translating your own weight and height into the C.G.S. System, and for answering such questions as: How many kilometres is the distance from here to London (or any other city)? What is the approximate height of this room in metres? What is the velocity of sound in C.G.S. units if it travels a mile in five seconds? How many centimetres per second is your ordinary rate of walking?

A fuller description of the derived units will be found in text-books on mechanics, properties of matter and heat.

* Reference may be made to the article Metrology in The Dictionary of Applied Physics for a comparison of the two systems and a reasoned argument on their relative advantages and disadvantages.

Learn how to convert a derived unit such as velocity or density from one system to another. Check your results by reference to the lists of conversion factors found in the tables of mathematical and physical constants.*

EXERCISES II

1. Explain why rounding off a half to the nearest even number will increase a measurement, in the long run, just as often as it will decrease it.

2. When you drew a line "just ten" centimetres long, was its length 10·0 cm.? Was it 10·00 cm.? 10·000 cm.? Measure it again, and explain why it is better to use one of these numbers in stating its length than to say "just" ten.

3. Does the sliding weight on a platform balance add its scale-indication to the right pan or to the left? How can it give correct results if the zero is not at the centre of its scale?

4. What disadvantage is there if the finger-tip that closes the top of the measuring pipette is not dry?

5. The *specific gravity* of a substance is defined as the ratio of its mass to the mass of an equal volume of water, *i.e.*, the *relative* density compared with water as a standard. (*a*) How do the specific gravity and the density of a substance compare if the density of water is 1 gm. per c.c.? (*b*) If it is a little less than one?

6. If *specific volume* u is defined for any substance as the volume per gram of mass, what will the equation be that shows the relationship between ρ (the density) and u?

7. If 10 cm. = 4 inches make a rapid mental calculation of the C.G.S. length of 12 inches. Of 40 inches; of 10 inches; of 3 feet; of 7 inches.

8. State the distance of 10 kilometres as the nearest whole number of miles. State 6 miles as the nearest whole number of kilometres.

9. Reduce (*a*) the pressure of 5 pounds per square in. to gm. per cm.^{2} (*b*) x lbs./ft.^{2} to gm./cm.^{2} (*c*) y miles/hour to cm./sec. (*d*) z pd./ft.^{3} to gm./cm.^{3}

10. Fairly accurate seconds can be counted off by repeating at ordinary conversational speed, "one thousand and one, one thousand and two, one thousand and three," etc. Try this by yourself and try it on your friends. Try it again out of doors and see if you can walk and keep step with it.

* Useful tables are given in Kaye and Laby's "Physical and Chemical Constants" (Longmans), and in J. B. Clark's "Mathematical and Physical Tables" (Oliver and Boyd).

CHAPTER III

ANGLES AND CIRCULAR FUNCTIONS

28. Unit of Angle.

When two straight lines intersect in such a way that each is perpendicular to the other the amount of their divergence is said to be ninety degrees or one right angle. A single degree is then a comparatively small difference in direction; two lines drawn from the observer toward the opposite edges of the sun's disc include an angle of about half a degree. For the sake of avoiding fractions at a time when decimals were not used, the degree (1°) was divided into sixty parts, each called one minute (1′), and a sixtieth of a minute was used as a still smaller unit, one second (1″). These three units are still in use all over the world for expressing the size of an angle as a "mixed" denominate number. For example, the circular arc which is of the same length as the radius corresponds to an angle of 57° 17′ 45″.

29. Circular Measure.

If the original right angle had been divided and re-divided into tenths or hundredths instead of into ninetieths and sixtieths the resultant units would have been much more convenient for purposes of calculation. The question suggests itself, however, as to why the right angle should be arbitrarily chosen for the quantity that is to be subdivided. Why not take the whole circumference or some other amount of angle? As a matter of fact, this arbitrariness is often avoided for scientific purposes by measuring an angle by an entirely different method: The vertex of the angle is taken as a centre, around which an arc of a circle is drawn extending from one side of the angle to the other (Fig. 5). The length of this arc

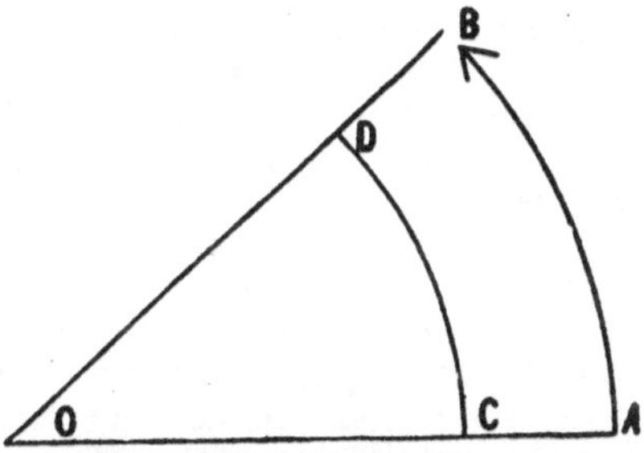

FIG. 5.—CIRCULAR MEASURE.

The ratio of any arc to its radius is taken as the numerical or circular measure of the angle at the centre. $DC/OC = BA/OA = 3/4$, approximately, for the angle represented here.

will depend not only upon the size of the angle but also upon the length of the radius that is used, but the *ratio* of length of arc to length of radius will depend only upon the size of the angle and so can be used as a measure of it. In the diagram the arc *AB* seems to be about $\frac{3}{4}$ as long as the radius *OA*, the arc *CD* is likewise $\frac{3}{4}$ as long as the radius *OC*, and the circular measure of this particular angle is accordingly $\frac{3}{4}$, or 0·75.

Draw an angle which is somewhat less than a right angle. Draw its arc, and measure the curved line as well as you can with an ordinary ruler and a piece of thread. Measure the radius also, and calculate the circular measure of the angle. It will probably turn out to be about 1·4. Is the circular measure of a right angle equal to just 1·5? State why.

30. Numerical Measure of an Angle.

Since the size of an angle is defined as the quotient *arc divided by radius*, it follows that this amount is not a number of centimetres or of any other arbitrary units but is a pure number. If the arc measures 6 centimetres and the radius is 3 centimetres the size of the angle is the abstract number 2, not 2 centimetres; and if both had been measured in inches the quotient would still have been merely the number 2, the arc being *two* times the radius, not *two centimetres* times the radius. The expression *numerical measure of an angle* has the same meaning as *circular measure of an angle*, and denotes the way in which the size of an angle is always expressed for theoretical purposes. One of the chief advantages of this method lies in the simplification which it causes. Just as the foot-pound system of measures makes the density of water about 62 lb./ft.^{3}, and hence makes specific gravity approximately equal to *density divided by* 62, instead of merely to *density* as in the C.G.S. System, so angular velocity would be represented by 57·28 v/r instead of v/r, and higher mathematical expressions would be more complicated if the angles were measured in degrees instead of in circular measure.

A suggestion of the reason why an angle ought to be expressed as an abstract number instead of in terms of a unit may be obtained by imagining a length of one centimetre and an angle of 45 degrees to be drawn on a sheet of paper and observed through a magnifying glass. The centimetre may appear to be enlarged to a length of two centimetres, but the angle of 45 degrees does not become an angle of 90 degrees; it remains exactly the same size as before. The same thing is of course true of an abstract number: with a very slight magnification two objects may both be made to look larger, but no amount of magnifying power will make them look like three.

31. The Angle π and the Unit Angle.

Draw an angle of 180° and its arc, viz. a semicircle. Obviously its numerical measure, semi-circumference divided by semi-diameter, is the same fraction as the ratio of the whole circumference to the whole diameter, which is denoted by the symbol π and is approximately equal to 3·1416. It is also clear that an angle which extends entirely around a point, that is, four right angles or 360 degrees, must have 2π for its numerical measure; one right angle must be equal to $\pi/2$, $\pi/4$ is the same as 45 degrees, etc. If $180° = \pi$, the numerical measure of one degree must be $\pi/180$ and the degree-measure of an angle that is numerically equal to one must be $180/\pi$.

Find the number of degrees in the unit angle by writing the value of π, carried out to at least eight or ten decimal places, and using the method of abridged division to find out how many times it is contained in 180. The latter number should be written 180·00000 . . . with as many ciphers as may be necessary.

Practise translating such numbers as the following into degrees until it can be done without any hesitation: $\frac{1}{2}\pi = ?$ $2\pi = ?$ $\frac{1}{4}\pi = ?$ $4\pi = ?$ $\pi/3 = ?$ $\frac{3}{2}\pi = ?$ $\pi/4 = ?$ $\pi/\pi = ?$ $\frac{1}{3}\pi = ?$ If an angle is numerically equal to three is it greater or less than 180°?

Practise translating the following numbers of degrees into circular measure until it can be done fluently: $90° = ?$ $360° = ?$ $45° = ?$ 3 right angles $= ?$ $180° = ?$ $30° = ?$ $60° = ?$ $1° = ?$ $270° = ?$ $57·296° = ?$

32. The Protractor.

A protractor is a scale of angles just as a graduated ruler is a scale of lengths. It consists essentially of a zero line on which is a point that represents the vertex of the angle, and a curved scale of

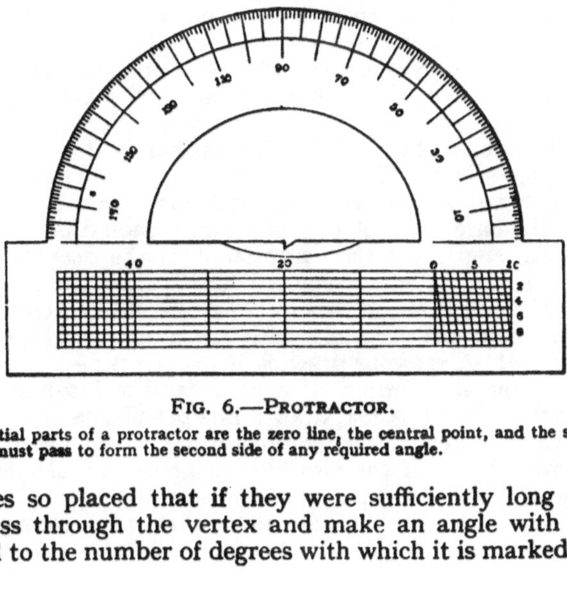

FIG. 6.—PROTRACTOR.

The essential parts of a protractor are the zero line, the central point, and the scale showing where a line must pass to form the second side of any required angle.

short lines so placed that if they were sufficiently long each one would pass through the vertex and make an angle with the base line equal to the number of degrees with which it is marked.

Examine the protractor and its scale of degrees. Notice that the centre toward which the slanting lines converge must be on the line joining 0° and 180° and hence must be at the *top* of the notch shown in the figure. It is always at the corner that is made rectangular, the other being obtuse or rounded. To draw a line which shall make a given angle with a given line at any particular point the protractor is placed so that its zero coincides with the given line and its centre with the given point. A fine line or dot is then made opposite the required number of degrees on the scale, the protractor is removed, and a ruler is used to draw a straight line through the dot and the particular point on the given line.

Use the protractor to draw a *triangle* of any convenient size and of such shape that its three angles are $\pi/2$, $\pi/3$, and $\pi/6$. Draw another so that its angles are $\pi/4$, $\pi/4$, and $\pi/2$.

33. The Diagonal Scale.

If the protractor is provided with what is known as a *diagonal scale*, notice that at the top of this scale there is a horizontal line which is divided into centimetres or inches and that one division at the end of the scale is divided into tenths. Decide for yourself how it is that any length, such as 3·4, within the limits of the total length of the scale, can be found already laid off as a single continuous stretch of the base line, the tenths being measured from the proper point to the junction of the tenths scale and the units scale, and the units then extending onward the required distance beyond the junction point. Notice that the tenths divisions are prolonged downward so as to cut diagonally across parallel horizontal lines and shift a single tenth to one side while dropping ten lines downward. This means that on the first level below the base line their shift will be only one hundredth, and on successive levels will be ·02, ·03, . . . 10, as the student can easily prove for himself by means of the principles of similar triangles. Accordingly, any number of units, tenths, and hundredths can be found marked off along the proper level. Thus a distance of 3·43 will be found on the third level between the same diagonal and vertical line as mark off 3·4 on the base line. A pair of dividers should be used to span an unknown length and then transfer it to the diagonal scale for measurement, or to take a required length from the scale for the purpose of laying it off on paper. They should be held rather flat against the scale, not perpendicularly, so as to avoid marring it.

Experiment.—Test your drawing instruments :—

(1) Test the straightness of a ruler (and a T-square). To do this place the ruler on paper, draw a fine pencil line along the edge, revolve the ruler about this line until it lies on the other side of it. Draw another line. If the two lines coincide the ruler is straight. If not, repeat until you get a ruler that is straight.

(2) Test the accuracy of the right angle of a set-square. To do this place a *straight* ruler on the paper and at a given point use

the square to draw a line at right angles to the ruler. Turn the square over until it lies flat on the other side of the line and draw another line at right angles to the ruler at the same point. If the two lines coincide the angle is a right angle. If not try another square. Repeat until you find one correct.

34. Measures of Inclination.

If a slanting line intersects a level line the inclination of the former may be measured either by the size of the angle between them or by the rise of the inclined line per unit of level distance (see § 14). The second of these two amounts is said to be the *tangent* of the first one; thus $2/5 = \tan 22°$. This relationship can easily be generalized so as to include cases in which neither side of the angle is horizontal. From any point on one side of an angle draw a straight line perpendicular to the other side; this will complete a right-angled triangle, and can be done in all cases where the angle is *acute* (*i.e.*, between 0° and 90° in size). The *tangent* of an acute angle of any right-angled triangle is defined as the ratio of the side opposite the angle to the side that is adjacent to the angle, the word "side" being used here to indicate either of the right-angled sides, not the hypotenuse.

In the diagram (Fig. 7) the angle G is an acute angle of each of three different right-angled triangles. Measure the height and the

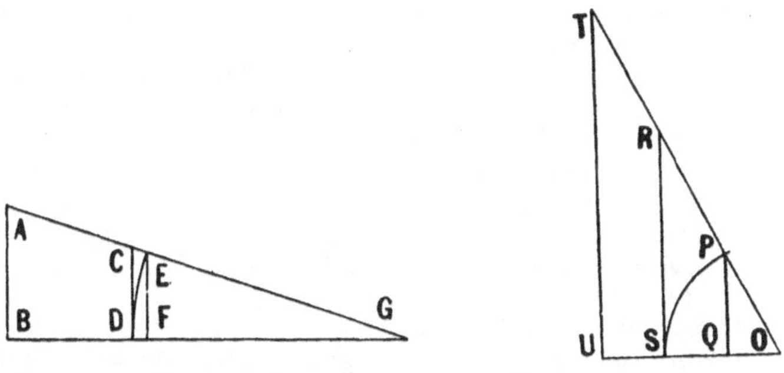

FIG. 7.—ANGLES AND THEIR TANGENTS.

The ratio of any vertical side to the corresponding base is a number which is called the *tangent* of the acute angle at the base. EF is the same fraction of GF as AB is of GB, namely about $1/3$. PQ is about twice OQ, and RS is about twice OS. Accordingly $\tan G = 1/3$ and $\tan O = 2$.

base of any one of these and calculate the tangent of the angle G. The reason for the name "tangent" may be understood by noticing that the line drawn between the two sides of the angle and *tangent* to one end of the arc will be numerically equal to the ratio in question if the radius GD or GE is of unit length, for then $CD/GD = CD/1 = CD$.

The inclination of two lines that form an acute angle of a right-angled triangle may also be measured by the ratio of the opposite side to the hypotenuse. This is called the *sine* of the angle; for example, in Fig. 7, the ratio of EF to GE, or, what amounts to the same thing, EF/GD, is the sine of the angle G, often abbreviated to "sin G." It is evident that these definitions of *sine* and *tangent* agree with those that have been previously given for the "gradient" and "per cent. slope" of an angle between a sloping line and a horizontal one.

35. Use of a Table of Tangents.

Close to the bottom of the next unused page of your notebook draw a fine horizontal line having a length of either 20, 25, or 50 of the squares made by the cross-lines of the paper. Draw it directly on one of the horizontally ruled lines, preferably so that it extends nearly across the page. Call its length unity (1); it may not be one foot or one decimetre, but it is to be considered as having a length of just one arbitrary unit, which need not be given any name. Mark a figure 1 under its right-hand end, a figure 0 at the left-hand end, and the scale of numbers 0·1, 0·2, 0·3, etc., at intervals of two, or two and a half, or five, squares, as may be required by the length of the line. From the right-hand end of this base line draw a fine line perpendicular to it, extending it as far as the top of the page. Beginning with zero at the junction with the base line lay off a similar scale along the vertical line. Do not number the successive squares of the paper 1, 2, 3, etc., but see that this scale indicates the same proportions as are given by the horizontal one. Turn to the table of circular functions in the appendix; look for the number 10 in the column headed DEG and notice that the number opposite it in the column TAN is 1763; this means that the tangent of 10° is 0·1763. On the vertical line just drawn make a small mark at a height of 0·1763 above the base line according to the vertical scale already made. In the same way lay off tan 20° on the same scale. The decimal points are omitted from the table; but the tangent of a large angle is obviously greater than that of a smaller angle, and the table shows that for successive degrees the tangent increases gradually and quite regularly. Notice that the tangent of 45° is 1·000000 . . . , and that angles and tangents larger than this must be sought for *above* the abbreviations DEG and TAN which are at the bottom of the page. If there is any trouble in understanding how the table is arranged use it to verify the following equations before proceeding further:

Tan 1°=0·0175; tan 2°=0·0349; tan 6°=0·1051; tan 44°=0·9657; tan 45°=1·000; tan 46°=1·036; tan 84°=9·514; tan 85°=11·43; tan 89°=57·29.

Lay off tan 30°, tan 40°, tan 50°, etc., as far as the length of the vertical line will allow; then draw slanting lines from each of these points to the left-hand end of the base line. With the protractor test the angles formed in order to make sure that they are accurately 10°, 20°, 30°, etc. If mistakes have been made repeat the construction on the next page; do not correct the first diagram by erasures.

36. Experimental Determination of Sines.

With the base line as a radius and its left-hand end as a centre draw an arc on the diagram that has just been made, extending it from 0° to 90°. Complete the series of angles as far as 90° by laying off successive ten-degree arcs or chords with a pair of dividers. Find the point where the line whose slope is 10° intersects the arc, and note carefully the vertical distance from the base line up to this point, but do not draw a vertical line. If the diagram has been carefully drawn with a sharp-pointed pencil the value should come out 0·174, being measured of course in terms of the figured scales, not in terms of the small ruled squares. Notice that this number is less than tan 10°, and that it corresponds to the ratio EF/GD in Fig. 7, and hence is the sine of the angle 10°. In the same way measure sin 20°, sin 30°, . . . sin 90°, or as many of them as may be directed by the instructor, and tabulate the results in the first two columns of a three-column table Then turn to the table of circular functions, find the true numerical values of the sines from the column headed SIN, and correct your measured sines by adding a third column as shown here; the sine of 40°, for example, appears to have been measured as 0·640 and then found to be actually 0·643, whence the correction of +0·003 in the third column. After your table of sines has been completed find the angle whose sine has the largest correction and divide this correction mentally by the true value of the sine in order to find the relative error of the measurement. Thus, if the quotient is about 3/600=1/200=0·005 your measurement has an error that amounts to three parts out of a total of 600, which is the same as 5 per thousand, or ½ per cent. Call your error "moderate" if it is anywhere between 0·3 per cent. and 1 per cent.; call it "large" if greater than 1 per cent., and "small" if less than 0·3 per cent.

angle	*sine*	
0°	0	
10°	0·175	− 0·001
20°	0·345	− 0·003
30°	0·500	+ 0·000
40°	0·640	+ 0·003

TABLE OF SINES.

The second column shows the measured value; the third column is the amount of change that must be made to obtain the true value.

37. Definition of Function.

Up to the present point it has been assumed that an angle of any size from 0 to $\pi/2$ or 90° will have a tangent which is a definite number for each definite size of angle. A quantity which can be assumed to have different sizes, whether restricted to a certain range or not, is called a *variable*; and a second quantity, which in general has a definite value for each particular value of the first, is said to be a *function* of that variable. For example x^2-3x+2 is called a function of x, because it has a definite value for any definite value that may be assigned to x. Similarly sin x, tan x,

$\sqrt{x}$, logarithm of x, x^n, a^x, are all functions of x, as is also any other algebraical formula which involves x. A function of x is often written $f(x)$.

38. The Cosine of an Angle.

Another function of an angle which is frequently useful is the *cosine*. In a right triangle, such as was used in defining the tangent and the sine of an angle, it is the ratio of the adjacent side to the hypotenuse; the ratio of GF to GE, in Fig. 7, is the cosine of the angle G, or, as it is usually abbreviated, $\cos G = GF/GE$.

39 Circular Functions.

The sine, cosine, and tangent are included under the general term *circular functions*, a phrase which includes also three other functions of an angle which are not so frequently used. These are the *cotangent*, the *secant*, and the *cosecant*, and may be defined by the equations $\cot x = 1/\tan x$, $\sec x = 1/\cos x$, and $\operatorname{cosec} x = 1/\sin x$. Another set of definitions, which are perhaps more interesting and easier to remember, may be obtained by the use of a circle diagram and extended so as to include angles greater than 90°. For the sake of uniformity the angle is so placed that one of its sides extends out horizontally to the right. A zero angle would have its second side coincident with the first, and larger angles may be supposed to have been generated by rotating the second side through the required angular distance from the first. The usual convention is that the direction of rotation shall be *counter-clockwise*, *i.e.*, in the opposite direction to that in which the hands of a clock turn. A circle whose radius is unity is drawn around the vertex of the angle, as in Fig. 8, which shows an angle of somewhat less than 45°. A line (DB) is drawn upward from the right-hand end (D) of the horizontal radius, and a perpendicular (AC) is dropped from the end of the inclined radius (A) to the base line (OD). Then the length of the tangent to the circle (DB) is the numerical tangent of the angle (O), the line (OB) that cuts the circle is the secant (Lat. *secans*, cutting), and the perpendicular (AC), which cuts off a rounded hollow of the figure, is called the sine (Lat. *sinus*, a bay). The cosine is

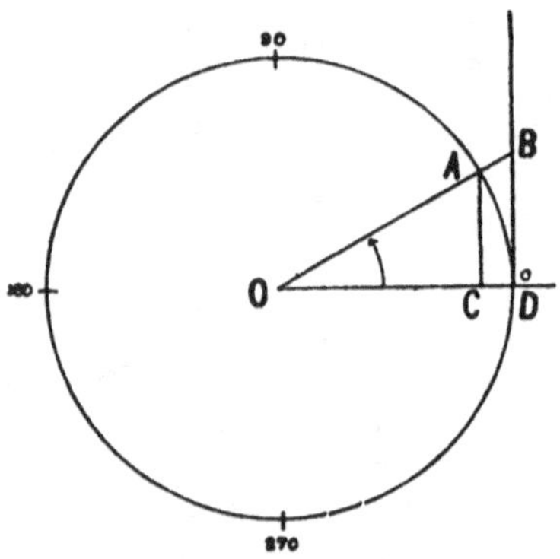

FIG. 8.—CIRCULAR FUNCTIONS.

$DB = \tan DOA$; $AC = \sin DOA$; $OC = \cos DOA$; $OB = \sec DOA$. The radius is supposed to be of unit length.

the sine of the complementary angle (*e.g.*, $\angle OAC$; if x is acute $\cos x = \sin(90° - x)$), and the cotangent and cosecant are similarly the tangent and secant respectively of the complementary angle, two angles being called *complementary* when their sum is $\pi/2$ or 90°. As OC is considered to be a positive amount when measured to the right of O it is only natural to consider the cosine as a negative number when C is to the left of O. Likewise CA or DB must be negative if below the base line instead of above. AO and DO are to be produced if necessary.

If $\tan 45° = +1$ and $\sin 45° = +0{\cdot}707$ what are the values of tan $(90° + 45°)$ and sin $(90° + 45°)$? Ans.: $\tan 135° = -1$; $\sin 135° = +1$.

What is the cosine of 135°? What are the tangent, sine, and cosine of $180° + 45°$? Of $270° + 45°$?

40. Generalized Idea of Angle.

For some purposes the term angle may be defined so as to include only angles less than $\pi/2$. For other purposes obtuse angles (*i.e.*, up to π or 180°) may need to be included. The circular functions have been already defined for angles of all sizes up to 2π (6·28, or 360°); but it is obviously unnecessary to stop at this figure. By supposing the line OA in Fig. 8 to have made more than a complete revolution it can be seen that the sine, cosine, tangent, etc., of $360° + 40°$ must all have the same values as the corresponding functions of 40°. In general, any circular function of any angle x is equal to the same function of $2\pi + x$, or of $4\pi + x$, or of $2n\pi + x$ if n is any whole number. A *negative angle* is of course one that is generated by a clockwise rotation from the position of the base line; thus the angle shown in Fig. 9 may be considered either as $+240°$ or as $-120°$, and obviously the circular functions of any negative angle $-x$ have the same values as those of the positive angle $2\pi - x$. Negative angles have to be considered occasionally, just as negative heights or lengths need to be. Angles larger than

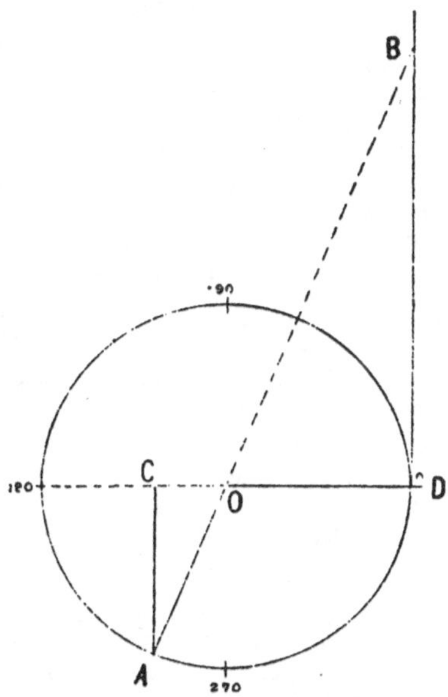

FIG. 9.—FUNCTIONS OF ANY ANGLES.

The cosine is negative if C is to the left of O; the sine and tangent are negative if A and B respectively are below the level of O; DO and AO are produced if necessary.

$+2\pi$ commonly come under consideration in connection with rotatory motion. Such objects as a spinning top, a fly-wheel, a planet, do not commonly move through an angle less than 360° and then stop, but their angular motion may be of almost any amount according to the extent of time occupied by it.

Exercises III

1. How can it be proved that the ratio which gives the numerical measure of an angle will have the same value whether the radius is long or short ?

2. Translate 135° into circular measure. Make an approximate mental calculation of the number of degrees in an angle whose numerical measure is 6.

3. After drawing a triangle whose angles were meant to be $\pi/2$, $\pi/3$, and $\pi/6$, what would you do if you found one of its angles inaccurate but the other two correct ?

4. The tangent of 80° is given in the table as 5671. Where do you place its decimal point and why ?

5. The *steepness* of a slope is the characteristic which corresponds to its sine ; a larger sine means a steeper slope. What characteristic can you name that will correspond to its cosine ?

6. Any radius of a rotating wheel describes an angle which increases steadily from 0 to (say) 40π. Explain how the sine of this variable angle behaves during the same time.

7. What is the approximate value of tan (−91°) ? Of tan (+89°) ? Of tan (−89°) ?

8. What is the approximate value of the tangent of the angle A in Fig. 7 ? How does it compare with the value of the tangent of the angle G ? What relationship is there between tan T and tan O ?

9. The *cotangent* of any angle, A, has been defined as the reciprocal of tan A, and also as the value of tan $(90°-A)$. Prove that these definitions are identical if $A<90°$ by drawing a right triangle having A for one of its acute angles.

10. Draw a right triangle and prove that, if $A<90°$, $\sin^2 A+\cos^2 A=1$ using the theorem of Pythagoras that the square of the hypotenuse is equal to the sum of the squares of the other two sides.*

11. Make A an acute angle of a right triangle and prove that $\tan A=\sin A/\cos A$.

12. Using a circle diagram prove that in general for an angle of any size, x,
   (a) $\tan x=\sin x/\cos x$
   (b) $\sin^2 x+\cos^2 x=1$
   (c) $\tan (90°-x)=1/\tan x$
   (d) $\cos x=\sin (90°-x)=\sin (90°+x)$

13. What synonyms have you already learned for the sine and the tangent of an angle ?

14. Find the number of seconds of arc in one radian.

15. Find sin 12″.

---

* $\sin^2 A$ is the conventional abbreviation of $(\sin A)^2$, not of sin (sin A) ; similarly $\sin^3 A$, etc. $\sin^{-1} A$, however, is always used to denote *the angle whose sine is A*, not $1/\sin A$.

CHAPTER IV

SIGNIFICANT FIGURES

41. Estimation of Tenths.

It sometimes happens that a measurement requires but a slight degree of accuracy, and time and trouble can be saved by making it only roughly. As a general principle, however, it is advantageous to make all measurements as accurately as possible. Thus, measurements of length made with the metre stick should be expressed not merely to the nearest centimetre but to the nearest millimetre or tenth of a centimetre. This is not the best that can be done, however. After noticing how each centimetre of the scale is divided in half by a long line and each half is subdivided into fifths by four short lines, thus indicating tenths of a centimetre, it is not difficult to imagine each of the millimetre intervals of the scale divided in the same way into tenths of a millimetre and to make a fairly good estimate of just how many of these parts are included in the length that is to be measured. Experienced observers even attempt to make a mental subdivision into hundredths of the smallest intervals on a graduated scale, and find that it is only occasionally that one man's estimate will differ from another's by more than one or two hundredths, but for the beginner even the estimation of tenths will be a rather uncertain process. To gain proficiency it will be found better to begin with larger subdivisions, such as a scale of centimetres that has no millimetre marks. The position of a mark placed at random on such a scale can be estimated mentally and the accuracy of such a determination can then be tested by actual measurement with a more finely divided scale.

42. Practice in Estimating Tenths.

Draw a short line at right angles to the edge of a card or slip of paper (Fig. 10) and hold this edge on a scale of centimetres and millimetres in such a way that the smallest graduations are hidden but the marks indicating centimetres and half-centimetres are visible. Notice the half-centimetre space in which the cross line comes and mentally divide it into five equal parts by four imaginary lines so as to make an estimate of the location of the line on the card. For example, in the figure the arrow seems to be either three-fifths or four-fifths of the way from the scale-division 19·5 to the division 20·0, making its position 19·8 or 19·9, but it may be difficult to decide which of these numbers is the nearer without actual measurement. In practice, however, the estimate should be written down and the card should then be allowed to slide carefully across

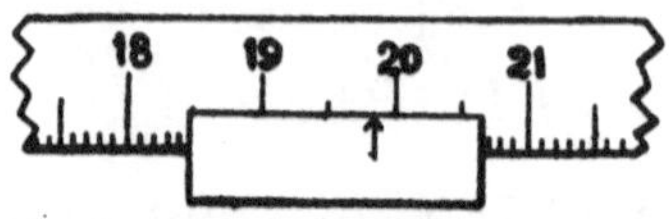

FIG. 10.—ESTIMATING TENTHS OF A CENTIMETRE.

The card is laid on the scale at random, but in such a way as to hide the small dividing lines. The location of the arrow, in millimetres, is then guessed, and afterwards verified by sliding the card downward enough to expose the whole scale.

the ruler until the millimetre scale is just exposed. The position on the scale which the line on the card occupies will then be ascertained and should be written in the notebook beside the previous estimate. The statement should be correct to the nearest millimetre, without any effort to decide upon fractions of a millimetre (see §§ 20 and 21).

Make ten such estimates with the card placed anywhere along the scale at random, and tabulate the determinations and the verifications in two parallel columns. Then hold the card a trifle higher, so as to hide the half-centimetre graduations as well as the millimetres and make twenty more determinations. These will also be estimates of the nearest millimetre, but will require the more difficult process of deciding upon tenths of a whole centimetre instead of fifths of a half-centimetre.

If the scale that is used in the foregoing exercise is on the lower edge of a metre stick there will probably be a duplicate scale along its upper edge, with the help of which it would be easy to make each estimate absolutely correct before verifying it. This furnishes a good illustration of the statement that the student should aim to *learn* rather than to *do*. Finding out the correct position of a line over a hidden scale is something that is utterly useless to him or to any one else; it is the learning to divide a centimetre into imagined tenths that will be valuable to him later when it becomes necessary for him to divide a millimetre into tenths without aid and without the possibility of later verification. In making any kind of a measurement care should be taken to avoid any extraneous influences, or bias or prejudice of any kind. In the present case, the upper scale on the metre stick should be kept covered in some way unless the student is sure of his ability to disregard it completely while making his estimations.

43. Mistakes in Estimating Tenths.

Omitting the preliminary estimate of fifths, examine your table of estimated tenths closely and find out what kind of error you are most apt to make. Some students find it hardest to estimate 0·3 and 0·7 correctly; others have almost a uniform tendency to read a position like 12·0 as either 11·9 or 12·1. The latter mistake is due to the fact that a minute deviation from the position of a visible graduation is very easily noticed and there is a tendency to consider it as a single tenth. Of course if it amounts to more than half of a tenth this is correct; but if it is less than half a tenth it should be considered as 0·0 instead of 0·1. The same bias may even cause a tendency to read 0·1 and 0·9 as 0·2 and 0·8. On the other hand, there may be just the opposite error if the graduated lines are rough or coarse, unless the observer is careful to estimate from the imaginary centre of such a line instead of from its margin.

If a definite kind of error is evident from a study of your table see if it can be overcome when making another short series of determinations. Then draw a second line on the card, place the latter in position as before, estimate both points, and find the distance between them by subtraction. This is the customary method of measuring a length, and is preferable to making one line coincide exactly with a scale division and estimating only the other one, in spite of an obvious additional source of error.

44. Value of π.

An experimental determination of the value of the constant, π, can be made by rolling the brass disc along a metre stick to find the length of its circumference, then measuring its diameter and calculating the ratio. Hold the disc loosely at its centre, using the thumb and forefinger only. Start it with its marked radius on some definite graduation of the scale and roll it in a straight line until the radius again comes vertically down on the scale. Read this second position, remembering not to be satisfied with the nearest millimetre (0·1 cm.), but to make as good an estimate as possible of tenths of a millimetre in order that the circumference may be correctly measured to a hundredth of a centimetre. Use the metre stick to measure the diameter of the disc with the greatest possible care, avoiding the end of the stick, which may be a little worn so that it does not represent precisely 0·00 cm. or 100·00 cm. Find the value of π by dividing circumference by diameter, and in this particular case using the unabridged method of division and carrying out the result until it has two or three figures that are different from the theoretical value of π, which is 3·14159265358979323846264338327950288419716939 9. Finally round off (§§ 20, 21) both your result and the true value to the same number of places, choosing that number so that your result will show just one incorrect figure; for example, $3\frac{1}{7}$ (or 3·1428571) will have just one wrong figure if it is rounded off to 3·143, because the theoretical value will round off to 3·142; while 3·14234567 should be rounded off to 3·1423 to compare with the correct value 3·1416.

45. Physical Measurement.

The operation of making a measurement is merely counting; it is the determination of how many units of a certain kind are required in order to be equal to a given quantity of the same kind. But while a count such as a census of the number of individuals in a town must give a perfectly definite whole number it usually happens that a physical measurement will not give a whole number, or even a commensurable number except as the result of an error, and successive repetitions of a measurement will give a number of different apparent values. (Try it; measure the circumference of the π-disc a second time.) Accordingly, any numerical statement of a measurement must be merely an approximation to an unknown true value, and so will be either indistinguishably correct or perceptibly incorrect according to how closely it can be examined.

46. Ideal Accuracy.

The average student is liable to have more or less difficulty in grasping the idea that accuracy is always a relative matter and absolute precision of measurement is an impossibility. This is usually because he has had very little practice in careful measurement and at the same time his previous study of arithmetic has emphasized a condition of infinite accuracy of numerical values. Such a number as 12·5 has been supposed not only to mean the same thing as 12·50 but also to be equal to 12·500000 . . . to an unlimited number of decimal places. This is quite proper and satisfactory as long as one realizes that he is dealing with imaginary quantities, or perhaps it would be better to speak of them as ideal quantities, perfections of measurement which have no more reality of existence than the point, line, plane, or cube, of geometry. The smoothest surface of a table does not come as near to being a plane as does the surface of an "optically worked" block of glass or a "Whitworth plane," and even the smoothest possible surface can be magnified so as to show that it contains irregularities everywhere. Perhaps if it were magnified enough we could see that its shape would not even remain constant, but individual molecules would be found swinging

FIG. 11.—DIAGRAMMATIC CROSS-SECTION.

A metal cube, greatly magnified, to show that there is no plane surface of contact between the metal and the air above it.

back and forth or possibly escaping from the surface (Fig. 11). A geometrical plane certainly corresponds to nothing in reality, and perfect accuracy of number is just as much an imaginary concept.

47. Decimal Accuracy.

If 12·5 cm., as a measurement, does not mean the ideal number 12·500000000 . . . to an infinite number of decimal places what does it mean? As different measurements are likely to be made with different degrees of accuracy the universally adopted convention is merely the common-sense one that *the statement of a measurement must be accurate as far as it goes; and it should go far enough to express the accuracy of the determination.* Thus "12·8 cm." means a length that is nearer to precisely 12·800 . . . than to precisely 12·7 or 12·9 cm., *i.e.*, that its "rounded-off" value would be 12·8 cm., not 12·7 or 12·9. If a length is written "12·80 cm.," however, this implies that the stated measurement is nearer to this same precise 12·8 or 12·80 or 12·800000 . . . than it is to either 12·79 or 12·81 cm., in other words, that it has been measured to hundredths of a centimetre and found to be between 12·795 and 12·805 cm., so that it can properly be rounded off to 12·80. The other description, "12·8 cm.," means between 12·75 and 12·85; it states nothing about hundredths of a centimetre, and can correctly represent any lengths between the limits just given; for example, 12·75, 12·76, 12·77, 12·78, 12·79, 12·80, 12·81, 12·82, 12·83, 12·84, or 12·85, for each one of these could be rounded off to 12·8. To write such a length of 12·8 cm. in the form "12·80 cm." would be to violate the rule that a statement should be accurate as far as it goes, for it would go as far as hundredths (stating that there were eighty of them), and the chances are ten to one that it would be one of the other numbers of hundredths given above. On the other hand, if an observer determined a length to be 12·80 cm., that is, if he measured the length as 12 cm.+8 tenths+0 hundredths—if he looked for hundredths and established the fact that there was none of them—then to state the measurement only as 12·8 cm. would not be doing justice to his own accuracy, for he would imply that the correct number of tenths was merely known to be nearer 8 than 7, namely greater than 7·5, whereas he had already found it to be nearer 8 than 7·9, namely greater than 7·95.

When a carpenter says "just 8 inches" he probably means "nearer to $8\frac{0}{8}$ than to $8\frac{1}{8}$ or $7\frac{7}{8}$ inches," a sixteenth of an inch one way or the other being unimportant. When a machinist says "just 8 inches," he may mean "nearer to $8\frac{0}{64}$ than to $7\frac{63}{64}$ or to $8\frac{1}{64}$," a half-sixty-fourth or hundred-and-twenty-eighth of an inch being negligible to him. When another person says "just 8 inches" we must know what kind of materials he works with before we can tell the meaning of his word "just." If decimal subdivisions were everywhere used the carpenter's eight inches would probably mean

8·0 while the machinist's would mean 8·00; for one man "8" would mean "between 7·950 and 8·050" while for the other it would mean "between 7·995 and 8·005." It is for the sake of avoiding such ambiguities that the scientist has adopted the rule that "8" means "between 7·500 and 8·500"; "8·0" means "between 7·950 and 8·050"; "8·00" means "between 7·995 and 8·005"; "8·000" means "between 7·999½ and 8·000½"; etc.; in other words: *no more figures are to be written down than are known to be correct; and, no figures that are known to be correct should be omitted.* This principle is simple enough when it has once been properly comprehended, and after that there is not much danger of the student's "rounding off" a carefully obtained measurement like 2·836 gm. to 2·84 gm. merely for the sake of doing some rounding off. There is a very decided likelihood, however, that he will often forget to write down a final significant zero; if two lengths are 147 mm. and 160 mm. the tendency when writing them in centimetres is to put down 14·7 and 16. If the zero is as important as the seven when writing millimetres the same is equally true when writing centimetres. Suppose the diameter of the π-disc is found to be "just 8" centimetres; the measurement should be stated as 8·00 cm. if tenths of a millimetre were estimated and none were found, but it should be given as 8·0 cm. if the student read the millimetres but was unable to make an estimate of smaller amounts. There is nothing to show which degree of accuracy was obtained if the diameter is put down as 8 cm. "because it came out just even."

48. Significant Figures.

In the expression 6·2 cm. both the figure 6 and the figure 2 mean something or are *significant.* In the expression 62 mm. there are likewise two significant figures. When the same length is written in the form ·062 metre there is no difference in what is signified, and although the number has three figures it is still said to have only two significant figures; the zero is present merely for the purpose of showing which decimal places are occupied by the six and the two, or, in other words, for fixing the location of the decimal point. If the same length is called 62 thousands of microns or 62,000 μ there are still only two significant figures, and again the ciphers serve only to show that the six is located in the tens-of-thousands' place, *i.e.*, to fix the position of the decimal point. *In arabic notation, the figures of which a number is composed, except for one or more consecutive ciphers placed at its beginning or end for the purpose of locating the decimal point, are called its significant figures.* In accordance with this definition it will be clear that only two of the three figures of 0·75 gm. are significant, and only one of the two figures of such an expression as 05 c.c., in which a superfluous zero is sometimes written. Non-significant ciphers occur sometimes on the left, as in the statement that a certain light-wave has a length of

·00005086 cm., and sometimes on the right, as when the sun is stated to be 93,000,000 miles from the earth; the first of these numbers has four significant figures, the second has only two. It will be noticed that a number like the last causes trouble in applying the rule of making it "accurate as far as it goes," for only the first two or three figures are known, but eight are needed in order to place the decimal point. A further source of trouble lies in the fact that the last figure which is significant may happen to be a cipher instead of some other digit. If this is to the right of the decimal point the zero is of course written when significant and omitted when not (as explained in § 47), but what is to be done when a similar case occurs in which the significant zero occurs to the left of the decimal point? If a building is said to be worth fourteen thousand pounds, how is any one to tell whether this means 14 thousands of pounds, *i.e.*, nearer to 14 than to 13 or 15 of these thousands, or whether it means exactly 14,000 pounds and no shillings or pence, or whether the number of significant figures is not intended to be either two or five but some intermediate number? Suppose the number of millions of miles from the earth to the sun at some particular time is found to be 93·00; we need two different symbols for our ciphers so that we can write 93,00o,ooo miles to show that the first two ciphers are significant while the last four are not. There is really no reason for using ciphers at all in the last four places, except that it is customary, and it would be better to use some other character, such as Ɛ, and write 93,00Ɛ,ƐƐƐ. Neither of these methods is ever used, however, but the same result is achieved by a notation that will be explained later on (§ 55).

The length of an inch has been determined to be between 25·39977 and 25·39978 mm. To state that it is 25·40 mm. is correct, because this value is accurate as far as it goes; but to say that 1 inch=25·39 mm. would be wrong, for the true number of hundredths is nearer to 40 than to 39. If it is desirable to use an approximate value, so that rounding off is permissible, how many of the following statements are correct and how many are positively wrong? 1 in.=25·4 mm.; 1 in.=25·40 mm.; 1 in.=25·400 mm.; 1 in.=25·4000 mm.

Which of the following values of π are correct, and which are incorrect? 3·141592; 3·141593; 3·141600; 3·14160; 3·1416; 3·1415; 3·142; 3·141; 3·15; 3·1; 3·142857 (=22/7); 3· (II Chronicles, chap. 4, v. 2); 3·16 ($=\sqrt{10}$).

At ordinary room temperature (20° C.) is the density of water equal to 1? Is it 1·0? 1·00? 1·000? To what point must its temperature be lowered in order to make its density 1·0000? (See tables, Appendix.)

49. Relative Accuracy.

The diagram (Fig. 12) shows two rectangles which are approximately squares. The difference between the height and the width of the larger one is just the same as the difference between the height and the width of the smaller, yet the small rectangle is obviously a less accurate approximation to the shape of a perfect square than is the large one. This may serve as an illustration of the important general statement that accuracy is a matter of relative amount rather than of absolute amount. A sixteenth of an inch has the same absolute value wherever it occurs, but it is a considerable part of a quarter-inch length while it is relatively insignificant in comparison with a whole inch.

The relative accuracy of a measurement accordingly depends upon two things: how much its absolute difference from the truth amounts to, and how large the measurement itself is. If two points on the earth's surface are found by careful surveying to be 10 miles apart the determination of distance may easily be in error by more than a foot, and even with the most extremely careful triangulation the error is likely to be as much as four inches. An error of a quarter of an inch, however, in measuring the thickness of a door could hardly be made even with the clumsiest of measuring apparatus. It would plainly be misleading to say that the clumsy measurement should be considered more accurate than the careful one on account of $\frac{1}{4}$ inch being less than 4 inches. The only consistent way of looking at the matter is to inquire how large a fraction of the total measurement the error amounts to. Suppose the thickness of the door is $1\frac{1}{2}$ inches; how large a part of this measurement is the error of $\frac{1}{4}$ inch? Obviously it is one-sixth of the total, or an error of more than 16 per cent.; while four inches out of a total of ten miles is not nearly a sixth, but is roughly an error of one out of a hundred and fifty thousand, or about six per million, or about 0·0006 of 1 per cent.

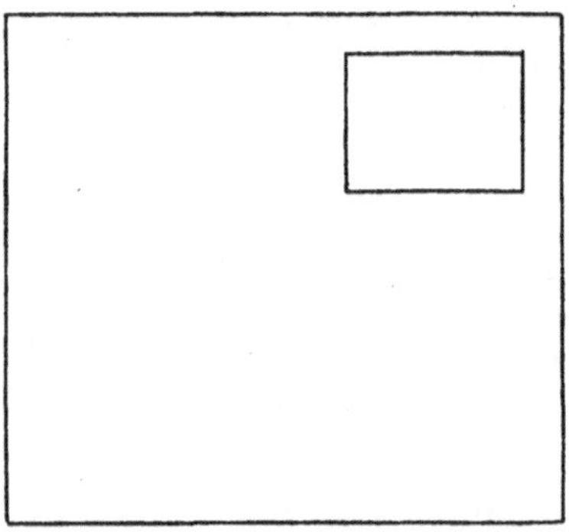

FIG. 12.—RELATIVE AND ABSOLUTE ACCURACY.

How large an error is an eighth of an inch when half of an inch is being measured (Fig. 12)? How large is a sixteenth of an inch out of a total of an inch and three-quarters?

If a measurement is stated to be 12·8 cm. when it is known to be between 12·750 and 12·850 cm., what is the greatest possible error of the statement? Answer: 0·05 out of a total of 12·75. This is the same as 05· out of 1275·, or 5 per 1275, or 1 per 255; 1/250 would be 4/1000 or 0·004, so 1/255 must be a little less than 0·004, *i.e.*, a little less than 0·4 per cent.

50. Calculation of Relative Errors.

The relative error of a measurement does not usually need to be calculated with any very great care. Where numbers are as different as 6 in tens-of-thousandths place (10-mile survey, above) and 16 in units place (thickness of door) the location of the decimal point is really more important than the size of the significant figure that occupies either place; to call the former number "a few ten-thousandths of a per cent." and the latter "some ten or twenty per cent." gives all the information that is needed. This means that a calculation of relative error never needs to be done on paper but can always be worked out as a rough mental calculation. Thus, in the illustration given above, 1 foot is 1/5280 of 1 mile, hence it is 1/52800 of 10 miles, or roughly about 1/50,000; and 4 inches, being $\frac{1}{3}$ of 1 foot, is $\frac{1}{3}$/50,000 of the whole distance, or 1/150,000. The denominator of this fraction is about a sixth of a million (since 15 is about $\frac{1}{6}$ of 100), so 1/150,000=6/1,000,000=0·000006 =0·0006 per cent.

Decide mentally what per cent. 0·01 is of 7·23. Ans.: 0·0015, or 0·15 per cent.

What per cent. of 94·07 is 0·01 ?

51. "Decimal Places" versus "Significant Figures."

If a length is stated as 174·2 cm. the inference is that it is nearer to that exact amount than to 174·1 or 174·3 cm., namely that its error certainly is not as much as 0·1 out of 174·2. This is the same as saying that it is not as much as 1 out of 1742, or 1 out of nearly 2000, or 5 per 10,000, or 0·0005, or 0·05 per cent.

In the following table the left-hand column contains five numbers, all of which are carried out to the same number of decimal places; namely, two. In the right-hand column notice that the same five numbers occur, but each one of them is carried out to the same number (3) of significant figures, no matter how many decimal places there may be. If the accuracy of each of these ten numbers should be worked out in the way that has just been explained, would the numbers in the left-hand column turn out to be all of approximately equal accuracy while the right-hand column showed great accuracy for one number and little for another, or would the numbers on the right be the ones that would be about equally accurate while those on the left fluctuated? In other words, is accuracy a matter of decimal places or of significant figures?

7·23	7·23
94·07	94·1
0·52	0·522
428·00	428·0
66·67	66·7

The table has been repeated on the next page and shows the answer to the question. Beside each of the ten numbers its accuracy has been written down as a percentage, and it will be seen that the numbers in the right-hand column all show about the same degree of accuracy, while those in the left-hand column differ widely. Hence the number of decimal places to which a measurement is carried out has nothing to do with its accuracy. It is the *number of significant figures* that determines the matter.

0·1 %	7·23	7·23	0·1 %
0·01 %	94·07	94·1	0·1 %
2·0 %	0·52	0·522	0·2 %
0·002 %	428·00	428·0	0·2 %
0·01 %	66·67	66·7	0·1 %

Turn to the table in § 138, where errors are classified according to their size, and write the appropriate word opposite each of the percentages given in the right-hand column of the table on this page. Then do the same way with those in the left-hand column.

52. Rule for the Relative Difference of Two Measurements.

The difference between 3·11 and π (=3·14) is 3 out of 314, and the difference between 3·17 and π is likewise 3/314, not 3/317; *i.e.*, the numerical error is to be divided by the true or theoretical value rather than by the experimental or erroneous value. It is often desirable, however, to compare two values which are equally good, according to one's available knowledge of them. When there is no standard and no reason for choosing one of the measurements rather than the other, the accepted procedure is to *divide the difference by the greater value*. For example, the numbers 4 and 5 would be said to differ from each other by 20 per cent., not 25 per cent., for the difference divided by the greater number is one-fifth, not one-fourth. In cases of fairly accurate measurements it is unimportant whether the larger number or the smaller one is taken for the divisor, but for the sake of uniformity it is customary to choose the larger one.

Apply this rule to your two measurements of an irregular area (§ 23), obtained by counting squares and by constructing geometrical figures. How much relative difference is there in the results of the two methods?

53. Accuracy of a Calculated Result.

Multiply 65·97 by 24·15, using the abridged method of multiplication. Compare your product with that of example (*c*), § 13. It will be noticed that changing the fourth figure of one factor has produced a change in the fourth figure of the product. This means that if only three figures of the factor had been known, the fourth being uncertain, no calculation could have given more than three trustworthy figures of the product, because the fourth figure would have depended upon the unknown fourth figure of one of the factors. Likewise, in division, if only five figures are known of either the divisor or the dividend there is nothing to be gained by keeping more than five figures in the other one; only five figures of the quotient will have any meaning, and if further figures are obtained by any process of calculation they will be unjustified and misleading. The general rule will be obvious: the result of a multiplication or division will have no greater accuracy than that of the least accurate of the data from which it is obtained.

54. Accuracy of the Abridged Methods.

Remembering that the accuracy with which a quantity is expressed depends not upon the number of decimal places but upon the number of significant figures, and keeping in mind the fact that the number of trustworthy figures in a product is the same as the number in its least accurate factor, turn back to your notes on §§ 12 and 13 and observe that the method of abridged division automatically gives just the number of figures in the quotient that are needed if no figures of the dividend are "brought down"; and that abridged multiplication always gives at least as many as are in the shortest factor. It will not give any superfluous figures if the longer factor is used as the multiplier, but will give as many as the longer factor contains if that is used as the multiplicand. Of course the best method is to round off the longer number before beginning the calculation, so that it has no more figures than the shorter one.

55. Standard Form.

To avoid a long string of figures when writing very large or very small numbers it is customary to divide a number into two factors, one of them being a power of ten. Thus, 0·00000017 and 632,000,000,000 are the same as 17×10^{-8} and 632×10^{9} respectively. This notation also makes it possible to write 93,000,000 unequivocally with either two significant figures or four, as may be desired (see § 48), for it can be put either in the form $9{\cdot}3 \times 10^{7}$ or in the form $9{\cdot}300 \times 10^{7}$. The same value and accuracy for $9{\cdot}300 \times 10^{7}$ would be retained just as well by writing $93{\cdot}00 \times 10^{6}$ or $930{\cdot}0 \times 10^{5}$, but it is customary to choose the power of ten so that the other factor shall have just one significant figure to the left of the decimal point. The number is then said to be written in *standard form*.

Write the following numbers in standard form: 2946·3; 632×10^{9} (ans. $6{\cdot}32 \times 10^{11}$); 17×10^{-8}; 25·39978; 0·0073; 0·007300; 666·6; 0·001; 0·0010; 107·42; 186,000; 2·5400; 3·1416; 9·9942; 2·5433. Prove that each result is correct by performing the indicated multiplication.

Write a definition of standard form in your own words.

EXERCISES IV

1. What precautions did you take in order to measure the diameter of the brass disc with the greatest possible accuracy ?

2. Write your measured value of π, carrying it out just far enough to show one wrong figure. How many significant figures of $3{\cdot}141625$ are correct ? Of $3{\cdot}1424$?

3. State some of the possible causes that make your determination of π incorrect. Would there be any advantage in taking the average of several measurements of the circumference ? In rolling the disc through two or more consecutive revolutions and measuring the total distance ?

4. Is there any difference in meaning between the italicized statement at the beginning of § 47 and the one near the end of the same section ? If so, what ?

5. How many significant figures do you think there are in the length of the earth's quadrant as given in § 16 ?

6. Show that the last statement in § 52 is correct by taking some measurement that you have made as an example.

7. If 1 gm. is equal to 15·432 grains, how much is 5 gm. ? If a weight of 5 gm. is the same as 77·16 grains, how much is 1 gm ? (The answers 77·16 grains and 15·432 grains are both wrong.)

8. How is it that a metre can be measured more accurately by simple methods (§ 16) than a centimetre ?

9. Each side of a square measures 82·5 mm. How many centimetres long is its entire periphery ? (" Just 33 cms." is not the correct answer.)

10. Use the abridged method for multiplying 12 by 13, and for multiplying 13 by 12. Why does the answer come out 16 each time ? How should it be pointed off ? How many of its figures are significant ? If you wanted to obtain three significant figures in the product, using the abridged method, what would you do ? (Answer: Use three significant figures for both factors, 12·0 and 13·0. Try it.)

11. In question 9 how many significant figures did you keep in the product of $82{\cdot}5 \times 4$? How many are you entitled to keep ? Does the 82·5 mean the same as 82·50 ? Does the 4 mean the same as 4·0 ? As 4·000 ?

12. Turn to the table in the Appendix where the density of water is given. Does water at a temperature of 4° C. have a density of 1 ? Of 1·000 ? Of 1·0000 ? Of 1·00000 ? Correct the following statement by crossing off the unjustifiable figures, but keep all that are correct : " at ordinary room temperatures water has density of 1·00000."

13. In § 31 why were you justified in adding ciphers *ad libitum* to the number 180 ?

14. Describe briefly the so-called " contracted methods " of arithmetic with simple examples ; illustrating among other things the fixing of the position of the decimal point. What are significant figures and why do we pay so much attention to them ?

15. What are meant by significant figures ? Explain the importance of them in your calculations. How many significant figures are there in : 426·25 gm.-cms. ; 42625×10^{-2} gm.-cms. ; 4262·5 gm.-mms. ; 0·0042625 gm.-kms. ?

16. The value of $\pi = 3{\cdot}14159265358979$ . . . Find the percentage error made by taking the following approximations :—

(*a*) $\pi = \frac{22}{7}$; (*b*) $\pi = \frac{355}{113}$; (*c*) $\pi = \sqrt{10}$; (*d*) $\pi^2 = 10$; (*e*) $\sqrt{\pi} = \frac{7}{4}$.

17. The value of e is given by the following series :—

$$e=1+1+\frac{1}{1.2}+\frac{1}{1.2.3}+\frac{1}{1.2.3.4}+\ldots (ad\ inf.)$$

Find what percentage error is made by taking (*a*) the first three terms only ; (*b*) the first four ; (*c*) the first five.

18. The resistance, R, of a platinum thermometer at 0° C. = 592·79 units, and at 100·04° C. 822·16 units. Assuming $R_t=R\,(1+at)$, where R_t is the resistance at t° C., find a. At what temperature is R = 640 units ?

19. The resistance, R, of a platinum thermometer is connected with the temperature, t, by the formula $R_t=R_0(1+at+bt^2)$, where a and b are constants.

R_0 = 347·1 units.
R at 99·9° C. = 451·6 units.
R at 444·2° C. = 784·8 units.
Find a and b.

CHAPTER V

LOGARITHMS

56. Definition.

In the equations in the accompanying table the constant 10 is called a "base" and any exponent is called the "logarithm" of the right-hand side of the equation; thus 3 is said to be the logarithm of 1000, or as it is usually abbreviated,

$$3 = \log 1000.$$

57. Properties of Logarithms.

Consider the equations

$$\begin{aligned} 10^6 &= 1000000 \\ 10^5 &= 100000 \\ 10^4 &= 10000 \\ 10^3 &= 1000 \\ 10^2 &= 100 \\ 10^1 &= 10 \\ 10^0 &= 1 \\ 10^{-1} &= 0{\cdot}1 \\ 10^{-2} &= 0{\cdot}01 \\ 10^{-3} &= 0{\cdot}001 \end{aligned}$$

Notice that $\log 100 + \log 1000 = 2 + 3 = 5 = \log 100000$. This suggests the general relationship

$$\log a + \log b = \log (ab) \quad . \quad . \quad . \quad . \quad . \quad (1)$$

Test also with numerical values the following equations:—

$$\log a - \log b = \log (a/b) \quad . \quad . \quad . \quad . \quad . \quad (2)$$

$$n \times \log a = \log (a^n) \quad . \quad . \quad . \quad . \quad . \quad (3)$$

$$(\log a)/n = \log \sqrt[n]{a} \quad . \quad . \quad . \quad . \quad . \quad (4)$$

These four equations express the fundamental principles that in the use of logarithms "addition takes the place of multiplication, subtraction of division, multiplication of raising to a power, and division of root extraction."

58. Common Logarithms.

The advantage of using 10 as a base is that

$$\begin{aligned} \log (10 \times a) &= \log 10 + \log a \qquad \text{(from equation 1)} \\ &= 1 + \log a \end{aligned}$$

and in general

$$\log(10^n \times a) = n \log 10 + \log a \qquad \text{(from equation 3)}$$
$$= n + \log a;$$

for example, log 365=log (10² ×3·65)=2+log 3·65. Accordingly tables of common logarithms are made out only for natural numbers between 1 and 10, the logarithms of all other numbers being self-evident from these.

nat. no.	*log.*
1	0·000
2	0·301
3	0·477
4	0·602
5	0·699
6	0·778
7	0·845
8	0·903
9	0·954
10	1·000

TABLE OF LOGARITHMS.

The logarithms of the natural numbers from 1 to 10 are given here as far as the first three decimal places.

If the logarithm of 3·65 is 0·562 what is the logarithm of 3650? Ans.: 3·562. What is log 365? Log 36·5? Log 0·365? Ans.: −1+0·562. Log 0·0000365? Ans.: −5+0·562. (Do not simplify these binomial forms. They are easier to use if left as they are.)

The logarithms of numbers other than powers of 10 are in general incommensurable and a logarithm (sometimes called the *mantissa*) is dependent only upon the arrangement of significant figures in the natural number; *e.g.*, log 36500=4·5623; log 0·0365=−2+0·5623. The integral part (sometimes called the *characteristic* of the logarithm), however, is determined only by the position of the first significant figure; for example, for any number beginning in the tens-of-thousands place it is 4, and for any number beginning in the hundredths place it is −2.* In order to save space a number like log 0·0365 is customarily written in the form $\bar{2}$·5623, the minus sign being written *over* the characteristic to indicate that it applies only to the whole number while the decimal part is always positive.

Write −2·60 in logarithmic form. Ans.: −2·60=−2−0·60 =−3+0·40=$\bar{3}$·40. Write −1·4377 with the decimal part positive. Ans.: $\bar{2}$·5623.

Write the characteristic of the logarithm of each of the following: 5441; 27; 79264; 264; 73; 0·73; 0·073; 0·000073. Make up a rule for finding the characteristic of the logarithm of any number, and write it in your notebook.

Write the logarithms of 984; 982; 981; 980; 98; 9·8; 0·98; 0·098; 7; 14. Add the last two and find their sum in the body of the table; see what number in the margin corresponds to it, and verify the result by multiplying 14 by 7.

*** In other words, for a number greater than unity the characteristic is one less than the number of figures (including any ciphers) to the left of the decimal point. For any number less than unity it is one more than the number of ciphers between the decimal point and the first significant figure and is negative.**

59. Use of a Table of Logarithms.

A four-place table of logarithms is in general satisfactory for obtaining the logarithm of a number that has four significant figures; for numbers of three significant figures it is not necessary to keep more than three decimal places of the logarithms; for five-figure accuracy a five-place table is needed, etc. A four-place table is generally made more compact by including only three-figure values for the natural numbers, and when the logarithm of a four-figure value is required it is found by a process called *interpolation*, in which it is assumed that small differences between logarithms are proportional to the corresponding differences in their antilogarithms. Turn to the table and notice that (log 633 − log 632) is exactly one-third as large as (log 635 − log 632); and of course the differences in the numbers 633 − 632 and 635 − 632 are in the same ratio.

Suppose the logarithm of 3·142 is required: The table gives log 3·14 and log 3·15. The required number 3·142 is larger than 3·140 but smaller than 3·150; in an orderly scale of numbers it would be located *just one-fifth of the way from 3·14 to 3·15*. The assumption is, accordingly, that its logarithm *likewise is situated one-fifth of the way* from log 3·14 to log 3·15. Log 3·14 = 0·4969; log 3·15 = 0·4983. One-fifth of the distance from 0·4969 to 0·4983 is obtained by first finding that distance—subtracting; 4983 − 4969 = 14;—then adding the required fraction of it to the lower logarithm—1/5 of 14 = 3 (the nearest whole number), and 4969 + 3 = 4972; ∴ log 3·142 = 0·4972.

In order to save work some logarithm tables contain on the right-hand side the proportional parts already worked out. Thus, if you look along the line 31 and under the column 2 you will find the number 3.

Suppose the number is required whose logarithm is 0·2753: Turn to the table and notice that the nearest logarithms are 2742 and 2765. Their difference (called the *tabular difference*) is 23, and the given logarithm 2753 is 11 larger than 2742; *i.e.*, it lies 11/23 of the way from 2742 to 2765. Accordingly the required anti-logarithm will be 11/23 of the way from one marginal number (188) to the other (189); that is, it will be $188\frac{11}{23}$, or 188·5. As the characteristic of the given logarithm is zero this should be pointed off as 1·885.

In order to save work tables of anti-logarithms have been compiled and these are also supplied with columns of proportional parts. Thus, if we find 0·275 in an antilog table we see 1884, and looking along the same line we find in the column under the number 3 at the top of the page the number 1. Adding this to 1884 we get 1885.

Exercise.—Find log 2·718. Ans. 0·4343. Find log 3·333; log 1·234; log 12·34; log 123·4; log 8888; log 0·4343; log 3449.

60. Common and Natural Logarithms.

Logarithms whose base is 10 are called "common" logarithms. For theoretical purposes "natural" logarithms are sometimes used. Their base is the number

$$2{\cdot}71828\ldots=1+1+\frac{1}{1.2}+\frac{1}{1.2.3}+\frac{1}{1.2.3.4}+\ldots \quad (\textit{ad inf.})$$

$$=\mathrm{Lt}_{n=\infty}\left(1+\frac{1}{n}\right)^{n}*$$

This number is denoted by e or ϵ (epsilon).

To avoid confusion the base is often written as a subscript; thus, $\log_{10} 100=2$ and $\log_e 100=4{\cdot}6052$ mean exactly the same thing as $10^2=100$ and $e^{4\cdot6052}=100$.

If the log of a number A to base 10 is y and to base e is x, then

$$A=10^y=e^x$$

Let the log of e to base 10 be μ, so that $10^\mu=e$.

Then $10^y=10^{\mu x}$ or $\mu x=y$, or $x=\dfrac{y}{\mu}$.

The tables give $\mu=0{\cdot}4343$; $\therefore\ x=y/0{\cdot}4343=2{\cdot}303y$.†

Therefore to get the log of a number to base e, multiply the log of the number to base 10 by $2{\cdot}303$.

Logarithm tables to base 10 are found in many text-books and tables, but logarithms to base e are not given so often. More often the value of e^x or e^{-x} are given for stated values of x. See tables at the end of the book.

61. The Probabilty Function.

In much of the student's later work it will be important to know how e^{-x^2} varies when x is given different numerical values. Before substituting any particular number for x it will be advisable to proceed as follows: Let the function e^{-x^2} be denoted by y; then

$$y=e^{-x^2};$$

taking logarithms to base 10 of each side this becomes

$$\log y=-x^2 \log e,$$

from which it follows that

$$-\log y=x^2 \log e$$

Taking logarithms of each side again

$$\log(-\log y)=\log(x^2)+\log(\log e)$$

or

$$\log(-\log y)=\log(x^2)+\bar{1}{\cdot}6378$$

* See a text-book of Algebra. † More precisely $0{\cdot}43429$ and $2{\cdot}30259$.

In the last equation it is easy to find the value of y that corresponds to any given value of x. The procedure is first to calculate x^2 and find its logarithm; then add $\bar{1}\cdot6378$. The result is stated by the equation to be the logarithm of an expression enclosed in parentheses, so the numerical value of that expression is easily found by the process of obtaining an antilogarithm. Then, when the value of $(-\log y)$ has been obtained it is easy to write the value of $(+\log y)$, and from the value of $\log y$ there is no difficulty in finding y itself.

When such a determination is to be made for several different values of x it is convenient to arrange the various calculated quantities in the form of a table. Using the last of the equations that was derived from $y=e^{-x^2}$, and the tables of squares and of logarithms in the Appendix, find in succession the values of x^2, $\log x^2$, $\log x^2+\bar{1}\cdot6378$, $\log(-\log y)$, $-\log y$, $\log y$, and y, for each one of the following values of x: 0, 0·2, 0·4, 0·6, 0·8, 1·0, 1·2, 1·4, 1·6, 2·0, 3·0, 4·0, 5·0. On the next unused left-hand page of your notebook tabulate the results in columns headed x, x^2, $\log x^2$, etc., carrying each line clear across the table, as shown in the illustration, before beginning the next line.

x	x^2	$\log x^2$	$\log(x^2) + \bar{1}\cdot6378$	$\log(-\log y)$	$-\log y$	$+\log y$	y
0·0	0·00	$-\infty$	$-\infty$	$-\infty$	0·0	0·0	0·1
0·2	0·04	$\bar{2}\cdot6021$	$\bar{2}\cdot2399$	$\bar{2}\cdot2399$	0·0174	$\bar{1}\cdot9826$	0·961
0·4	0·16	$\bar{1}\cdot2041$	$\bar{2}\cdot8419$	$\bar{2}\cdot8419$	0·0695	1·9305	0·852
0·6	0·36	$\bar{1}\cdot5563$	$\bar{1}\cdot1941$	$\bar{1}\cdot1941$	—	—	0·698
0·8	0·64	—	—	—	—	—	—
—	—	—	—	—	—	—	—
—	—	—	—	—	—	—	—

TABLE CONSTRUCTED WHEN FINDING THE VALUES of e^{-x^2}.—It is important that the table be filled out line by line, not column by column.

Leave the opposite right-hand page of the notebook vacant until directions for using it are reached in another chapter.

Check the last column by reference to a book of tables * which tabulates e^{-x^2}.

EXERCISES V

1. Find the numerical values of $(1+\frac{1}{10})^{10}$; $(1+\frac{1}{100})^{100}$; and $(1+\frac{1}{1000})^{1000}$.
2. If $\log 10001 = 4\cdot00004343$ find the value of $(1+\frac{1}{10000})^{10000}$.
3. Find the natural logarithm of π.
4. Use the algebraical fact that $10^x 10^y = 10^{x+y}$ to prove equation 1 of § 57, which was merely stated without proof.

---

* *E.g.*, The Appendix, also the Smithsonian tables.

5. Turn back to the numbers that you have written in standard form (§ 55) and write the characteristic of the logarithm of each of them. What relationship can be observed ?

6. Write down any number that has three significant figures. Find its logarithm from the table, subtract it mentally from 0 (=log 1), and find the antilogarithm in order to obtain the reciprocal of the number that was written down. The mantissa is all that is needed for each logarithm, as the answer can be pointed off by inspection. For example, suppose the reciprocal of π is required : log 3·14=497 ; working from left to right, subtract each figure of the logarithm from 9 except the last one, which is to be subtracted from 10 ; result, 503 ; antilog 503=318 ; pointed off, 1/3·14=0·318.

Find 1÷269 by using the logarithm table, but without writing down any figures. The answer must be approximately 0·04.

Find the reciprocal of a four-figure number (for example, π or e) in the same way.

7. Work out some of the problems given in the Miscellaneous Exercises at the end of the book until you become familiarized with logarithms, and have convinced yourself that their use effects a great saving of time.

8. Calculate by continued extraction of the square root each to four significant figures the values of $10^{\frac{1}{2}}$, $10^{\frac{1}{4}}$, $10^{\frac{1}{8}}$, $10^{\frac{1}{16}}$. Combine these by ordinary arithmetic to get $10^{\frac{3}{16}}$, $10^{\frac{5}{16}}$, $10^{\frac{3}{8}}$, $10^{\frac{7}{16}}$. Plot a curve between the values of these powers of 10 as ordinates and the indices as abscissæ. Calling the indices the logarithms of the corresponding numbers use the curve to find the logarithms of 1, 2, 3, 4, 5, 6, 7, 8, 9, 10. (Check with logarithmic tables.)

9. From $\epsilon^x = 1 + x + \frac{x^2}{\lfloor 2} + \frac{x^3}{\lfloor 3} + \frac{x^4}{\lfloor 4} + \ldots$ calculate to 5 significant figures the values of $\epsilon^{\frac{1}{10}}$, $\epsilon^{\frac{1}{5}}$, $\epsilon^{\frac{1}{4}}$, $\epsilon^{\frac{1}{2}}$, $\epsilon^{\frac{3}{4}}$, $\epsilon^{\frac{4}{5}}$, ϵ^1. Tabulate your answers. Do the same for negative exponents. Check your answers by testing if $\epsilon^{0\cdot 2} \times \epsilon^{0\cdot 8} = \epsilon^1 = \epsilon^{\frac{1}{4}} \times \epsilon^{\frac{3}{4}}$ and whether $\epsilon \div \epsilon^{0\cdot 5} = \epsilon^{0\cdot 5}$.

Put $y = \epsilon^x$, then $x = \log_\epsilon y$. Plot x horizontally and y vertically. Read off the logarithms to base ϵ of 1·5, 2, 2·5, 2·7, and compare with the values in the mathematical tables.

10. Radium decays according to an equation $M = M_0 e^{-\lambda t}$, where M_0=initial amount, and M=mass left after a time t. For radium $\lambda = 1{\cdot}1 \times 10^{-11}$ sec.$^{-1}$. How many years must elapse before the mass is only one-half its initial value ? If T is this time, what is the relation between T and λ ?

11. The growth of radium emanation in a radium solution originally exhausted of the emanation obeys the law $Q = Q_0(1 - e^{-\lambda t})$, where Q_0 is the constant amount which the value of Q finally reaches (show that this is so by putting $t = \infty$), and Q is the amount grown in a time t. If $\lambda = 2{\cdot}08 \times 10^{-6}$ sec.$^{-1}$, find the time in days for the emanation to grow to half its final amount ?

CHAPTER VI

SMALL MAGNITUDES

62. Negligible Magnitudes.

Suppose a ruler graduated in centimetres and millimetres is used to measure the side of a square, and by estimating tenths of a millimetre the length is found to be 2·87 cm. The mathematical square of this quantity is 8·2369 cm.^{2}, but it has already been seen (§ 53) that only three figures of this area can be trusted, because nothing is known about the fourth figure of the measurement from which it is derived. Accordingly, the square is said to have an area of 8·24 cm.^{2} Similarly, if one side of a square measures 1·03 cm., the measurement being correct to tenths of a millimetre but nothing being known about hundredths of a millimetre, then its area will be correctly expressed by the quantity 1·06 cm.^{2} and not 1·0609 cm.^{2}, and the volume of a cube that has this square for one of its sides will be 1·09 cm.^{3} (and not 109·2781 cm.^{3}). It should be noticed (*a*) that with an ordinary ruler it is impossible to measure a length of a few centimetres with an accuracy greater than is expressed by three significant figures ; (*b*) that the area or volume calculated from such data cannot be trusted further than its third significant figure ; (*c*) that the example just given suggests a remarkably simplified process of calculation where some quantity which is to be squared or cubed is *a little greater than unity*, namely : (*one*+*small amount*)^2= *one* + *twice small amount*, and (*one* + *small*)^3=1+3 (*small*).

Decide mentally, by induction, the value of (1·—0·02)^2 ; then prove the result in two ways : (*a*) expanding in accordance with the binomial theorem ; (*b*) squaring 0·98 by abridged multiplication.

The justification of the simplified process of raising a number which is approximately unity to a power will perhaps be made more evident by the following example :—

Suppose that a metal cube has been constructed accurately enough to measure 1·00000 cm. along each edge. If it should be brought from a cold room into a warm room a delicate measuring instrument might show that the change of temperature had increased each dimension to 1·00012 cm. and by unabridged multiplication it would be easy to prove that the area of each side was 1·0002400144 cm.^{2} and that the volume had become 1·000360043201728 cm.^{3}. If the most careful measurements make it just possible to distinguish units in the fifth decimal place then tenths of those units (represented by the sixth decimal place) would be impossible to measure, and the attempt to state not only tenths, but hundredths and thousandths of those units would be absurd. By noticing that the number 1·0002400144 differs from the value obtained by abridged multiplication (1·00024) by only a few thousandths of the smallest measurable amount we can see clearly why the area of a 1·00012-cm. square is and must be 1·00024 cm.^{2}. Similarly, the volume of the cube is neither more nor less than 1·00036 cm.^{3}, and the string of figures running out ten decimal places further is absolutely meaningless.

It will be noticed that the number 1·0002400144 is in the same form as $1+2x+x^2$, the square of $(1+x)$, where $x=0·00012$; also that 1·000360043201728 corresponds to the cube, $1+3x+3x^2+x^3$. This is an illustration of the fact that when dealing with objects of the real world which is evident to our senses, it may happen that a measured amount is so small that its higher powers, algebraically speaking, are relatively minute beyond all perceptive ability. Of course this must not be understood as meaning that the cube of a measurable length can ever be an impalpable volume; the cube of $1+x$ is even a larger number than $1+3x$; it is the *difference* in size between this "physical" value, $(1+x)^3=1+3x$, and the true mathematical value, $(1+x)^3=1+3x+3x^2+x^3$, that eludes perception on account of x^2 and x^3 being extremely small *in comparison with* x, which is itself minute.

63. Formula for Powers.

The examples that have been given above suggest that, if x is small enough, $(1+x)^2=1+2x$, $(1+x)^3=1+3x$, and in general $(1+x)^n=1+nx$. The matter can be tested by making use of the binomial theorem:

$$(1\pm x)^n=1\pm nx+\frac{n(n-1)}{1.2}x^2\pm\frac{n(n-1)(n-2)}{1.2.3}x^3+\ldots$$

This shows that $(1\pm x)^n$ is equal to $1\pm nx$ if x is so small that x^2, x^3, etc., are negligible, the only possible exception being in case n should be so large that it could counterbalance the small size of x and prevent the terms

$$\frac{n(n-1)}{1.2}x^2, \text{ etc.}$$

from becoming negligible. In the practical use of the formula, however, n is rarely larger than 2 or 3 while x is at most only a few hundredths and is usually very much smaller.

64. Properties of Deltas.

The small quantities which have been considered in this chapter are usually symbolized by the Greek letter δ (delta). It has been seen that an important property of deltas is given by the equa-

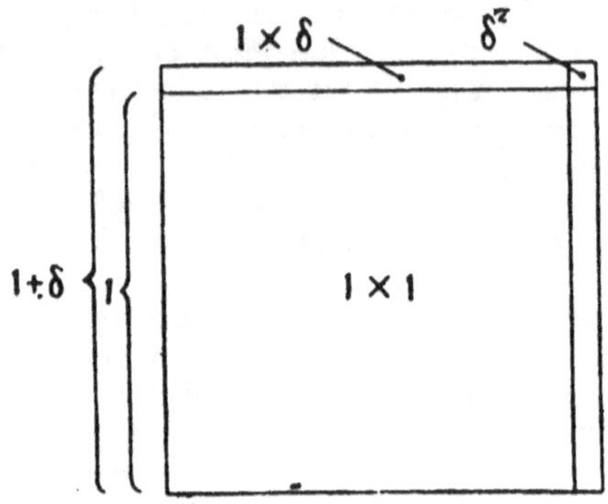

FIG. 13.—THE SQUARE OF $1+\delta$.

The fact that δ is small means that δ^2 must be very small. The error produced by taking the area $1+2\delta$ for the square of $1+\delta$ is only a minute fraction of the total area.

tion $(1\pm\delta)^n=1\pm n\delta$ (see Fig. 13), but the student is advised to learn this in the form

$$(1+\delta)^n=1+n\delta \quad . \quad . \quad . \quad . \quad . \quad . \quad . \quad (1)$$

with the understanding that δ may have either a positive or a negative value; for example, the square of $1+(-0\cdot03)$ is equal to $1+2(-0\cdot03)$ or $1-0\cdot06$, or $0\cdot94$.

Find the value of each of the following expressions mentally: $0\cdot99^2$; $0\cdot98^2$; $0\cdot98^3$; $0\cdot97^2$; $1\cdot00012^2$; $1\cdot00012^3$; $(1-0\cdot008)^2$; $0\cdot992^3$.

The ordinary process of algebraical division shows that $1/(1+x)$ $=1-x+x^2-x^3+x^4-x^5+\ldots$ From this it follows that

$$\frac{1}{1+\delta}=1-\delta \quad . \quad . \quad . \quad . \quad . \quad . \quad (2)$$

Divide 1 by $0\cdot997$. Ans.: $0\cdot997=1-0\cdot003$; $1/(1-0\cdot003)$ $=1+0\cdot003=1\cdot003$.

Find mentally the reciprocal of $1\cdot00012$.

Find $1/(1\cdot00012)^2$ mentally by using first formula (1) to simplify the denominator and then formula (2) to clear of fractions.

Find $(1\cdot00012)^{\frac{1}{2}}$ and complete the following formula for yourself: $\sqrt{1+\delta}=\ldots$

If a number is so small that its square is negligible it will similarly be true that the product of two such numbers will be negligible. For example, if a number that is carried out to thousandths place is $1\cdot007$ its square will not differ from $1\cdot014$ by a single thousandth:

$1·007 \times 1·007 = 1·014049$; likewise $1·007 \times 1·006$ will not differ by a thousandth from 1·013, its "exact" value being 1 013042, as the student can easily prove by considering it to be a special case of $(1+x)(1+y)=1+x+y+xy$. Accordingly,

$$(1+\delta_1)(1+\delta_2)=1+\delta_1+\delta_2 \quad . \quad . \quad . \quad . \quad (3)$$

This equation shows the advantage of keeping the signs positive and allowing the deltas to be either positive or negative. If the deltas were restricted to positive values there would be a liability to error unless the equation could be written $(1 \pm \delta_1)(1 \pm \delta_2)=1 \pm \delta_1 \pm \delta_2$.

Find mentally $1·04 \times 0·98$. Ans.: $0·98=1·-0·02$; $(1+0·04)(1-0·02)=1·02$.

Find mentally $1·03 \times 0·98$; also $1·00012 \times 0·99890$, and prove the latter by abridged multiplication.

In your notebook work out neatly the value of $1·0021 \times 1·0037$ in three different ways: (*a*) by unabridged multiplication; (*b*) by abridged multiplication; (*c*) by the use of deltas, writing the two numbers one under the other and adding merely the deltas on paper in the customary manner. Compare the three processes carefully and decide which process you will use in future.

Remembering that a/b is the same as $a \times (1/b)$ write the formula for $(1+\delta_1)/(1+\delta_2)$.

When $x<1$ your text-book of algebra proves that

$$\log_e(1+x)=x-\frac{x^2}{2}+\frac{x^3}{3}-\frac{x^4}{4}+\ . \ . \ .$$

When x is very small this reduces to $\log_e(1+x)=x$, or $\log_{10}(1+x)=0·4343x$. Look up your logarithm tables and confirm this by reading off the logarithms to base 10 of 1·001, and 1·002.

65. Deltas with nearly Equal Expressions.

If two quantities are nearly equal to each other the square root of their product (or geometrical mean) can be obtained by taking their average (or arithmetical mean). If the two quantities are denoted by a and $a+\delta$ (to indicate that their difference is a small magnitude) this statement can be proved as follows: The geometrical mean

$$=\sqrt{a(a+\delta)}=\sqrt{a \cdot a(1+\delta/a)}=a\sqrt{1+\delta/a}=a(1+\delta/2a)$$

since δ/a is also a small magnitude. But

$$a(1+\delta/2a)=a+\frac{\delta}{2}=\frac{a+(a+\delta)}{2}$$

$$=\text{the arithmetical mean} \qquad (4)$$

This fact is useful when making weight measurements.

Suppose an object appears to weigh m_1 when placed on one of the scale pans of a balance, and m_2 when on the other pan. Let x=its true weight and let the ratio of the arms of the balance be 1 to

$1+\delta$. Then by the upper figure of Fig. 14, $x \times 1 = m_1 \times (1+\delta)$, and by the lower figure $m_2 \times 1 = x \times (1+\delta)$. Eliminating $(1+\delta)$ we get $x/m_2 = m_1/x$, or $x = \sqrt{m_1 m_2}$. For example, a metal block weighed 15·19 and 15·23 gm. on the two sides of a balance. Its true weight is accordingly $\sqrt{15 \cdot 23 \times 15 \cdot 19}$ gm. Equation (4) shows that this is the same as 15·21 gm.

Weigh a set (say 16 ounces) of avoirdupois weights, considered as an unknown mass on a platform balance against the brass gram

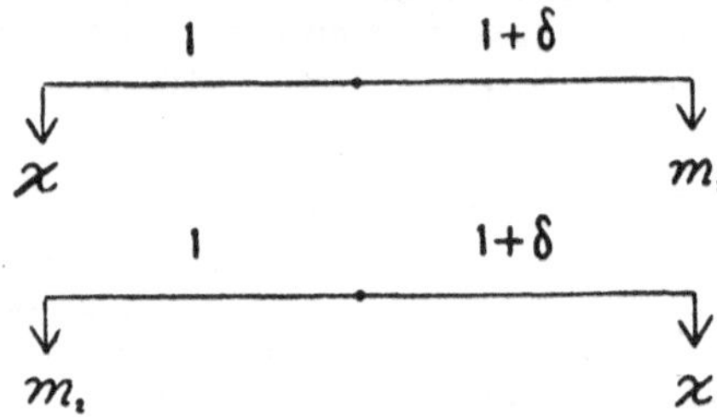

FIG. 14.—DIAGRAM OF A BALANCE OF UNEQUAL AREAS.

weights. Repeat the process on the other pan of the balance, remembering that the reading of the sliding weight on the arm of the balance is additive in one case and subtractive in the other. Find the true weight, in grams, of the avoirdupois set.

66. Properties of "Deltas" with Angles.

It will be almost self-evident from Fig. 15 that in the limit when δ is very very small

$$\frac{\tan \delta}{\delta} = 1 \quad \text{and} \quad \frac{\sin \delta}{\delta} = 1 \quad . \quad . \quad . \quad . \quad (5)$$

if it is remembered that an angle is measured by the ratio of arc to radius.

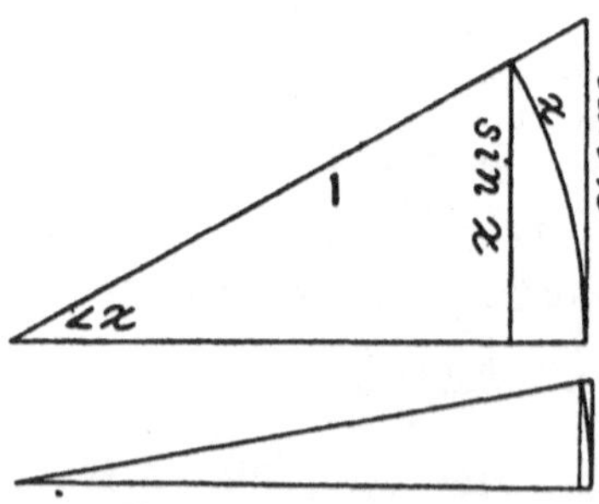

FIG. 15.—FUNCTIONS OF A SMALL ANGLE.

The tangent is always larger than the angle and the sine is always smaller, but the ratio of either function to the angle comes nearer and nearer to unity as the angle is made smaller and smaller. A formal proof of the fact that the limit of $\sin x/x$ = limit of x = limit of $\tan x/x$ when x becomes very very small can easily be worked out if two equal angles are juxtaposed so that their sine lines complete the chord of the double arc.

By consulting the table of circular functions find the largest whole number of degrees for which $\tan\delta=\delta$ to four decimal places. The numerical measure of each angle will be found in line with the number of degrees, but in the column headed RAD, an abbreviation of *radian measure*, which is another synonym for circular measure.

If an accuracy of three decimal places is all that is needed how large can δ be without differing from $\tan\delta$? Ans.: 6°; but not 7°. How large if δ is to equal $\sin\delta$?

67. Transformation of Operands.

It will be noticed that most of the rules that have been given for operating upon small quantities involve functions of $1+\delta$, not of δ alone. This means that factorizing, or some other process, is often necessary in order to put an expression into the form of $1+\delta$. For example, $\sqrt{50}$ is equal to $\sqrt{7^2+1}$, an expression which is not in the form of a function of $1+\delta$ but can be made so by dividing the binomial by an assumed factor 49: $\sqrt{49+1}=\sqrt{49(1+1/49)}$ $=7\sqrt{1+1/49}=7(1+1/98)$. Now it will be noticed that 98 is a little less than 100, and so

$$7\left(1+\frac{1}{98}\right) = 7\left(1+\frac{1}{100-2}\right) = 7\left\{1+\frac{1}{100(1{\cdot}0-0{\cdot}02)}\right\}$$

$$= 7\left\{1+\frac{1}{100}(1+0{\cdot}02)\right\} = 7(1{\cdot}0102) = 7{\cdot}0714$$

The first four figures of this result are correct. Extreme accuracy cannot be expected where a delta is as large as 1 or 2 per cent.; in most physical calculations it is much smaller than this.

68. Recapitulation.

The formulæ for calculation with small quantities are collected here for reference. Notice that the second and fifth are special cases of the first, and the first formula, *for n equal to a whole number only*, is a special case of the fourth,

$$(1+\delta)^n=1+n\delta$$

$$1/(1+\delta)=1-\delta$$

$$\frac{1+\delta_1}{1+\delta_2}=1+\delta_1-\delta_2$$

$$(1+\delta_1)(1+\delta_2)(1+\delta_3)\ \ldots\ =1+\delta_1+\delta_2+\delta_3\ \ldots$$

$$\sqrt{1+\delta}=1+\tfrac{1}{2}\delta$$

$$\sqrt{a(a+\delta)}=\frac{a+(a+\delta)}{2}=a+\delta/2$$

$$\frac{\tan\delta}{\delta}=\frac{\sin\delta}{\delta}=1$$

69. Rate of Increase.

If y is some function of x and x increases by a small amount y will also increase. If the increase of x be δx,* and the corresponding increase of y be δy, the value of $\frac{\delta y}{\delta x}$ is called the rate of increase of y with respect to x. More strictly this last expression is the value of $\left(\frac{\delta y}{\delta x}\right)$ when δx is made vanishingly small

(1) Suppose $$y=2x+4$$
Then $$y+\delta y=2(x+\delta x)+4$$
$$\therefore\ \delta y=2\delta x$$
or $$\frac{\delta y}{\delta x}=2$$

or the rate of increase of y with respect to x is always equal to 2.

(2) Suppose $$y=4x^2$$
Then $$y+\delta y=4(x+\delta x)^2$$
$$=4\{x^2+2x\delta x+(\delta x)^2\}$$
$$=4x^2+8x\delta x \text{ nearly}$$
$$\therefore\ \delta y=8x\delta x \text{ nearly}$$
or $$\frac{\delta y}{\delta x}=8x$$

or the rate of increase of y with respect to x is always eight times the value of x.

(3) Suppose a distance s feet is related to the time t seconds by the formula

$$s=40t+16t^2$$

Find $\frac{\delta s}{\delta t}$.

We have

$$s+\delta s=40(t+\delta t)+16(t+\delta t)^2$$
$$=40t+40\delta t+16t^2+32t\delta t+16(\delta t)^2$$
$$=40t+40\delta t+16t^2+32t\delta t \text{ nearly}$$
$$\therefore\ \delta s=40\delta t+32t\delta t$$
or $$\frac{\delta s}{\delta t}=40+32t$$

Students of mechanics will recognize $s=40t+16t^2$ as the formula for a body shot downwards with a velocity of 40 feet per second, and will also recognize $40+32t$ as the velocity of this body, whence they may draw a conclusion as to the meaning of $\frac{\delta s}{\delta t}$.

* In this notation δx does not mean $\delta\times x$, but simply a little increase in x.

(4) Suppose $y = \sin x$

then
$$\begin{aligned} y+\delta y &= \sin(x+\delta x) \\ &= \sin x \cos \delta x + \cos x \sin \delta x \end{aligned}$$

But if δx is a small angle $\cos \delta x = 1$ and $\sin \delta x = \delta x$ (x is supposed measured in radians)

$$\therefore y+\delta y = \sin x + \cos x \times \delta x$$
$$\therefore \delta y = \cos x \times \delta x$$

or
$$\frac{\delta y}{\delta x} = \cos x$$

or the rate of increase of $\sin x$ with respect to x is equal to the cosine of the angle x.

(5) Suppose $y = \log x$, the logarithm being natural, then $y+\delta y$ $= \log(x+\delta x) = \log\left\{x\left(1+\frac{\delta x}{x}\right)\right\} = \log x + \log\left(1+\frac{\delta x}{x}\right) = \log x + \frac{\delta x}{x}$ (cf. § 64).

$\therefore \delta y = \frac{\delta x}{x}$ or $\frac{\delta y}{\delta x} = \frac{1}{x}$, or the rate of increase of $\log x$ with respect to x is equal to the reciprocal of x.

Similarly the rate of increase can be found in other cases when the value of y is known in terms of x. Chapter VIII will give a geometrical method of treating the same problem.

EXERCISES VI

1. Find the value of $\frac{400}{797}$. (Suggestion: divide both numerator and denominator by 800 in order to convert the latter into $1-\delta$.)

2. Find the value of $0{\cdot}504 \times 0{\cdot}498$. (Suggestion: the first of these is a little more than $1/2$, *i.e.* is equal to $1/2$ multiplied by a little more than one; likewise, the second is a little less than $1/2$.)

3. The water equivalent of a calorimeter is its weight (w) multiplied by the specific heat (s) of the metal of which it is made. If $w = 264{\cdot}3$ gm. and $s = 0{\cdot}095$, find mentally the water equivalent to the nearest tenth of a gram.

4. Write the general formula for the value of the expression $(1+\delta_1)^m(1+\delta_2)^n(1+\delta_3)^p(1+\delta_4)^q \ldots$

5. In the equations of § 65 eliminate x instead of δ and show that $1+\delta = \sqrt{m_2/m_1}$. Then suppose that $m_2 = m_1 + \Delta$ and show that $1+\delta = 1+\Delta/2m$, and hence that $2\delta = \Delta/m$. If the two weighings of your avoirdupois pound (§ 65) differed by $0{\cdot}3$ per cent. how much do the two arms of the balance differ in length?

6. If $\sqrt{10} = 3{\cdot}162$ find the value of π^2. (Suggestion: $\pi = 3{\cdot}142$ $= 3{\cdot}162(1-2/316)$).

7. Use the δ notation to find,

(a) $\sqrt{50}$ to four significant figures.

(b) $\dfrac{1+0{\cdot}000018t}{1+0{\cdot}000182t}$ when $t=20$.

(c) $\sqrt[3]{8009}$ to five significant figures.

8. Find the value of :—

(*a*) The side of a square whose area, measured directly, is 4·00072 sq. cms.

(*b*) $\frac{18}{2{\cdot}9985}$.

9. When warmed up to ordinary room temperature the brass scale of a barometer is too long; as a result its reading falls short of the correct number of centimetres and needs to be multiplied by 1·00037. The mercury, however, expands at a different rate from the brass and tends to make the reading too great; to correct this error alone the reading must be divided by 1·00364. The correct height of the mercury will accordingly be obtained if the observed height is multiplied by 1·00037/1·00364. What is the percentage of difference between the corrected reading and the original reading? (Suggestion: call the observed height unity.)

10. If the observed height in Q. 9 is 754·3 mm. find the corrected height. (Don't multiply 754·3 by the decimal fraction containing several nines. Put the correction factor in the form $(1-\delta)$ and then subtract δ times 754·3 (which can often be done mentally) from 754·3.)

11. Work out the following calculations which have occurred in actual laboratory experience:

(*a*) $763{\cdot}3\ (1-0{\cdot}000167\times 12)$.

(*b*) Reduction of a barometer to 0° C.

$$H_0=764{\cdot}7\left(\frac{1+0{\cdot}000027\times 20}{1+0{\cdot}000182\times 20}\right)\text{ mms.}$$

(*c*) Reduction of a barometer:

$$H_0=30{\cdot}242\left(\frac{1+0{\cdot}000018\times 20}{1+0{\cdot}000182\times 20}\right)\text{ ins.}$$

(*d*) Calculation of the height of a barometer

$$H=761{\cdot}05\left(\frac{1+0{\cdot}000019\times 15}{1+0{\cdot}000182\times 15}\right)\text{ mms.}$$

(*e*) Calculation of a length

$$L=200{\cdot}34(1+0{\cdot}000017\times 90)\text{ cms.}$$

(*f*) Calculation of a coefficient of expansion

$$1+60c=\frac{19{\cdot}087}{18{\cdot}606}.\quad\text{Find } c.$$

(*g*) Deduction of coefficient of volume expansion, c, from the densities at two known temperatures.

$$\frac{2{\cdot}6001}{2{\cdot}5967}=1+50c.$$

12. The formula for the true mass of a body weighed in air

$$=w_1+w_1a\left(\frac{1}{D}-\frac{1}{d}\right)$$

where a, D, d are the densities of air, the body, and the weights respectively, and w_1=apparent "weight" of the body. If $w_1=17{\cdot}2625$, $a=0{\cdot}0012$, $D=2{\cdot}5$, $d=8{\cdot}4$, all in C.G.S. units, find the true mass.

13. The formula for the density of a body by an application of the principle of Archimedes is

$$D=\frac{w_1\Delta}{w_1-w_2}-\frac{w_2a}{w_1-w_2}$$

where w_1=weight in air whose density is a and w_2=weight in water whose density is Δ. For a penny $w_1=9{\cdot}464$ gms. and $w_2=8{\cdot}404$ gms., and $a=0{\cdot}00117$, $\Delta=0{\cdot}99823$ gm. per c.c. Find D.

14. Find the rate of increase of the area of a rectangle with respect to one side.

15. Find the rate of increase of

(1) The circumference of a circle;
(2) The area of a circle;
(3) The volume of a sphere;

with respect to the radius.

16. Find the rate of increase of $4x^2+3x+6$, $\sqrt{x}$, $\cos x$, $\dfrac{2x+3}{x+4}$ with respect to x.

17. If $s=128t-16t^2$, find the rate of increase of s with respect to t. If s is in feet and t in seconds, find when the rate of increase becomes zero.

18. Plot the values of $y_1=\theta-\sin\theta$, $y_2=\tan\theta-\theta$, against θ for angles rising by steps of 5° from 0° up to 40°. The abscissæ must be in radians. Find how far we can go with increasing values of θ before the values y_1 and y_2 reach 5 per cent. of θ. (Hint: Plot also $y=\frac{1}{20}\theta$.) (This exercise may be left until Chap. VIII. has been read.)

19. Between 0° C. and 30° C. the density ($=\rho$) of water varies with the temperature ($=\theta$° C.) according to the following equation:

$$\rho=0{\cdot}999868+0{\cdot}0000552\theta-0{\cdot}0000069\theta^2$$

Find the density at 0°, 10°, 20°, 30° C.

20. Without looking up a logarithm table find the logarithm of $\frac{1864}{1863}$ to base e and also to base 10.

21. If a exceeds b by a very small amount prove that $\log_e \dfrac{a}{b}=\dfrac{a-b}{b}$.

CHAPTER VII

THE SLIDE RULE

70. Addition by Means of Two Scales.

If two scales of centimetres should be laid parallel so that the zero of the second one coincided with the seventh division of the first (Fig. 16) it would be evident that the fifth division of the second would be opposite the twelfth division of the first. In other words, when a length of five has been added in this manner to a length of seven the result can be seen at a glance to be equal to twelve. Notice also that the arithmetical difference between scale divisions that lie opposite each other is everywhere the same, and that it is equal to the number on the first scale which is opposite the beginning of the second.

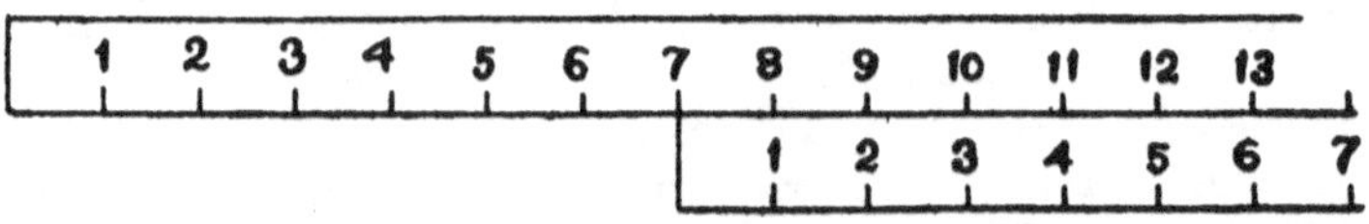

FIG. 16.—ADDITION WITH UNIFORM SCALES.

By starting afresh at 7 of the original scale and measuring 5 (or any other number, n) on the new scale we reach the distance of $7+5$ (or $7+n$, as the case may be) on the original scale. It should be noticed that this illustrates subtraction as well as addition: $12-5=7$, also the difference between any other pair of opposite scale numbers is equal to 7.

71. Multiplication with Logarithmic Scales.

If the same experiment is tried with two scales whose divisions are not a succession of whole numbers but are the logarithms of such a series the result will be different, for adding logarithms corresponds to multiplying natural numbers. Accordingly if we start at log 7 (Fig. 17) and measure a further distance of log 5 we

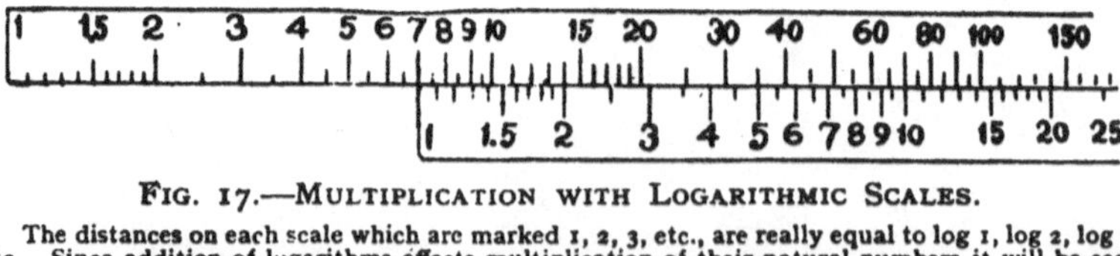

FIG. 17.—MULTIPLICATION WITH LOGARITHMIC SCALES.

The distances on each scale which are marked 1, 2, 3, etc., are really equal to log 1, log 2, log 3, etc. Since addition of logarithms effects multiplication of their natural numbers it will be seen that the number above the 5 of the lower scale is not $7+5$ as in Fig. 16, but is equal to 7×5, or 35. Since subtraction of logarithms corresponds to division it will be evident that the opposite pairs of numbers have their quotients equal to seven, instead of their differences.

shall come out with log 7+log 5, which is not equal to log (7+5) but to log (7×5). Notice not only that 5 on the second scale comes opposite 35 on the first, but also that quotients of corresponding numbers are everywhere equal to 7, just as differences are in Fig. 16; and furthermore, the upper scale in Fig. 17 forms a multiplication table (a seven times table in this case) for the numbers on the lower scale just as it did an addition table in Fig. 16.

72. The Slide Rule.

The apparatus called a *slide rule* is essentially a ruler containing a groove in which is a movable slide. Logarithmic scales are so marked that one of them (the "*A* scale") can be held stationary while another (the "*B* scale") can be placed in any required position below it (Fig. 18).

Two scales (*C* and *D*) along the lower edge of the slide can be used in the same way and can be read a little more accurately on account of their subdivisions being larger, but the two upper scales (*A* and *B*) will be found the most convenient for general use. Each one of them is really two complete logarithmic scales, as will readily be seen, but its right-hand half is often marked 10 to 100 instead of 1 to 10. In calculating with the aid of the slide rule it is advisable to ignore all decimal points while using the apparatus, and to point off the required answer by making a rough preliminary calculation mentally, as illustrated in § 11. It is then possible to use either the 7, for example, in the left-hand part of the *A* scale, or the 70 (or 7) in the right-hand part, whichever may happen to be the more convenient, in order to represent 7000, or 700, or 0·07, or any other number whose significant figure is 7.

By consulting Fig. 17 find the value of 70×20. Ans.: 1400. Find $3\frac{1}{2}\div5$. Ans.: 0·7, because it is evident that the quotient must be somewhat less than 1. Find the approximate value of 0·0007×1·4. Ans.: nearly 0·0010. Find 70×45. Ans.: as nearly as the value can be read from the diagram, it seems to be 32; since 70×40= 2800 the required product must be 3200, the ciphers not being significant. Two-figure accuracy is the highest that can be obtained from Fig. 17.

73. Reading a Logarithmic Scale.

The ordinary slide rule is usually constructed so carefully that three-figure accuracy is always attainable, although it requires the "estimation" (§ 41) of halves or fifths or even tenths of the graduated intervals. This is sufficient for the great majority of practical calculations in everyday life; if four-figure accuracy or greater is required it is usually more convenient to use logarithm tables, but a slide-rule makes a 3-figure logarithm table superfluous because it can be used with greater rapidity.

It will be observed, both in Fig. 18 and on the slide rule itself, that the distances from log 1 to log 2, log 2 to log 3, log 3 to log 4, etc., grow progressively smaller and smaller, so that those near the left-hand end of the scale can be more finely subdivided than those

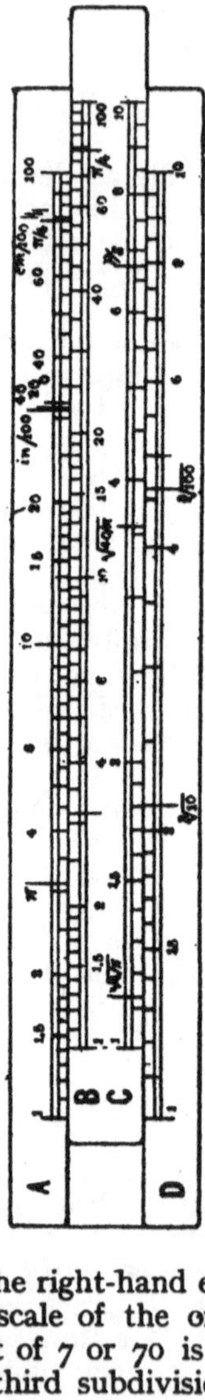

FIG. 18.—THE SLIDE RULE.

The essentials of the slide rule are four logarithmic scales arranged in two pairs and called *A*, *B*, *C*, and *D*, as shown in the diagrams even if not lettered on the apparatus.

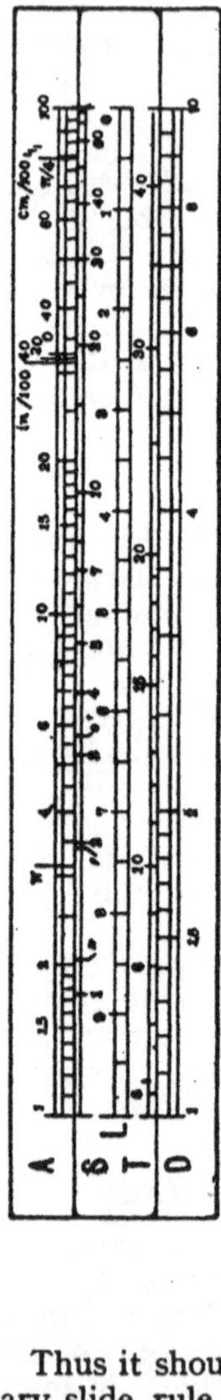

FIG. 19.—THE SLIDE RULE, SHOWING THE REVERSE SIDE OF THE SLIDE.

Opposite any angle on *S* is found its sine (on *A*); opposite any angle on *T* is found its tangent (on *D*). The numbers on *L* are the logarithms of those on *D*, but arranged backward so that they can be used when the *A*, *B*, *C*, and *D* scales are in their usual position.

toward the right-hand end. Thus it should be noticed that on the *A* or *B* scale of the ordinary slide rule the third subdivision to the right of 7 or 70 is 73, while on the *C* or *D* scale it is 715; but the third subdivision to the right of 1 represents 106 on the upper scales and 103 on the lower ones. To avoid mistakes in picking out the position of a given number it is best to find both the next larger single figure and the next smaller one; then look for the long 0·5 graduation which is about halfway between them.

Find 365 on the *A* scale of your slide rule by locating first 3, then 4, then 3·5, then 3·6 and 3·7, and finally 365. Find 235 in the same way; then by estimation set the transparent runner with its vertical line as nearly as possible at 238. Move the slide so that the end (1) of the *B* scale comes under this 238, and see if 210 comes just under 500. Set the transparent runner at 128; then at 106, and 109. Repeat the work of this paragraph using the *D* and *C* scales *respectively* instead of the *A* and *B* scales.

In reading a given position on the scale it is likewise best to read the next larger single figure and the next smaller one, also the halfway point between them.

Set the left-hand 1 of the *B* scale at random somewhere between 1 and 2 of the *A* scale. Then read its position accurately; also the positions of 2, 3, 4, 5, 6, 7, 8, and 9. On some slide rules the subdivisions between 1 and 2 of the *C* and *D* scales are marked 1, 2, 3, . . . instead of 1·1, 1·2, 1·3, etc. If this is the case with your slide rule it will be necessary, when looking for a number such as 7, to see that it belongs in the series that runs 7, 8, 9, 10, out to the right-hand end of the rule, and not in the series that runs merely from 1·0 to 2·0.

74. Multiplication.

To find the product of two numbers with the slide rule either end of the *B* scale should be set opposite one factor on the *A* scale, then the other factor on the *B* scale will be opposite the required product on the *A* scale. *Try this process with small numbers* by setting 1 on *B* opposite 3 on *A* and observing that the position of 2 on *B* gives the product 6 on *A*.

Multiply 2×4; 2×5; 2×6; 3×9; 7×8. Find 7×13, and 7×16, remembering that the two halves of the *B* scale (and also those of the *A* scale) are identical, so that if 16 (or 1·6) when taken on the right half of the *B* scale falls beyond the end of the *A* scale the same result can be read over 16 (or 1·6) on the left half.

Multiply 1·5 by 2; $1{\cdot}8\times2$; $1{\cdot}7\times2{\cdot}1$ (estimate the third figure of the product); 17·8 by 2; 1·79 by 2·53.

Find $0{\cdot}6\times183$ (caution: not $1{\cdot}6\times183$; see end of § 73); $7{\cdot}3\times1{\cdot}09$; $325\cdot\times106{\cdot}5$; $0{\cdot}073\times0{\cdot}0016$. If you have any difficulty in pointing off the last product master it before going any further. The use of " standard form " often makes the preliminary checking easier; thus in the last two examples given 3×10^2 multiplied by 1×10^2 gives 3×10^4 or 3000; and $(7+)\times10^{-2}$ multiplied by $(2-)\times10^{-3}$ gives 14×10^{-5}, or 0·00014.

75. Division.

In Fig. 16, above, notice that a length of twelve may be considered as being laid off from left to right, and then, beginning at the point 12, a length of 5 is laid off to the left, or subtracted from the original 12, giving a final result of 7. Similarly, in Fig. 17, log 35 minus log 5 equals log 7. The rule for division is accordingly: place the divisor on *B* under the dividend on *A* and read the answer on *A* over either end (or the middle) of the *B* scale. Such directions should never be memorized by the student, but he should practise the process until he becomes familiar with it, and in case of doubt should *experiment with small numbers* where the answer is obvious beforehand.

Divide 35 by 5; 30 by 5; 30 by 4; 11 by 4; 11 by 7; 11 by 10; 11 by 11; 11 by 12; 11·8 by 99·0; 114·0 by 3·4; 3·4 by 114.

76. Ratio and Proportion.

In Fig. 16 it will be seen that $8-1=9-2=10-3=7$. Since subtraction of logarithms corresponds to division of natural numbers Fig. 17 shows that $14\div2=21/3=28:4=35/5$, etc. That is, with the slide set in a given position any two opposite numbers are in the same ratio as any other two. *Test this on the slide rule with small numbers*; for example, set 6 under 2 and notice that 15 is under 5, for 6 : 2 : : 15 : 5. Solve the following proportions by setting the rule so that the answer is always found on the *A* scale.*

$$6 : 2 :: 15 : x,$$
$$2 : 6 :: 15 : x,$$
$$3 : 2 :: 9 : x :: 12 : y :: 10 : z$$

Solve 31 : 750 :: 0·005 : x; first notice that 750 is about twenty times as large as 31. Solve 2,300 : 0·36 :: 990 : x. If 26 inches=66 centimetres solve the following equation as accurately as possible: 26 : 66 :: 1 : x. What is the length in centimetres of 1 inch? From the slide rule find the *approximate* length of 4 inches. How many centimetres in 41 inches?

Set 1 precisely under the special mark that indicates π (=3·142) and notice that 7 is nearly under 22 because 22/7 is approximately equal to π. Look along the scales for another ratio which is equal to π, and try to find one that is not numerically equal to 22/7 or to 3·142857. . . . Then find its decimal value by *unabridged* division, carrying out the work until the figures become different from those of the correct value of π, as was done in § 44.

* To avoid uncertainty about which scale should be read it is advisable to get into the habit of setting the slide rule so that the answer will always be found on one of the fixed scales, *A* or *D*; not on either of the movable ones, *B* or *C*.

77. Equivalent Measures.

Turn to the back of the slide rule and find a statement about centimetres and inches. It will usually be given in tabular form, such as cm. : in. : : 26 : 66, together with various other data; if not, use the table given here. Find a statement of the relation

26 in. = 66 cm.	30 atmo. = 31 k/cm.^{2}
292 ft. = 89 m.	128 lb./in.^{2} = 9 k./cm.^{2}
35 yd. = 32 m.	500 lb./in.^{2} = 34 atmo.
87 mi. = 140 km.	
	340 ft.-lb. = 47 kgm.-m.
31 in.^{2} = 200 cm.^{2}	134 hph. = 100 kwh.
140 ft.^{2} = 13 m.^{2}	1 kwh. = 860 Cal.
19 f℥ (U.K.) = 540 cm.^{3}	$\pi = 355/113$
23 f℥ (U.S.A.) = 680 cm.^{3}	$\sqrt{\pi} = 39/22$
36 in.^{3} = 590 cm.^{3}	
22 gal. (U.K.) = 100 l.	$\angle 1 = 57° 17' 45''$
14 gal. (U.S.A.) = 53 l.	$= 3437'{\cdot}7$
	$= 206264''{\cdot}8$
108 gr. = 7 gm.	$1° = 1{\cdot}74533 \times 10^{-2}$
1940 av. oz. = 55 kgm.	$1' = 2{\cdot}90888 \times 10^{-4}$
22 lb. = 10 kgm.	$1'' = 4{\cdot}84814 \times 10^{-6}$

TABLE OF EQUIVALENTS FOR USE WITH THE SLIDE RULE.

The particular numbers are chosen so as to be accurate as far as three significant figures, at least. Thus not only does 26 in. equal 66 cm., but also 26·0 in. = 66·0 cm. The value of π is not only 355/133, but also 3550/1330, and 35500/13300, even to 355000/133000, as may easily be verified by the process of abridged division.

between pounds and kilograms and calculate your own weight in kilograms, remembering to set the slide rule so that the answer will be found on the *A* scale.

78. Reciprocals.

Set 4 or 3 or any other small number on *B* under 1 on *A*, and notice that above 1 (or 10 or 100) of the *B* scale will be found 0·250 or 0·333 or the reciprocal of whatever small number was used. Verify $1 \div 7 = 0{\cdot}142857$, as accurately as the apparatus will allow.

79. C and D Scales.

Experiment with any small number, using the *C* and *D* scales, and show where it must be set in order to find its reciprocal on the (fixed) *D* scale. Notice that if 8 on *C* is set over 10 on *D* the value of 1/8 will be found under 1, but if 2 is set over 1 the value of 1/2 will be found under the 10 instead.

Try multiplication, division, proportion, and equivalents on the *C* and *D* scales, remembering to set the slide so that the answer always comes on the stationary (*D*) scale. If the method for any of these processes has been forgotten *experiment with small numbers* so that you know the required answer beforehand. In multiplication if the answer runs off the end of the rule set 10 instead of 1 on C opposite the first factor, and in division read the result under 10 instead of under 1 when necessary. If the fourth term of a proportion cannot be read the runner must be placed over the end of the C scale and the slide then moved so that the other end of the C scale is under the runner. As an example, set 26 on C over 66 on D and show that the number of centimetres in 5 inches is 12·7.

80. Squares and Square Roots.

Set the slide so that the ends of the B and C scales coincide with the ends of the A and D scales. Then move the runner so that its vertical line falls on 9 of the C and D scales and notice that it also comes on 9^2 or 81 of the A and B scales. Set it at 8, 7, 6, 5, etc., on the lower scales and notice that the number on the upper scales just over n on the lower ones is always n^2. For a slide rule that has no runner set 1 on C over n on D and 1 on B will indicate n^2 on A. Read the square of 12; of 13; of 19. Find the (four-figure) square of 43, knowing that the last figure must be 9.

Under any number on the A (or B) scale will be found its square root on the D (or C) scale, but it must be remembered that any given arrangement of significant figures has two different square roots, according to how it is pointed off. For example, $\sqrt{1500}=40-$; $\sqrt{150}=12+$; $\sqrt{15}=4-$; $\sqrt{1{\cdot}5}\doteq 1{\cdot}2+$; $\sqrt{0{\cdot}15}=0{\cdot}4-$; $\sqrt{0{\cdot}015}=0{\cdot}12+$; etc. Notice that under 150 of the left-hand half of the A scale one of these roots, 122, is found; while the other root, 387, occurs under the 150 of the right-hand half. Since one of these is about three times as large as the other the simple precaution of making mentally a rough preliminary calculation of the root will avoid the possibility of obtaining the wrong number. For example, is the square root of 0·036 given by the significant figures 19 or 60, and how should they be pointed off?

81. Compound Operations.

A quotient such as m/n is found by setting n on the B scale under m on the A scale and reading the value opposite the end of B. (If this is not perfectly obvious *try it with small numbers.*) If m/n is further to be multiplied by some other number, say x, notice that the slide rule does not need to be re-set as it is already arranged so that the required product will be found over x of the B scale. Find two-sevenths of thirteen in this way and notice particularly that it is not necessary to read the value of the fraction 2/7; merely set 7 under 2 and find the required answer over 13.

In continued operations you will find it best to move the runner when multiplying and the slide when dividing. If the expression contains a number of factors in the numerator and denominator alternately divide by a factor in the denominator and multiply by a factor in the numerator instead of doing all the multiplications first, and finally the divisions.

Similarly, m^2n will be found on A opposite n of the B scale if 1 on C is set at m on D; and $\sqrt{mn}$ will be found on D under n on B if 1 on B is set to m on A.

A series of fractions like a/m, b/m, c/m, d/m, . . ., can be read off merely by setting the slide rule for $1/m$ and looking opposite a, b, c, d, . . . This is useful for finding percentages of the same maximum of a large set of figures.

There is usually a special mark for π on the left-hand half of the A and B scales, and for $\pi/4$ (or 0·7854) on the right-hand half. By placing the mark of the B scale opposite the end of the A scale the area of any circle will be found on B opposite the diameter on C.

82. Determination of Circular Functions.

In most slide rules the back of the sliding part is provided with three scales, which are named, and sometimes marked, S, L, and T, from above downward. If the slide be reversed and placed so that the ends of the S and T scales coincide with the ends of the A and D scales respectively (Fig. 18), the sine of any angle from 1° to 90° will be found on A opposite the number of degree and minutes on S; and the tangent of any angle from 6° to 45° will be found on D opposite the number of degrees and minutes on T. The decimal point is located by recalling the facts that $\sin 90°=1$, $\tan 45°=1$, and if two angles differ only slightly their sines (or tangents) will also be only slightly different.

By using the slide rule show that $\sin 70°=0·940$; write the values of $\sin 50°$; $\sin 30°$; $\tan 30°$; $\sin 11° 30'$; $\tan 11° 30'$; $\sin 6°$; $\sin 5°$; $\sin 1°$.

Since $\tan(45°+a)$ and $\tan(45°-a)$ are reciprocals of each other, tangents of angles greater than 45° are easily obtained from the slide rule. For example, to find $\tan 49°$, which is $\tan(45°+4°)$, set 41°, which is $45°-4°$, opposite 10 on D and read the required value, 1·15, on D opposite the left-hand of the T scale.

For sines less than 0·01 and tangents less than 0·1 different types of slide rule employ different methods, usually based on the formulæ $\sin\delta=\tan\delta=\delta$ (§ 66). If no special marks for angles are found on the C scale or the S scale use the equations for 1°, 1′, and 1″ in § 77.

83. Determination of Logarithms and Antilogarithms.

Set 1 on C opposite any number, n, on D, and $\log n$ will be found on the L scale opposite a special mark on the *back* of the slide rule. *Try this with small numbers* whose logarithms are already known; *e.g.* $\log 3=0·477$.

84. Cube Roots and Higher Roots.

To find the nth root of a number set 1 on the C scale opposite the number of the D scale. Turn the rule over and the logarithm of this number will appear on the L scale opposite the scratch at the right end of the scale. Prefix the characteristic, if any, divide the new number by n and shift the L scale until the mantissa of the quotient comes opposite the scratch. The number will then be on the D scale opposite the 1 on the C scale.

85. Other Exercises.

Many other exercises that can be done with the slide rule, such as the solution of triangles and of quadratic equations, will be found described in special manuals on the slide rule.

The slide rule is of no use unless the student continually practises with it. Now that logarithmic and trigonometrical tables are so common the chief use of the slide rule is to save time. Four-figure accuracy can be obtained on the C and D scales, but only three-figure accuracy on the A and B scales. The rule is very convenient if a large number of simple calculations of the same type have to be made. For example, in the calculation of percentages of a long series of numbers based on the same maximum, one setting of the rule enables one person to read off percentages almost as fast as another person can call them out and write the results down.

Try some of the problems in the Miscellaneous Exercises at the end of the book with the slide rule and compare the time taken with the times by ordinary arithmetic and bv logarithms.

Exercises VII

1. If decimal points are disregarded, one square root of a given arrangement of significant figures is stated (§ 80) to be about three times as large as the other. What is the exact value? Why?

2. Explain how it is that the process in § 82 for finding tan 49° really gives the *reciprocal* of tan 41.

3. Set 1 on the C scale over a tentative cube root of n on D and see whether $\sqrt[3]{n}$ on B comes under n on A. Practise this method of finding cube roots, *e.g.* to find $\sqrt[3]{64}$, take 64 on the right-hand scale of A, move the slider along until the same number is on B opposite 64 on A as is on D opposite the 1 on C. This number is the $\sqrt[3]{64}$. In what way would the marks $\sqrt[3]{10}$ and $\sqrt[3]{100}$ in Fig. 18 be helpful?

4. Discuss the relative merits of ordinary arithmetic, contracted arithmetic, 4-figure logarithm tables, and the slide rule in working out the numerical results from experimental data.

5. Describe the slide rule. Explain its method of working, its degree of accuracy and its limitations. When would you use in your calculations
(1) the slide rule,
(2) four-figure logarithms,
(3) seven-figure logarithms,
(4) the ordinary processes of arithmetic?
Give illustrative examples (these need not be worked out).

6. (*a*) Set the value of π on your slide rule (*A* and *B* scales) and looking along the rule find coincidences or "near" coincidences and so get a number of fractions whose values are nearly equal to π.

(*b*) Do the same with epsilon ($\epsilon = 2.71828$).

(*c*) Do the same with $\sqrt{2}$, $\sqrt{3}$, $\sqrt{5}$, $\sqrt{6}$, $\sqrt{7}$.

7. Find the

(*a*) squares of 15, 7.1, 132, 0.08.
(*b*) cubes of 15, 7.1, 132, 0.08.
(*c*) square roots of 784, 78.4, 7.84.
(*d*) cube roots of 243, 24.3, 2.43, 0.243.
(*e*) reciprocals of 4.2, π, 0.00367.
(*f*) the twelfth root of 2.

8. Work out with the slide rule:

(*a*) 3.45×5.67.
(*b*) 0.108×43.9.
(*c*) $108 \div 720$.
(*d*) $0.056 \div 0.0022$.
(*e*) $\frac{4}{9} \times \frac{18}{13} \times \frac{63}{37}$.

9. Find the percentage composition of potassium chlorate ($KClO_3$). Take K = 39, Cl = 35.4, O = 16. Try other chemicals. Check with a book on chemistry.

10. Work out with the slide rule the area of a circle whose diameter is 6.25 yards.

11. Work out with the slide rule the weight of an iron ball of diameter 6.0 inches (1 cub. ft. of iron weighs 450 lbs.).

12. Work out with the *slide rule* the following calculations (logarithms or ordinary arithmetic must not be used. Fix the decimal point by approximate calculation).

(*a*) $22.32 \times \dfrac{273+182}{273} \times \dfrac{960}{950}$

(*b*) $2 \times \dfrac{0.1066}{54.30} \times \dfrac{374}{273} \times \dfrac{760}{298} \times \dfrac{1}{0.0000899}$

(*c*) $0.054/(18.0 \times \frac{273}{294} \times \frac{737}{760} \times 0.0000899)$

(*d*) $J = \dfrac{3768 \times 552 \times 47.65 \times 981}{389 \times 5.94}$

(*e*) $\eta = \dfrac{\pi \times 981 \times 14.1 \times (0.051)^4}{8 \times 20.5 \times 83.5}$

(*f*) $t = 2\pi\sqrt{\dfrac{39.14}{32.2 \times 12}}$

13. (*a*) Out of a possible 175 marks the students obtained the following respectively, 21, 142, 37, 137, 49, 78, 125, 152, 163, 7, 141, 129, 135, 98, 159. Find, with the slide rule, their percentages to the nearest unit.

(*b*) Twelve students in a class get, out of a total of 179, the following marks: 162, 85, 98, 143, 37, 156, 14, 111, 133, 175, 27, 59. Find their percentages to the nearest unit.

(*c*) Six students obtained 173, 111, 206, 256, 96, and 101 marks respectively out of a total of 273. Find their percentages with the slide rule.

14. The volume of a cube is 269 cubic inches. Find its radius $\left(V = \frac{4}{3}\pi r^3 \quad \therefore \quad r = \sqrt[3]{\frac{3V}{4\pi}}\right)$.

CHAPTER VIII

GRAPHIC REPRESENTATION

A. Graphic Diagrams

86. A Graphic Diagram.

A *graphic diagram* consists of two lines of reference, called axes, a numerical scale along each of them and a point or assemblage of points or a line which corresponds to the numerical values which are to be represented (Fig. 21). The vertical distance of any point from the horizontal line or x-axis is called the ordinate of that point. It is denoted by y. The horizontal distance from the vertical line or y-axis is called the abscissa of the point. It is denoted by x.

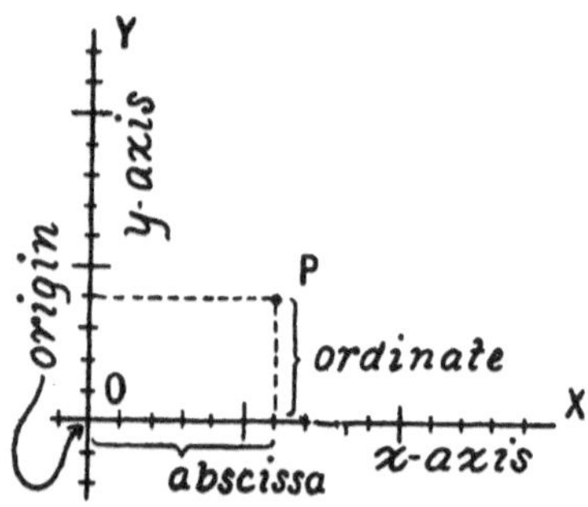

Fig. 20.

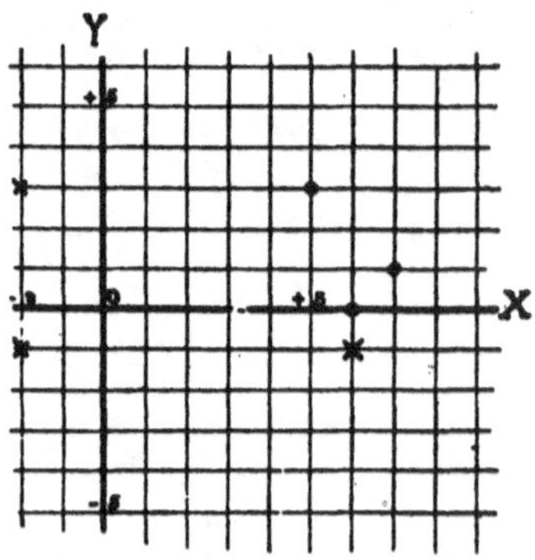

Fig. 21.—Points on a Graphic Diagram.

Considered collectively the two distances are called the co-ordinates of the point, and the point whose abscissa is x and whose co-ordinate is y is called the point (xy). The point whose co-ordinates are both zero, viz. the intersection of the two axes, is called the origin of the diagram.

In Fig. 20 the abscissa of P is 6, the ordinate is 4, and P is called the point (6, 4).

The co-ordinates may be positive or negative. The abscissa is positive if the point is to the right of the y-axis, negative if to the left. The ordinate is positive if the point is above the x-axis, negative if below.

In Fig. 21 the points represented by small dots have positive co-ordinates, those shown by crosses have negative values for one or both of their co-ordinates.

Exercise.—Show that the distance between the points (3, 4) and (7, 5) is equal to $\sqrt{(3-7)^2+(4-5)^2}$, and write a general formula for finding the distance between any two points, such as (x_1, y_1) and (x_2, y_2).

87. Choice of Scales.

When only a few points are to be plotted and the range of extreme values is not very great, it is advisable to use a normal scale of one

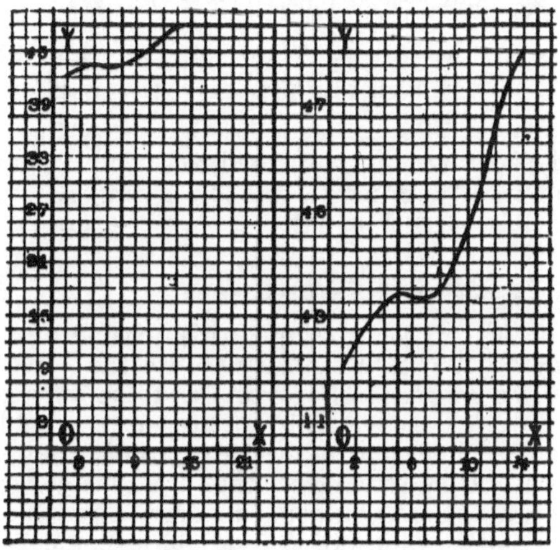

FIG. 22. FIG. 23.

FIG. 22.—GRAPHIC DIAGRAM WITH CONDENSED SCALES.

Notice that the x-values have been so condensed that the curve does not extend far to the right; and the y-values have been condensed to the same extent, making the curve relatively flat.

FIG. 23.—GRAPHIC DIAGRAM.

Notice how both scales have been arranged so that the same table as was illustrated by Fig. 22 now has its values much better displayed.

unit for each square of the ruled paper. Condensed scales are most useful for large numerical values, including data given in round numbers, and for keeping a diagram within the limits of a definite space. Expanded scales are necessary for minute values and those which need accurate representation. For example, when the width of a square is not less than one-tenth of an inch or two millimetres it is possible to divide it into tenths by estimation and to show a perceptible difference in position to correspond to a difference of one-tenth in numerical values, but values differing only by a number of hundredths or thousandths need an enlarged scale to make their variations show in the graphic diagram. In some cases the importance of the relation or ratio of x-value to y-value makes it necessary that both x-scale and y-scale shall be the same, whether normal or condensed or extended. In other cases the two sets of values represent measurements which are mutually incommensurable (as in the case of time and temperature), so that it is theoretically of no importance whether the two scales correspond or not. It is also not necessary to start either scale from zero. Where the slant of the curve, or of some important part of it, is fairly uniform, however, it is often more satisfactory to choose the scales in such a way as to make the slope approximately 45 degrees, as in Fig. 23.

Figs. 22, 23, show graphs plotted from the data:

x: 1, 2, 3, 4, 5, 6, 7, 8, 9, 10, 11, 12, 13, 14
y: 42, $42\frac{1}{2}$, 43, $43\frac{1}{4}$, $43\frac{1}{2}$, $43\frac{1}{3}$, $43\frac{1}{3}$, $43\frac{1}{2}$, 44, $44\frac{3}{4}$, $45\frac{1}{2}$, $46\frac{1}{2}$, $47\frac{1}{2}$, 48

Both graphs in Figs. 22, 23 are poor, in that the heavy lines of the squared paper are not numbered, and simple numbers are not used. It is far better in most cases to use simple sequences such as 0, 1, 2, 3, 4, . . ., or 0, 2, 4, 6, 8, . . ., or 0, 5, 10, 15, 20 . . .

88. General Principles of Plotting.

When one of two variable quantities changes as the result of changing the other it is customary to use the horizontal scale of x-values for the *independent* variable and the vertical scale of y-values for the *dependent* variable. Thus, for a graphic diagram of temperature and time, it is natural to consider the temperature as being the result of the time rather than the time as depending upon the temperature. Before starting to draw a graphic diagram the largest and smallest values of the variables should be observed, so that the axes can be located in a satisfactory position on the paper The next step should invariably be to *lay off a scale along each axis before plotting a single point*, and that *equal numerical intervals are always represented by equal distances* on any one scale, the (positive) x-values *always increasing from left to right*, and the y-values *from below upward*.

89. Representation of Tabular Values.

After the points corresponding to a set of tabular values have been located on a graphic diagram it is customary to draw a straight line from each point to the adjacent point on the left or right, making a broken line for the " curve " that shows the fluctuations in the value of the dependent variable. If all the points lie on a smooth curve it is advisable to draw such a curve as evenly as possible, but the experimental values given in a table are apt to show little irregularities which make it impossible to draw a smoothly flowing curve through their graphic points. In such cases the broken line serves the purpose of visually assembling the points in a series, but is not supposed to indicate that intermediate points, if obtainable, would lie exactly along the straight lines.

Exercise.—Draw a graphic diagram in which the horizontal scale represents the 24 hours of a single day, beginning at midnight and running through twelve o'clock noon, to the next midnight. For the vertical ordinates use the temperatures given in the adjoining table. Use a horizontal distance of one square to represent one hour and a vertical distance of one square to represent one-tenth of a degree. Connect each point with the next by means of a straight line, being careful not to omit any point. Notice how much more striking and "graphic" the diagram is than the table; how it shows at a glance facts that could be gleaned from the table only with much greater effort.

hour	*temperature* A.M.	P.M.
12	36·9°C.	37·4
1	36·8	37·4
2	36·8	37·6
3	36·6	37·5
4	36·4	37·5
5	36·5	37·6
6	36·6	37·6
7	36·8	37·7
8	36·9	37·6
9	37·1	37·4
10	37·2	37·4
11	37·2	37·2
12	37·4	36·9

NORMAL TEMPERATURE OF THE HUMAN BODY.

Exercise.—Study the different curves and graphs you see in the newspapers, Government bulletins and other pamphlets. Notice how the scales are chosen to emphasize the points at issue.

90. Smoothing of a Graphic Curve.

In the temperature diagram just made it is probable that the little irregularities of temperature are due to accidental causes and would not be exactly repeated in taking the temperature of another individual or even of the same individual on another day. In such cases a more typical picture is given by drawing a smooth, flowing curve in such a way that it passes through the midst of the scattering points and follows their general upward and downward trend without necessarily passing through most of them or even any one of them. To make a good curve use a long pencil and keep the wrist on the concave side of the curve so that the forearm may function as a radius.

It ought not to pass below most of the points, nor above most of them, but should leave them distributed, some above and some below, as evenly as possible, subject to the general condition that it must be a smooth curve.

Exercises.—1. In another place on the squared paper of the notebook plot merely the *points* as was done for § 89; then draw a *smooth* curve along their general course, outlining it tentatively with a light pencil mark, erasing and correcting this until it is satisfactory, and then tracing it plainly with ink. It should show not more than one downward loop and one larger upward swing.

2. What is the most probable value of the average temperature of a healthy individual at 4 p.m.? Would you prefer to decide the matter from the graphic curve of § 89 or from that of § 90?

This process, which is called *smoothing* a graphic diagram, is advisable only when one can be reasonably certain that the small fluctuations that are eliminated by the process represent unavoidable experimental errors or chance variations, or else that they are due to causes which are negligible in the case that is under consideration. Instances have occurred in which even able scientists have missed the discovery of important facts on account of the "smoothing out" of what have seemed to be only accidental irregularities.*

B. Curves and Equations

91. Graphic Representation of a Natural Law.

The graphic diagram which is obtained from a table of measurements or experimental data usually shows irregularities, which are sometimes retained by the use of a broken line and are sometimes eliminated by the process known as "smoothing" (§ 90). Neither

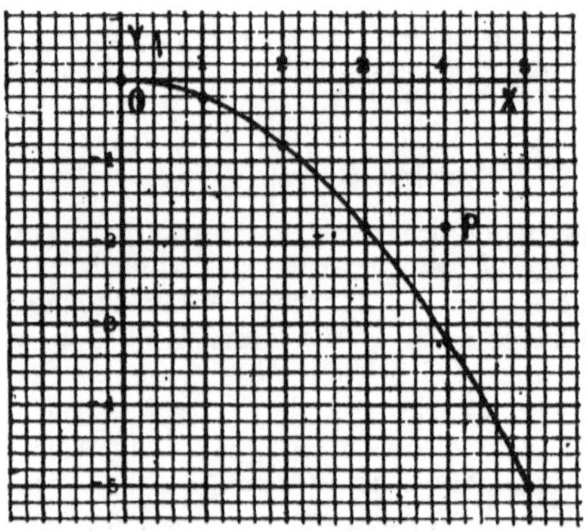

Fig. 24.—Parabolic Path of a Falling Body.

The total vertical movement is proportional to the square of the horizontal movement.

of these procedures is necessary if the variables follow some definite natural law in regard to their changes or if their relationship can be expressed by an equation. In such cases the plotted points in general will lie precisely along a smooth curve without showing

* Regnault, the great French experimenter in heat, thought that the vapour-pressure curve of ice was continuous with that of water and made one smooth curve link up all the readings. It is now known that at 0° C. the two curves should intersect at slightly different slopes, and an investigation of Regnault's diagram shows that his experiments had actually indicated this fact, but he had smoothed it out with his curve (see text-books of Heat under the heading, "Vapour Pressure").

any irregularities whatever. For example, a ball that is thrown horizontally will travel, under the influence of gravity, along a curved path in such a way that its progress in a horizontal direction during 1, 2, 3, 4, . . . n seconds will be proportional to the numbers 1, 2, 3, 4, . . . n, while its downward movement will be proportional to the numbers 1, 4, 9, 16, . . . n^2.* If points are so located on a graphic diagram that the vertical distance of each below the x-axis is proportional to the square of its distance to the right of the y-axis (Fig. 24) it will be found that a smooth curve can be drawn so as to pass exactly through all of them, and the relationship between the x-value and the y-value of each and every point will of course be given by the equation $y=-kx^2$. Such a curve is shown more or less steadily by the surface of water that forms a waterfall or by a jet of water that issues from a hose pipe, and is known as a *parabola*.

It will be seen that the scales have been so chosen in Fig. 24 that $k=1/5$, in other words the equation $y=-kx^2$ has become $y=-\frac{1}{5}x^2$. Test the diagram by assuming that x has the value of $3/2$, finding what the corresponding value of y must be by solving the equation, and then locating the point $(3/2, -9/20)$ which has these two numbers for its x-value and y-value.

92. Graph of an Equation.

A single equation that contains two unknown quantities has an infinite number of solutions, for any value whatever may be assigned to one unknown quantity and the corresponding value can always be calculated for the other. Any such solution of an equation that involves x and y will consist of an x-value and a y-value, and so can be represented by the position of a point. All of the infinite number of solutions of a given equation of this sort can theoretically be represented by an infinite number of points on a graphic diagram; thus every point that lies on the curve of Fig. 24 corresponds to two numbers, x-value and y-value, which are a solution of the equation $y=-\frac{1}{5}x^2$. In general, all of the points that represent the solutions of a given equation will be found to lie along a smooth curved or straight line, which is accordingly called the *locus* or " *curve* " *of the equation*. If the locus of an equation extends to an infinite distance, as is the case with the curve of $y=-\frac{1}{5}x^2$, its distant portions are usually more or less flat or straight and uninteresting, so that there is no objection to omitting them from the graphic diagram.

It should be carefully kept in mind that the relationship between an equation and its " curve " or locus means that *every solution of the equation gives a point which is located on the curve*, and that *every point on the curve furnishes a solution of the equation*. From these two statements it is evident that the co-ordinates of a point which is not on the curve will not satisfy the equation (try the point $(4, -9/5)$

* For the horizontal motion $x=ut$. For the vertical motion $y=-\frac{1}{2}gt^2$.

shown by the letter P in Fig. 24), and that an incorrect solution (try $x=2$, $y=+4/5$) will give a point which does not lie on the curve.

It is obviously impossible to calculate the position of an infinite number of points, but since an equation is known to give a smooth curve it is only necessary to plot enough points to prevent uncertainty as to the shape and position of any part of it.

Make a graphic diagram for the equation $y=x^3-x$ by the following process: Assume that x has various small integral values, positive and negative, and calculate the corresponding values of

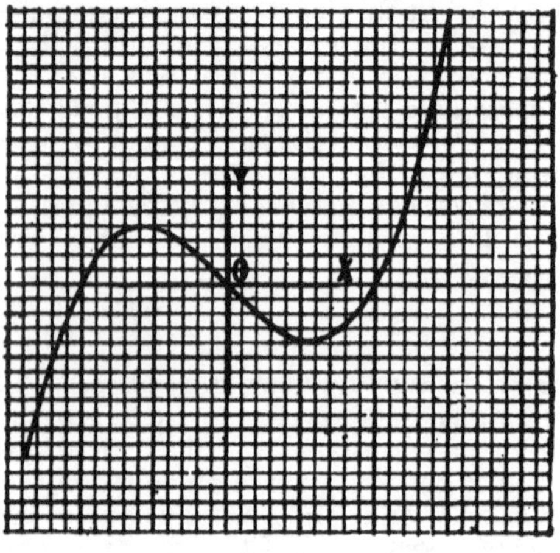

FIG. 25.—GRAPH OF THE EQUATION $y=x^3-x$.

x	y
0	0
$\frac{1}{2}$	$-\frac{3}{8}$
1	0
2	6
3	24
4	60
5	120
$-\frac{1}{2}$	$+\frac{3}{8}$
-1	0
-2	-6
-3	-24

TABLE OF SOLUTIONS OF THE EQUATION $y=x^3-x$.
The y-values increase so rapidly beyond $x=5$ that the curve becomes nearly straight.

y, arranging them in tabular form, as shown in the adjacent table. The x-values have not been carried beyond $+5$ on account of the y-values increasing so rapidly. Negative x-values are seen to have the same numerical y-values as positive ones but with the sign changed. The next step is to choose suitable axes and scales. Plot the points given by the table before proceeding further. It is not probable that the line has a straight portion between (1, 0) and (−1, 0) while the rest of it is curved, so it is necessary to calculate at least two more points on the curve, *e.g.* $x=1/2$ and $x=-1/2$; and these will be found sufficient to determine a smoothly flowing curved outline (Fig. 25).

98. General Procedure.

In plotting the curve of an equation the general procedure is to calculate the table of values, substituting successive positive and negative integral values of x (and fractional values if necessary) and solving for y; then consider the available space on the paper and draw the axes in a suitable location. Before locating the points that correspond to the tabular values a scale of numbers should always be marked along the x-axis and another along the y-axis. The two scales need not be the same, but along each axis, considered by itself, equal distances must always correspond to equal numerical differences, and x-values must always increase to the right and y-values increase upward. If a condensed scale is used it is always advisable to let the dimensions of a small square of the ruled paper represent a simple round number; if the first tabular value is 9,200 do not number the successive squares 9,200, 18,400, 27,600, etc., but use the simpler round numbers 10,000, 20,000, 30,000. Where the relative value of the x-unit as compared with the y-unit is unimportant the scales can often be advantageously chosen so as to give the chief part of the curve a slope of about 45°. Intersections of the graph with the lines of the squared paper can be read most accurately if this is done.

94. The Straight Line.

The equation $y=2x+3$ represents a sloping straight line. For such a simple equation it is hardly necessary to construct a table of values. Locate the x-axis and y-axis in your notebook where there is room for the y-values to extend approximately from $+10$ to -10 and for the x-values to extend at least to ± 5. Plot four or five points from solutions of $y=2x+3$ obtained mentally, and use a ruler to draw a straight line passing through them and extending a little distance beyond them in each direction. Label this line by writing the equation $y=2x+3$ alongside it, close to one end (Fig. 26). Without making a new diagram use the same axes and scales for plotting the following straight lines: $y=3-2x$; $y=3+\frac{1}{2}x$;* $y=3+0x$; $y=2x$; $y=-1+x$; $y=-1+\frac{1}{2}x$; $y=-1-2x$; $y=-1-x$ (Fig. 26). Label each of these in the same way. Examine the lines whose equations have 3 for the numerical term. Do they all pass through the point (0, 3)? Do the lines of the equations that have -1 for a numerical term all pass through the point (0, -1)? Which line passes through the origin (0, 0)? Does it seem probable that *the equation $y=a+bx$ gives a line that passes through the point* (0, a)? Test the point (0, a) by ordinary algebra in order to determine whether it lies on the line $y=a+bx$.

Test the lines you have plotted for a proof of the statement that if the equation of the line is $y=a+bx$ the coefficient of x (namely b) gives the slope of the line.

NOTE.—If the point P is (x, y) and an adjacent point Q is $(x+\Delta x, y+\Delta y)$, where Δx, Δy, are the small increases in x and y going from P to Q it follows that the slope of the line is $\Delta y/\Delta x$. This is obvious

* Notice that this equation, as it stands, is an explicit statement of the value of y; "clearing of fractions" before starting to substitute would only complicate the work.

from a diagram. It can also be shown by the method of § 69 that if $y=a+bx$, $b=\Delta y/\Delta x$.

The equation $y=a+bx$ is the general equation for all straight lines except those that run parallel to the y-axis. The latter are

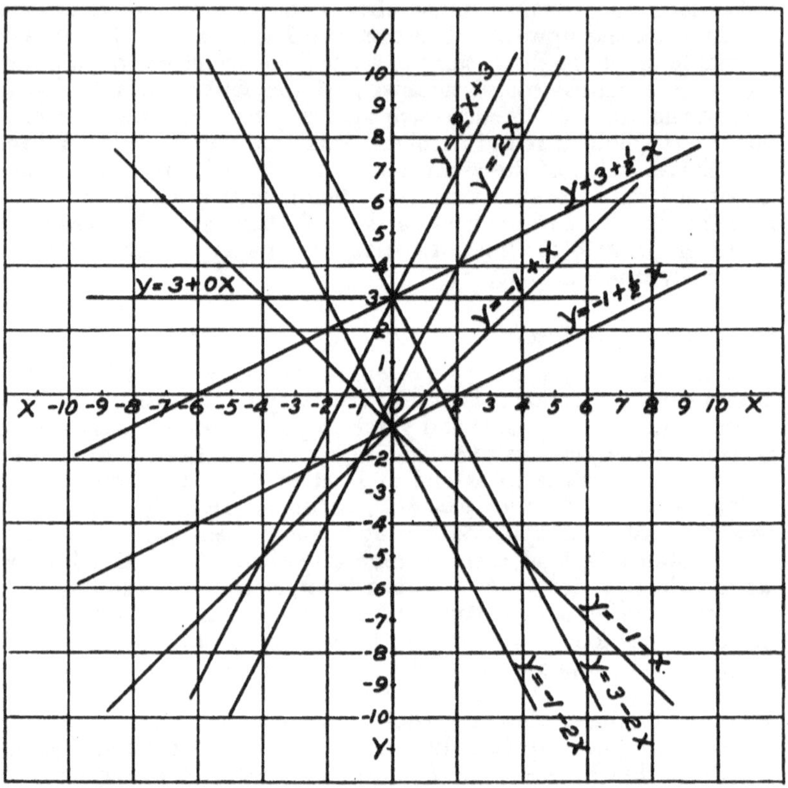

FIG. 26.

given by the equation $x=k$. The most general equation of the straight line is $Ax+By+C=0$. This can be reduced to the form $y=a+bx$ if B is different from zero, and to the form $x=k$ if B is zero.

If we take $Ax+By+C=0$ and divide through by C, we can get the equation in the form

$$\frac{x}{-C/A}+\frac{y}{-C/B}=1$$

which may be written:

$$\frac{x}{l}+\frac{y}{m}=1$$

This is called the intercept form of the equation, for l and m are the intercepts made by the line on the axes $OX . OY$. If the line is parallel to OY, m is infinitely great, and $\therefore x=l$ is the equation of such a line, as is otherwise obvious.

Exercise.—Throw the equation of the line $y=3-2x$ into the intercept form. Verify from the graph in Fig. 26.

95. The Parabola.

On a single graphic diagram draw and label the curves $y=x^2$, $y=x^2+12$, and $y=x^2/10$ (Fig. 27). These should be drawn from

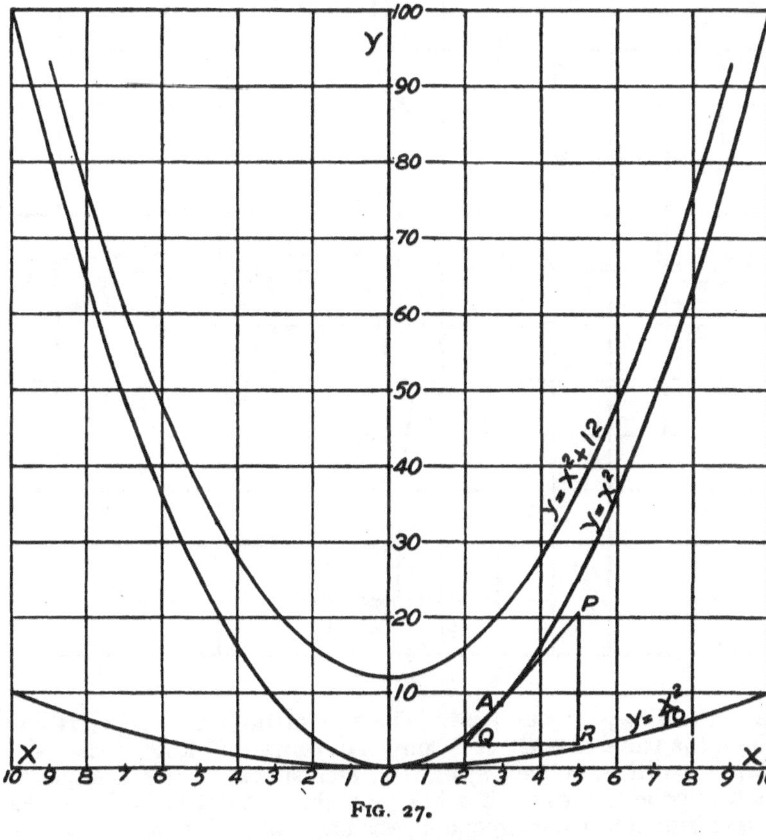

FIG. 27.

about $x=-10$ to $x=+10$. They are all similar figures and are called *parabolas*; the curves $y=x^2$ and $y=x^2/10$ differ in size but the portion of the latter that extends from -10 to $+10$ is of exactly the same shape as the part of the former that is included between $x=-1$ and $x=+1$. Notice that the curve $y=x^2+12$ differs only in position from $y=x^2$; it appears to be the same curve moved twelve units higher up, and the equations make it clear that for each x-value the y-value of the first curve is greater by twelve than that of the second.

On another graphic diagram draw and label the three curves $y=x^2-3x$, $y=x^2-6x$, and $y=x^2-7x$. Notice that they are not only of the same shape but of the same size as well.

The general equation of all parabolas whose axis of symmetry

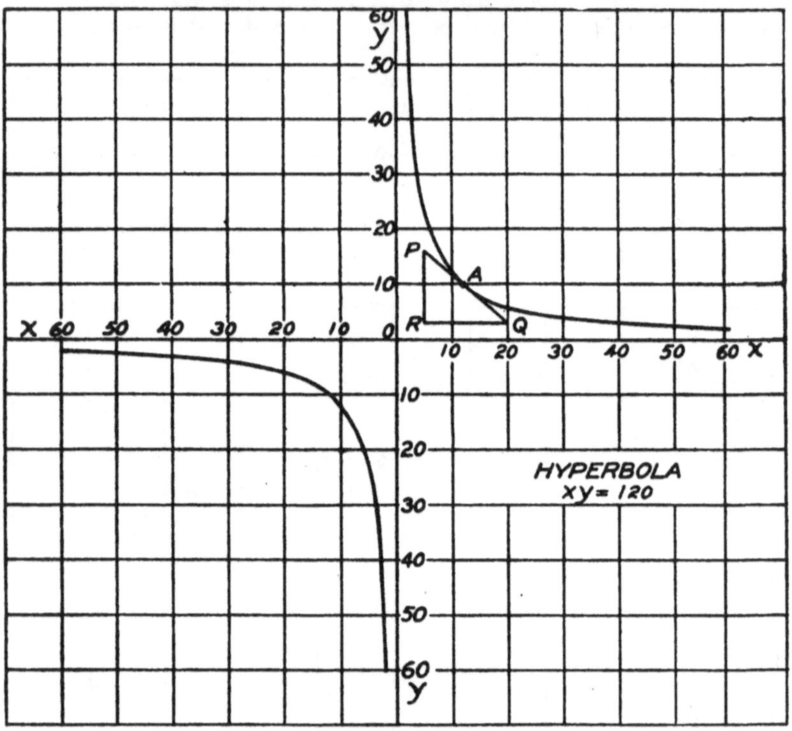

FIG. 28.

is vertical is $y=a+bx+cx^2$. The size of the curve depends only upon the value of c and it is convex downward (" festoon-shaped ") if c is positive, convex upward (" arch-shaped ") if c is negative. A change in the value of a has been shown to raise (or lower) the curve through a corresponding distance, and obviously it always cuts the y-axis at a height of a. Since any curve cuts the x-axis at the points where $y=0$ it follows that the parabola $y=a+bx+cx^2$ must cross the x-axis at points whose abscissæ are the two roots of the quadratic equation $0=a+bx+cx^2$; these roots of course depend upon the values of all three of the parameters a, b, and c.

Exercise 1.—Show by the method of § 69 that the slope of the parabola $y=kx^2$ or $y=kx^2+a$ at the point (x_1, y_1) is equal to $2kx_1$. Draw a tangent to your curve and test this. (See tangent at A, Fig. 27.)

Exercise 2.—Show also that the slope of the parabola $y=a+bx+cx^2$ at the point (x_1, y_1) is equal to $b+2cx_1$.

96. The Rectangular Hyperbola.

Plot the curve xy=a constant. Take, for example, $xy=120$ and plot it from $x=+2$ to $x=+60$ and $x=-2$ to $x=-60$. The curve (Fig. 28) consists of two separate parts ("branches") each of which extends to an infinite distance and approaches two fixed lines,

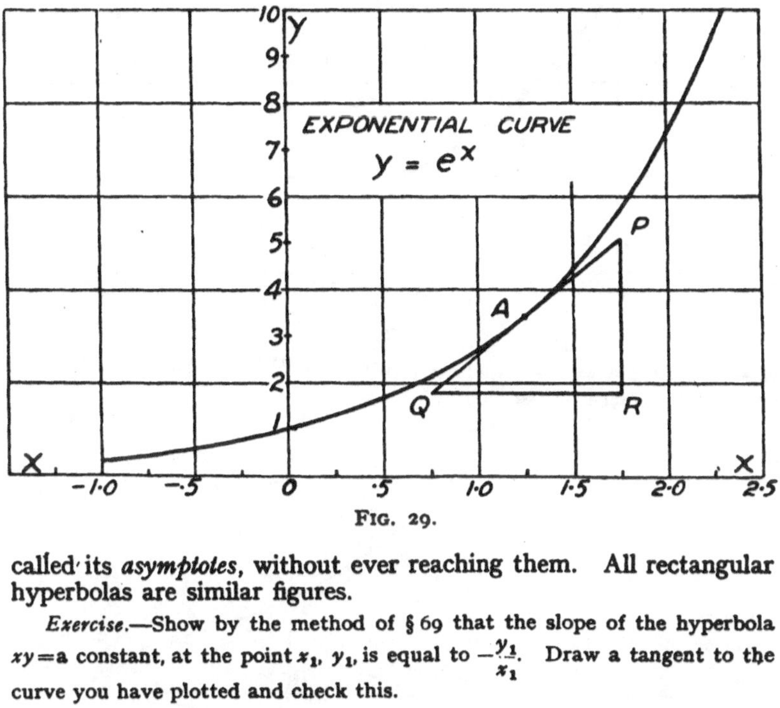

FIG. 29.

called its *asymptotes*, without ever reaching them. All rectangular hyperbolas are similar figures.

Exercise.—Show by the method of § 69 that the slope of the hyperbola xy=a constant, at the point x_1, y_1, is equal to $-\frac{y_1}{x_1}$. Draw a tangent to the curve you have plotted and check this.

97. The Exponential Curve.

Plot $y=\epsilon^x$ (Fig. 29) and $y=\epsilon^{-x}$ (Fig. 30). Get the values of y in terms of x from the tables and plot from $x=0$ to x=about 2·5 or 3. A remarkable property of the curves is that their slope at every point is numerically equal to the ordinate of the point. Check this by the method of § 69, and also by drawing tangents to the curve. In the more general equation $y=\epsilon^{mx}$ the slope is proportional to the ordinate, so that the curve may be used to represent a relationship like *Newton's Law of Cooling*, viz. the rate at which the temperature of a cooling body decreases is proportional to its excess temperature over its surroundings. Curves for $m=1$, $m=2$,

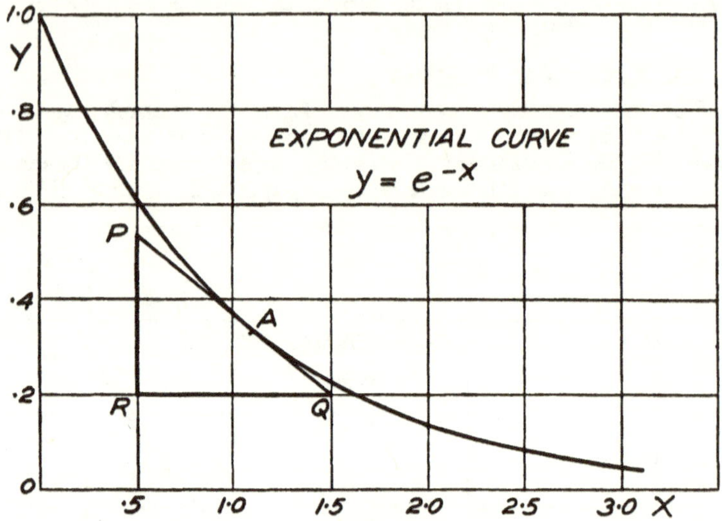

FIG. 30.

$m=-1$, and $m=-\frac{1}{2}$ are shown in Fig 31, the first by a full line and the rest by dotted lines.

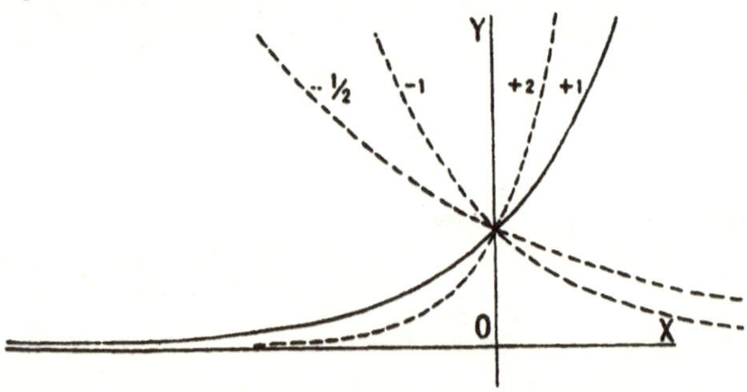

FIG. 31.

98. The Logarithmic Curve.

$y=\log x$ (Figs. 32, 33). Plot this curve from $x=1$ to $x=10$, using logarithms to base 10 (Fig. 32). Also plot the curve $y=\log_e x$ or $x=\epsilon^y$ (Fig. 33), using the table of natural logarithms at the end of the book. This latter form of the equation shows the similarity between this curve and the exponential curve just plotted. Find out how to hold your sheet of paper to make the curves you have drawn for $y=\epsilon^x$ and $x=\epsilon^y$ appear identical.

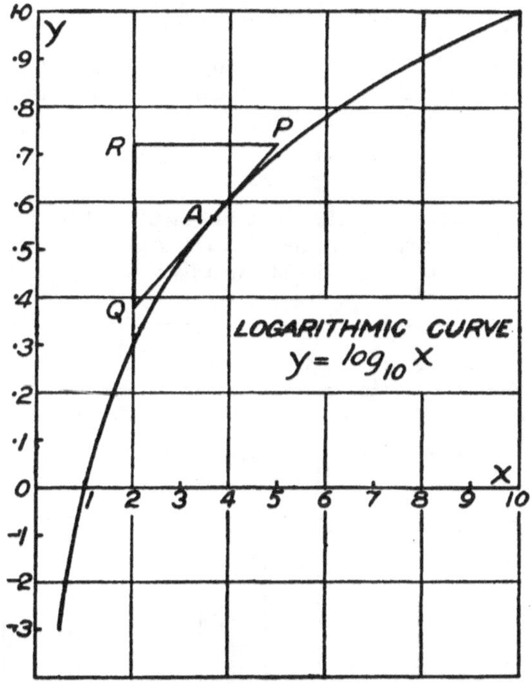
Y
X
R
P
A
Q
LOGARITHMIC CURVE
y = log10 x
1·0
·9
·8
·7
·6
·5
·4
·3
·2
·1
0
-·1
-·2
-·3
1
2
3
4
5
6
7
8
9
10

FIG. 32.

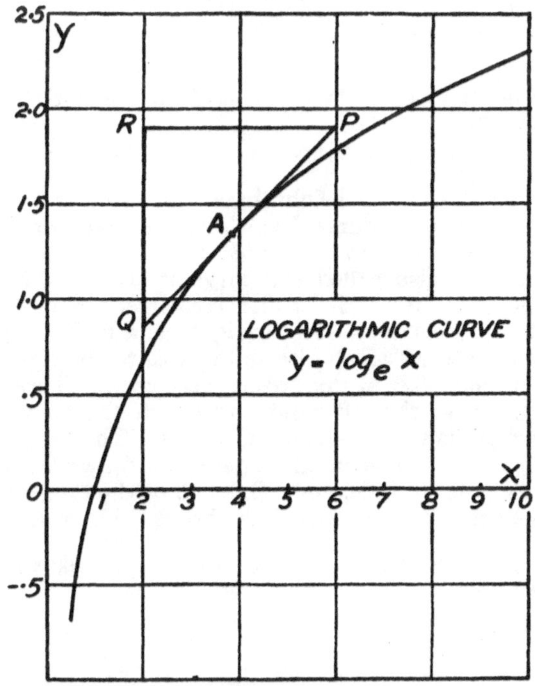
Y
X
R
P
A
Q
LOGARITHMIC CURVE
y = loge x
2·5
2·0
1·5
1·0
·5
0
-·5
1
2
3
4
5
6
7
8
9
10

FIG. 33.

A feature of the curve $y = \log_e x$ is that the slope of the curve at every point is numerically equal to the reciprocal of the abscissa of that point. Check this from your drawing.

99. The Probability Curve or Curve of Errors.

This curve may be plotted from the data obtained in § 61 for the equation $y = e^{-x^2}$. The greatest value of $y = 1$ and the curve slopes down symmetrically on the two sides of the axis of y (Fig. 34).*

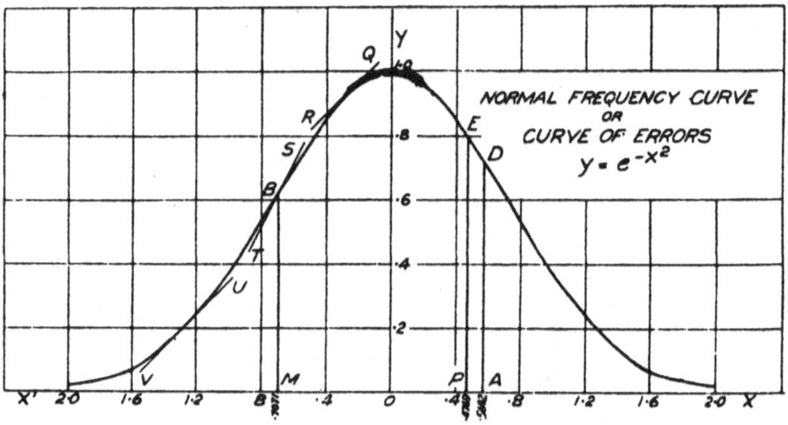

FIG. 34.

Fig. 35 shows the forms of the curves plotted from the slightly more complicated equation

$$y = b\epsilon^{-(ax)^2}$$

The three curves having the same height have the same b but different values of a. The lower curve has the same value of a as the uppermost curve, but a smaller value of b.

The probability curve is also called the curve of errors. The reason for this will be seen later. It is also convenient now to point out three important values of x. On the right hand of Fig. 34 an ordinate PE has been erected which bisects the area to the right of the axis of x. The x-value, OP, of this ordinate is 0·4769. Three tangents have been drawn to the left-hand portion of the curve. The uppermost tangent QR is wholly above the curve. The lowest tangent UV is wholly below. The middle one ST is drawn at a point B which is called a point of inflexion. One half of the tangent is above the curve, the other half below. The abscissa OM of this

* In Fig. 34 the curve has been drawn a little incorrectly. It ought to go through the point $x = 0$, $y = 1$.

point is $1/\sqrt{2}$ or 0·7071. A third ordinate AD has been drawn at 0·5642. The importance of this ordinate will be seen later. If the

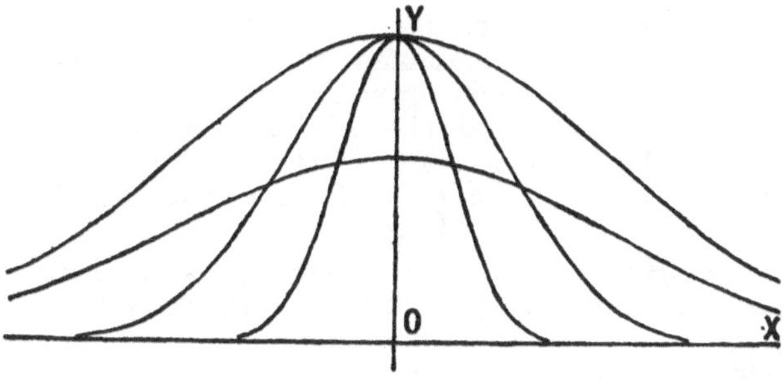

FIG. 35.

equation of the curve is $y=\epsilon^{-x^2/a^2}$ the x values of the three ordinates would be 0·4769a, 0·7071a, 0·5642a.

100. Equation of a Graph.

The process of finding the curve that corresponds to a given equation is not usually difficult, but the reverse operation may be much harder. If the given "curve" is a straight line that cuts the y-axis its equation will be of the form $y=a+bx$, and the values of a and b can be determined according to § 94.

Draw a line through the two points (1, 3) and (2, 1), and find its equation. Ans.: $y=5-2x$.

Draw a line through (0, 2) and (3, 0), and find its equation.

Find the equation of a line drawn through (0, −3) and (5, 0).

See if your last two equations can be transformed into $x/3+y/2=1$ and $x/5+y/(-3)=1$, respectively. What is the apparent significance of the denominators m and n in the equation $x/m+y/n=1$?

If the given curve is a parabola symmetrical about the y-axis it has been seen that its equation will be $y=mx^2$ *if the origin is at the vertex of the curve.* Turn to your plot of $y=x^2-6x$ (§ 95) and draw a new pair of axes through the point (3, −9), which is the vertex of the curve. If the co-ordinates of points on the curve, referred to these new axes, are denoted by the capital letters X and Y, notice that the same curve corresponds exactly to the relationship $Y=1X^2$. From this equation it is possible to deduce the original equation as follows: Let P be *any* point on the curve; notice that its X-value is 3 less than its x-value, also that its Y-value is 9 more than its y-value. That is, for every point on the curve the equations $X=x-3$ and $Y=y+9$ hold good, as well as $Y=1X^2$. Substituting in the last equation the values of X and Y given by the other two, $y+9=(x-3)^2$ or $y=x^2-6x$.

With a pencil compass draw carefully a circle whose radius is 5 and whose centre is the point (4, 3) on a graphic diagram. The curve will be seen to pass exactly through the points (0, 0), (1, −1), (4, −2), and three corresponding integral points in each quadrant. Mark any point P on the circumference, and considering the centre of the circle as a new origin draw the ordinate Y of the point P and from its base draw the abscissa X along the new X-axis. From the theorem of Pythagoras and the fact that all radii are equal the co-ordinates of the point P must satisfy the relationship $X^2+Y^2=5^2$; and since $X=x-4$ and $Y=y-3$ this relationship becomes $(x-4)^2+(y-3)^2=5^2$, or $y=3+\sqrt{25-(x-4)^2}$, which is the equation of the given circle with respect to the original axes.

101. Change of Scales.

Lay off an x-scale in which three squares of the ruled paper correspond to each unit and a y-scale of two squares per unit, and plot the circle $x^2+y^2=25$ from the following twelve points $(0, \pm 5)$, $(\pm 3, \pm 4)$, $(\pm 4, \pm 3)$, $(\pm 5, 0)$, drawing as smooth a curve as possible. The result is an *ellipse*, or a " strained " (*i.e.* distorted) circle.

To find the equation of such an ellipse, *referred to uniform scales*, let each square of the ruled paper be considered as equal to unity. Then each point on the ellipse will be twice as high or low and three times as far to the right or left as a corresponding point on the true circle $x^2+y^2=25$. That is, if the co-ordinates of a point on the ellipse are called X and Y, then $X=3x$ and $Y=2y$; in other words, $\frac{1}{3}$ of the X-value and $\frac{1}{2}$ of the Y-value will satisfy the relationship for the circle, so that $(X/3)^2+(Y/2)^2=25$. This is an illustration of the general fact that re-writing an equation with x/a and y/b in place of the original x and y will stretch out the length and height of the diagram to a and b times the original dimensions, respectively.

On one graphic diagram plot the loci of $x+y=1$ and $x/3+y/5=1$.

102. Definitions of Circular Functions.

The use of a graphic diagram makes it possible to give very concise definitions of the circular functions of an angle of any size: Let a line that coincides with OX (Fig. 36) rotate counterclockwise through an angle a to a new position OP. If the co-ordinates of the point P are denoted by x and y, and its distance from the origin by $+r$, then $\sin a=y/r$, $\cos a=x/r$, $\tan a=y/x$, $\cot a=x/y$, $\sec a=r/x$, and $\operatorname{cosec} a=r/y$. These relationships hold true whether the angle is larger or smaller than 360°, and for negative angles when described by a clockwise rotation. They furnish the easiest method of determining the *sign* of a circular function; in Fig. 36, right-hand figure ($\angle POX=103°$), the value of y is positive but x is negative; r is always considered positive, so it is immediately obvious that the sine of an angle "in the second quadrant" (between 90° and 180°) is positive while the cosine is negative.

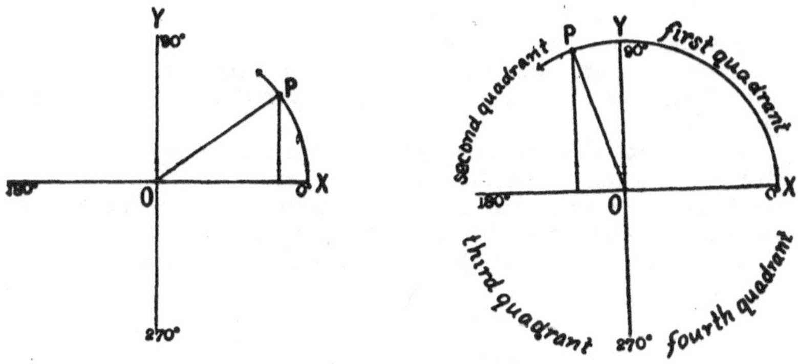

FIG. 36.—FUNCTIONS OF AN ANGLE.

For an angle of any magnitude, described counterclockwise from OX, the sine, cosine, tangent, cotangent, secant, and cosecant are respectively y/r, x/r, y/x, x/y, r/x, and r/y, where r is the distance from the point (x, y) on the rotating line to the origin.

103. Rate of Increase, Rate of Growth.

Let OX and OY (Fig. 37) be axes of x and y, the scales being

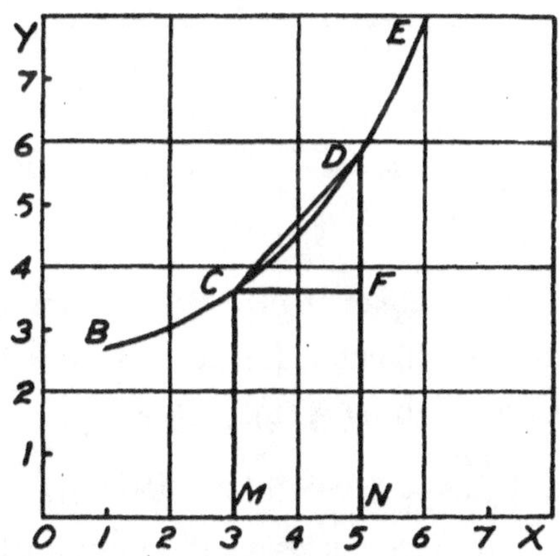

FIG. 37.

equal. Let BE be a smooth curve such that when x increases y also increases.

Take two points C and D fairly close together and draw the straight line CD, also the perpendiculars CM, DN, CF.

As x increases from OM to ON, y increases from CM to DN.

The increase of x is MN, the increase of y is DF. The ratio of the increase of y to the increase of $x = DF/CF =$ tangent of the angle DCF. This is the average value of the rate of increase of y with respect to x over the region CD.

Suppose now the increases of x and y are small, so that C and D are close together. The increases may be called δx and δy, and the rate of increase of y with respect to x over the region $CD = \delta y/\delta x = \tan DCF$.

We see that as δx diminishes C and D get closer together and that ultimately when δx diminishes very nearly to zero C and D practically coincide, and the line CD becomes coincident with the tangent to the curve at the point C (or D).

This is shown in Fig. 38, where C and D have coalesced at A.

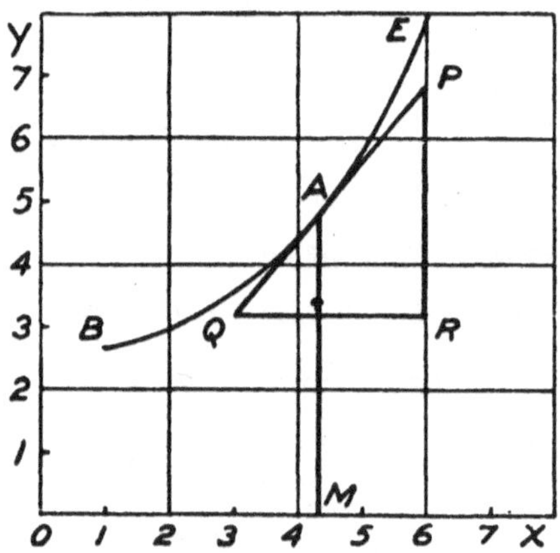

FIG. 38.

The "instantaneous" rate of change of y with respect to x is called $\frac{dy}{dx}$, this being the notation used for the value of $\left(\frac{\delta y}{\delta x}\right)$ when dx has become vanishingly small (*cf.* § 69). Therefore:

$$\left(\frac{dy}{dx}\right)_{\substack{\text{when}\\ x=OM}} = \text{tangent } \angle PQR = \frac{PR}{QR}$$

PR and QR are the sides enclosing the right angle of any right-angled triangle drawn under the tangent PQ in the manner shown. It is obvious that the ratio of PR to QR does not depend on the size of the triangle PQR.

The instantaneous rate of change of y with respect to x is sometimes called the rate of growth of y with respect to x. For example, y may represent the height of a plant and x the age of the plant. If we are given the curve of growth and asked to find the rate of growth at any given value of x or y, all that we have to do is to mark the position of the corresponding point on the curve, draw the tangent at that point, complete the triangle PQR and find the ratio of PR to QR.

The scales of x and y being equal the rate of growth is numerically equal to the tangent of the angle PQR, so that if this angle were measured with a protractor the tangent could be obtained from the trigonometrical tables.

But if the scales are not equal we must express PR, QR in terms of the units of y and x respectively, and then obtain the quotient (an example will be given in the next section).

This method of finding rates of increase or rates of growth is very useful when the equation of the curve is unknown, as is commonly the case when the curve is the result of experiment. Its accuracy depends upon the accuracy with which tangents can be drawn to a curve. If the curve is clean and smooth and a transparent ruler is used there is usually no difficulty. If the ruler is opaque, be careful to use it on the convex side of the curve so that all the curve is visible.

If the equation of the curve is known, *e.g.* $y=$ some known function of x (*i.e.* $y=f(x)$), the algebraical method of finding the rate of growth of y with respect to x is to proceed as in § 69 or to *differentiate* y with respect to x. Reference may be made to any elementary book on the differential calculus to see how this is done. For example, if $y=2x+3x^2$, $\dfrac{dy}{dx}$ may at once be written down by the calculus as $2+6x$, and its value can then be deduced for any given value of x.

104. Space-Time Curves, Velocity-Time Curves, Acceleration.

Considering only motion along a straight line, the instantaneous rate of change with time of the distance of the moving particle from any fixed point on the line is called the velocity. Hence, if we are given the space-time graph (Fig. 39) of such a moving particle it is comparatively easy by drawing tangents to find the velocities at given instants.

For example, it is evident that the velocity at the 10″ second

$$=\frac{PR\text{, expressed in feet}}{QR\text{, expressed in seconds}}=\tfrac{19}{8}=2\tfrac{3}{8}\text{ ft./sec.}$$

Having got the velocities at a number of instants, a velocity-time graph may be plotted.* By drawing tangents to this curve at selected instants it is possible to find the rates of increase of the velocity with the time, *i.e.* the accelerations at these instants, and in this way to find the characteristics of the motion.

* A curve obtained by plotting the slope of the initial curve against the abscissæ is said to be the *first derived* curve of that curve. The velocity-time curve is the first derived curve of the space-time curve.

Thus, in the case of a body slowly rolling down an inclined plane the times may be noted as the body passes marked points, the space-time curve plotted, the velocity-time curve deduced, and finally

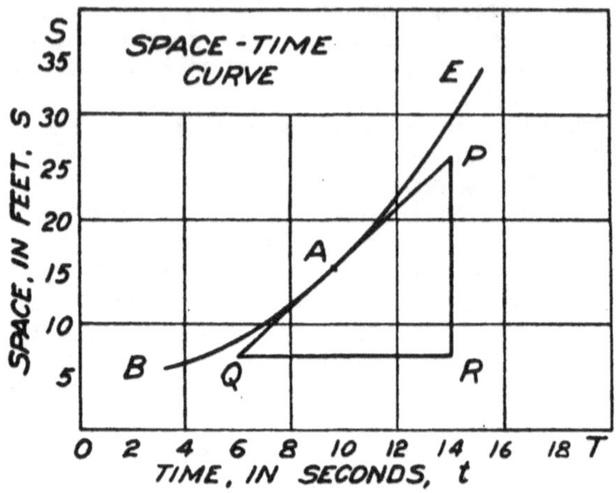

FIG. 39.

the acceleration evaluated. In this case it will be seen that the acceleration-graph is a straight line parallel to the axes of time, or in other words, that the acceleration is constant.

Galileo tried such an experiment by rolling a ball down a groove in an inclined plane; but a more convenient method is to take a large disc, supported on a small axle, and let this roll down between the two straight vertical sides of a framework like a long narrow

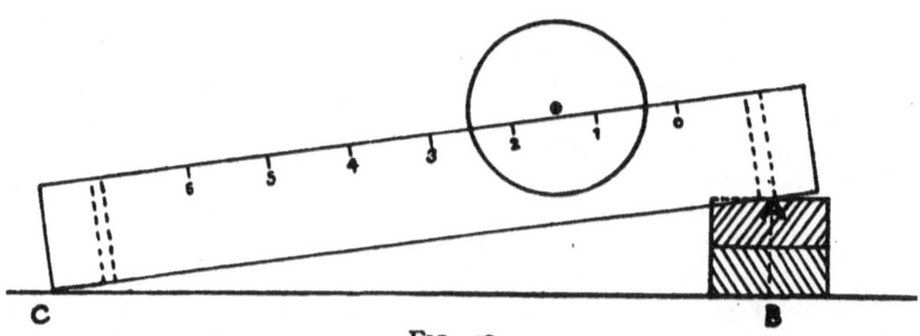

FIG. 40.

box without a top (Fig. 40). Times are taken by a watch as the axle passes the distance marks placed on the side of the box. By putting blocks of wood under one end of the box different tilts may be given to the frame and the relation of the acceleration to the slope may be investigated.

Fletcher's trolley (described in text-books on Mechanics such as Eggar's or Allen and Moore's) is another still more convenient apparatus for the same purpose.

Exercises VIII

1. If one point on the curve of $y=x^3-x$ (Fig. 25) is (a, b) prove that another one is $(-a, -b)$.

2. Knowing the shape of the graph of $y=a+bx+cx^2$, what facts can you deduce in respect to the graph of $x=a+by+cy^2$?

3. What effect will be produced if the equation of a curve is re-written with $x+p$ and $y+q$ substituted everywhere for the original x and y? (Try the parabola, $y=x^2$, and $p=0$, $q=3$, if there is any uncertainty.)

4. What effect will be produced if the equation of a curve is re-written with mx substituted everywhere for the original x? If my for the original y?

5. Make a graphic diagram of the equation $y=30/x$. Plot enough points to show clearly the form of both parts of the curve. On the same diagram plot $y=1/x$. Each of these curves is a (*rectangular*) *hyperbola*.

6. Plot the curve of $y=1/x^2$ carefully. How would the appearance of the curve $y=30/x^2$ differ?

7. Plot the curves of one or more of the following equations: (*a*) the finite curve $y^2=x(10-x)^2$, using a condensed scale along the y-axis; (*b*) the curve $y=x \log x$; (*c*) $x^2-y^2=0$; (*d*) $x^2+y^2=0$.

8. Instead of being located by latitude and longitude (*rectilinear co-ordinates*, referred to an x-axis and a y-axis), the position of a point may be located by its distance and direction, *i.e.* the length r and the angle a of § 102 (its *polar co-ordinates*, referred to a *pole* and an *initial line*). Obviously, $\tan a=y/x$ and $r^2=x^2+y^2$; or $x=r \cos a$ and $y=r \sin a$. The distance and angle are called *polar co-ordinates* and are usually indicated by the letters r (radius vector) and θ (angle).

Make a rough graphic diagram of the curve of the equation $x^2+(y-h)^2=h^2$ (a circle that passes through the points (0, 0), (0, $2h$), and $(\pm h, +h)$; compare § 100); then substitute $x=r \cos \theta$, $y=r \sin \theta$ and show that the *polar equation* of this circle is $r=2h \sin \theta$.

9. In the spiral of Archimedes equal increases in vectorial angle and radius vector accompany each other. Draw such a spiral of two convolutions given that the radius vector is initially $\frac{1}{8}$-in. and increases $\frac{1}{8}$-in. for every 30° increase of vectorial angle. Write down its equation. Give an example of an Archimedean spiral.

10. In the Logarithmic Spiral as the vectorial angle increases in arithmetical progression the radius vector increases in geometrical progression. Draw such a spiral of two convolutions, the initial radius vector being $\frac{1}{4}$ in. and the increase of radius vector for an increase of vectorial angle of $22\frac{1}{2}°$ being in the ratio 1 to 1·08. Write down its equation. Give an example of a logarithmic spiral found in Nature.

11. (*a*) The co-ordinates of A and B are: (3, 1), (−2, 5) respectively. Plot these points and obtain from your diagram the equation of the straight line AB.

(*b*) The equation of the straight line that passes through the two points (x_1, y_1) and (x_2, y_2) is

$$\frac{y_2-y_1}{x_2-x_1}=\frac{y-y_1}{x-x_1}$$

Draw a diagram of the line and the two points, and explain the significance of the subtracting, the dividing, and the equating in the above formula. Apply this formula to the line in (*a*).

12. Plot (1) $y=\sin x$, (2) $y=\cos x$, (3) $y=\tan x$, from $x=0$ to $x=720°$ (*i.e.* $x=4\pi$).

13. Plot (1) $y=\sinh x$, (2) $y=\cosh x$, (3) $y=\tanh x$ from $x=0$ to $x=5$.

14. Plot (1) $y=e^x$, (2) $y=e^{-x}$ from $x=0$ to $x=5$.

15. Plot the parabolic curve $y=x^2$ from $x=0$ to $x=8$. Draw tangents at the points $x=1, 2, 3, 3, 5, 7$. Find the slopes of these tangents. From the slopes get the values of $\frac{dy}{dx}$. Plot a curve between the values of $\frac{dy}{dx}$ and x. Draw the smoothed curve and find its equation.

16. Do the same for $y=\frac{1}{x}$. Plot from $x=0\cdot1$ to $x=1\cdot0$ and draw tangents at $x=0\cdot2$, $0\cdot3$, $0\cdot5$, $0\cdot7$.

17. Do the same for $y=\sqrt{x}$. Select as values of x, $x=0\cdot1111$, $0\cdot25$, 1, $2\cdot25$, 4, 9. Draw tangents at these points.

18. Do the same for $y=e^x$. (Let x range from 0 to 2.)

19. Do the same for $y=e^{-x}$. (Let x range from 0 to 4.)

20. Do the same for $y=\log_e x$. (Let x range from 0 to 3·0.)

21. Do the same for $y=\log_{10} x$. (Let x range from 0 to 10.)

22. Do the same for $y=\sin x$. (x must be in radians.)

23. Do the same for $y=\cos x$.

24. The average temperature on the first of the month for fifty years at Greenwich has the values :—

	Degrees F.		Degrees F.
January 1st	38·4	July 1st	61·3
February 1st	38·5	August 1st	62·8
March 1st	40·1	September 1st	59·7
April 1st	44·1	October 1st	54·0
May 1st	49·5	November 1st	46·6
June 1st	56·5	December 1st	41·1

Plot the temperature curve from January 1st to January 1st. Find the average temperature for the year. (To do this, find the height of the rectangle which has the same area as the area under the smoothed temperature curve. Count the squares to get the area.)

25. The average price of wheat per bushel as received by the farmers in Canada from 1913–1922 was as follows :—

	$		$
1913	0·67	1918	2·02
1914	1·22	1919	2·37
1915	0·83	1920	1·62
1916	1·31	1921	0·81
1917	1·94	1922	0·85

Plot and comment.

26. The following observations were made on the growth of a plant. The time is reckoned in days from the first observation.

Days	0	1	2	3	4	5	6	7	8
Height in inches	0·75	1·20	1·75	2·50	3·45	4·70	6·20	8·25	11·50

(*a*) Plot the curve of growth (make a smooth curve).

(*b*) By drawing tangents find the rates of growth at (1) 3 days, (2) 7 days from the start.

27. Observations on a growing plant gave the following readings:

Days from first observation	0	1	2	3	4	5	6	7	8
Height in inches	0·51	0·68	1·04	1·55	2·18	2·90	3·70	4·53	5·40

(*a*) Plot the curve of growth.
(*b*) Find the heights at $1\frac{1}{2}$, $3\frac{1}{2}$, $5\frac{1}{2}$ days from the start.
(*c*) Find the rates of growth at the end of the third and seventh days respectively.

28. The actual run of the non-stop Royal train to the West of England on the G.W.R. line in July, 1903, was as follows:—

Station.	Mileage from London.	Time. h. m.
Paddington	0	10.40
Bristol	$118\frac{1}{2}$	12.4
Taunton	$162\frac{3}{4}$	$12.35\frac{1}{2}$
Exeter	$193\frac{1}{2}$	1.2
Newton Abbot	209	1.21
Plymouth	$246\frac{1}{2}$	$2.33\frac{1}{2}$

Plot the space-time curve. Between which two stations was the speed greatest? between which least? Find the average speed for the whole journey.

29. Plot a curve between x and y from the following data:

x	0	0·5	1·0	1·5	2·0	2·5	3·0	3·5	4·0
y	0	0·165	0·272	0·448	0·739	1·218	2·009	3·312	5·460

(*a*) Draw tangents at the points whose abscissæ are: 0·5, 1·5, 2·5, 3·5. From their slopes find $\frac{dy}{dx}$ (*i.e.* the rate of increase of y with respect to x) at these points. Tabulate.

(*b*) Plot a curve between these values of $\frac{dy}{dx}$ as ordinates and the values of x as abscissæ. The two curves are to be drawn on the same paper with same axes and scales.
(*c*) Comment on the relative shapes of the two curves.

30. Plot a curve from the following corresponding values of x and y:

x	y	x	y
0·0	1·000	1·0	2·718
0·2	1·221	1·2	3·320
0·4	1·492	1·4	4·055
0·6	1·822	1·6	4·953
0·8	2·226	1·8	6·050

Find by drawing tangents the values of $\frac{dy}{dx}$ at the points $x=0·4$, $0·8$, $1·2$, $1·6$ respectively. Plot the first derived curve, that is the curve between x and $\frac{dy}{dx}$. Find if you can the equations of both curves.

31. The equation of the velocity-time curve of an ascending body is $v=1600-32t$, where v is in feet per second and t is given in seconds, measured from the instant the body left the ground. Plot the velocity-time curve and show what part of the graph respresents the time of ascent. By integrating the above formula or counting squares deduce the equation between space and height and calculate the height of ascent.

32. Plot the curve whose equation is $s=16t^2$, where s is in feet and t in seconds.
(*a*) Draw tangents at the points $t=1$, 2, 3, 5, 7, and deduce the velocities at these instants. Tabulate. Name the unit of velocity taken.
(*b*) Plot a curve between velocities and times. Find its equation.

(*c*) Draw tangents to the second curve and find the acceleration at the same instants. Tabulate. State the unit.

(*d*) Plot a curve between acclerations and times. Find its equation.

33. Plot the space-time curve from the following data:

t (secs.)	0	1	2	3	4	5	6	7	8
s (feet)	6	13	24	39	58	81	109	139	174

Plot the velocity-time and the acceleration-time curves, and get their equations. Try to find the equation of the space-time curve.

34. A ball is shot upwards. Its equation of motion is $s=128t-16t^2$, where s is in feet and t is in seconds. Plot the curve between s and t from $t=0$ to $t=8$ seconds. Find graphically the velocity when $t=1\frac{1}{2}$, $3\frac{1}{2}$, 4, $4\frac{1}{2}$, and 8 seconds, respectively.

35. The space-time equation of a point moving along a straight line is

$$s=0{\cdot}4023\ t+0{\cdot}000137t^2$$

s is in feet, t is in seconds.

Find the velocity and acceleration at the instant $t=10$ seconds.

36. Find the rate of increase of the area of a circle with respect to the radius. Confirm graphically. Work out its numerical value in the case of a circle of radius 4·50 inches.

37. The densities of aqueous solutions of common salt are related to the concentrations as follows:—

Gm. salt per 1000 gm. water.	Density at 20° C.
0·0	0·9982
5·1	1·0017
10·8	1·0060
20·1	1·0128
46·9	1·0329

Plot a curve between density and concentration and find the relation between density and concentration for very dilute solutions. State it in the forms $C=k(D-0{\cdot}998)$ and $D=k'C+0{\cdot}998$.

38. If V=volume of free gas (free means if measured at atmospheric pressure) contained in a cylinder of unit volume and P=the pressure of the gas in the cylinder in atmospheres, the relation between V and P is approximately $V=1{\cdot}06P$ for oxygen and $V=0{\cdot}88P$ for hydrogen. Plot these lines. After the war a company bought a large number of gas cylinders from the Government. These were warranted to hold 176 cubic feet of hydrogen (measured at atmospheric pressure) when pumped up to 200 lb. per square inch pressure. The company used them with oxygen pumped up to the same pressure, and, being ignorant of the above relationships, charged their customers for 176 cubic feet of gas. What error did they make per cylinder? What per cent. error?

39. Between 0° C. and 30° C. the density of water ($=\rho$) varies with the temperature C. ($=\theta$) according to the following equation:

$$\rho=0{\cdot}999868+0{\cdot}0000552\theta-0{\cdot}0000069\theta^2$$

Find the density at 0, 10, 20, 30° C. Plot a curve between ρ and θ, and find the mean coefficients of expansion between 0° and 10°, 10° and 20°, 20° and 30°. (Coefficient of expansion at temperature $\theta=\frac{1}{\rho_\theta}\times$ (rate of diminution of ρ with increase of θ).)

40. The following data were obtained in an experiment designed to study the motion of a moving particle:

Time, secs.	Distance from origin, cms.	Time, secs.	Distance from origin, cms.
0	0	9	170·71
1	3·61	10	186·60
2	13·40	11	196·59
3	29·29	12	200·00
4	50·00	13	196·59
5	74·12	14	186·60
6	100·00	15	170·71
7	125·88	16	150·00
8	150·00	17	125·88

Plot the space-time curve and obtain from it the velocity-time and acceleration-time curves. Plot also the acceleration-space curve. What is the characteristic of the motion?

41. Plot the curve $y=x+\frac{1}{x}$. Show that this curve has for ordinates the sum of the ordinates of the curves $y=x$ and $y=\frac{1}{x}$. For what value of x is y least? Answer now Question 11 of Exercises I. Find this value of y.

42. Plot the curve $y=x(11-x)$. Find the greatest value of y and the corresponding value of x. (See Exercise I, Question 8.)

43. Plot the curve $y=x+\frac{5}{x}$. Find the value of x which makes y a minimum. (See Exercise I, Question 9.)

44. Check the following values of $\frac{dy}{dx}$ at the point A for the curves mentioned:

Curve.	Equation.	Point selected.	$\frac{dy}{dx}$
Fig. 27 . . .	$y=x^2$	$x=2·9$	5·8
Fig. 28 . . .	$xy=120$	$x=12$	−·84
Fig 29 . . .	$y=e^x$	$y=3·3$	3·3
Fig. 30 . . .	$y=e^{-x}$	$y=3·1$	−·32
Fig. 32 . . .	$y=\log_{10}x$	$x=3·7$	·117
Fig. 33 . . .	$y=\log_e x$	$x=3·8$	·26

CHAPTER IX

GRAPHIC ANALYSIS

105. Interpretation of Equations.

Turn back to the diagrams that were drawn to represent the equations $y=2x$ and $y=2x+3$ (§ 94) and notice that for every point of the first there is another point just three units higher on the second of the two straight lines. The equation $y=2x+3$ makes the statement that the value of y is as much as the $2x$ of the first equation with three more added. Another way of looking at the same equation can be shown by writing the terms in reversed order, $y=3+2x$. This can be considered as making the statement that, for every value of x, y is equal to the amount 3, to which there is further added the amount $2x$ (Fig. 41). It is worth while to form the habit of always investigating the meaning of an equation as far as possible, especially for the student who intends to proceed further with the study of any of the practical applications of mathematics. For example, the equation, $s=ut+\frac{1}{2}at^2$, of uniformly accelerated motion means that the space traversed is equal to the space, ut, that would have been traversed by an object moving at an initial velocity of u without acceleration, plus the $\frac{1}{2}at^2$ of distance that a stationary body would have been made to describe if acted upon by the accelerating force alone—an illustration of the addition of vectors, which in this case are directed along the same straight line. The student of physics will find no difficulty in further analysing each of these two separate terms.

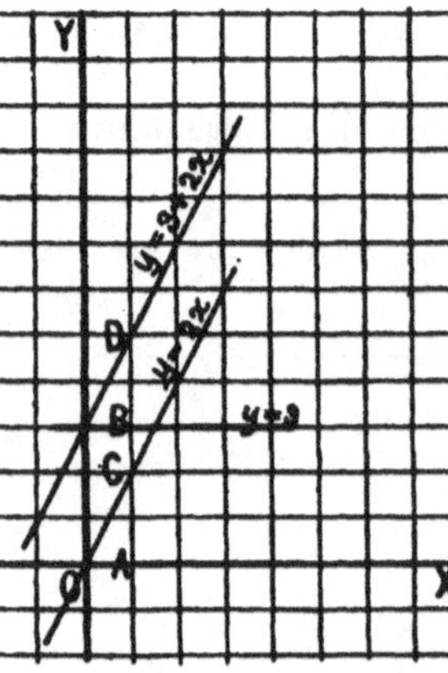

FIG. 41.—GRAPH OF THE EQUATION $y=3+2x$.

The ordinate of any point (D) on the line may be considered as being either 3 more than $2x$ (CD added on to AC) or $2x$ more than 3 (BD added on to AB).

106. The Graph of $y=a+bx$.

It has already been seen (§ 94) that the equation $y=a+bx$ has a straight line for its "curve," and that this line cuts the y-axis at a height of a and has a slope that is numerically equal to b. For this reason it is often spoken of as a *linear equation*, and the law of variation which it illustrates is known as the "*straight line*" *law*. It should be noticed that the values of x and y are not proportional except for those cases in which the straight line passes through the origin.

Draw the loci of the following equations *without calculating* the values of x and y for any point. For each one rule a straight line in such a position that it intersects the y-axis at the required point and has the required gradient. If there is any doubt as to the meaning of a negative value for the b of the equation $y=a+bx$ a few points may be calculated from the equation (5) in the usual manner. (1) $y=2+2x$. (2) $y=2+x$. (3) $y=2+\frac{1}{2}x$. (4) $y=2+0x$. (5) $y=2-\frac{1}{2}x$. (6) $y=4-2x$. (7) $y=4-x$. (8) $y=0-\frac{1}{2}x$. (9) $y=\frac{1}{2}x$. (10) $y=-2-x$. (11) $y=-2+x$. (12) $y=-2$.

Lay the ruler on the squared paper at random in any position and draw a straight line. Mark the axes along any convenient ruled lines of the paper and determine the equation of the line that was drawn, expressing the numerical term and the coefficient of x as decimal fractions.

107. The Straight Line Law.

If a homogeneous metal bar is heated it may be expected to expand in such a way that its length is always proportional to its thickness, and if these two variables are denoted by x and y the relation between them will be given by the equation $y=kx$, in which k has a constant value. If we compare length and temperature, however, there is no proportionality between them. The bar is not made twice as long by heating it twice as hot, but it is not difficult to show experimentally that there is a certain law of relationship, namely, the *change* in length is proportional to the *change* in temperature. If a bar whose length is 10 at a temperature of 15° C. expands to 12 at 20° C., then its length at 30° C. will be 16.* In other words, 12–10 : 16–10 : : 20–15 : 30–15. Draw a graphic diagram to illustrate this example, plotting temperature horizontally and length vertically, and notice that the points (15, 10), (20, 12), and (30, 16) lie in the same straight line. The variables x and y are not proportional; it is only their respective changes or differences that show proportionality. Using the notation of § 94, we can write $\Delta y : \Delta x :: \Delta' y : \Delta' x$, or $\Delta y/\Delta x=k=2/5$ for the graph (draw these Δ's on your diagram), and for the general physical law $\Delta(length)/\Delta(temp.)=k$. When two variables show proportional

* These numbers are very greatly exaggerated as compared with those for ordinary materials. Metals expand only a few hundred-thousandths of any dimension when heated a few degrees. Furthermore the law is only approximate; careful experiments on a given bar will show that its length at t° C. will not be exactly expressible in the form $L=L(1+kt)$, but will need a more complicated equation, $L=L_0(1+k_1t+k_2t^2)$ or even $L=L_0(1+k_1t+k_2t^2+k_3t^3)$.

changes they are said to follow a *linear law*, or a *straight line law*.

Prove (*a*) that proportionality is always in accordance with the straight line law, but (*b*) that a straight line law does not mean proportionality between the two variables except when the straight line *passes through the origin*.

108. The "Black Thread" Method.

If a set of experimental measurements, such as those of temperature and length of a metal bar, are found to correspond approximately to a straight-line law they may be plotted as the x's and y's of a graphic diagram, and their irregularities may be eliminated ("smoothed") by drawing the straight line that appears to come closest to all of the points. This is called the *black thread method* because the position of the line is decided by making use of a stretched thread instead of a ruler; the thread and the points can all be seen at the same time, while a ruler would hide half of the points if properly placed.

Even better than the black thread is a strip of transparent celluloid, about fourteen inches long and an inch and a half wide with a fine straight line scratched down the middle of one face.

x	y
1	9·8
2	8·5
3	8·0
4	7·2
5	6·7
6	6·5
7	6·2
8	5·5
9	5·0
10	4·1
11	3·9
12	3·2
13	2·3

EXPERIMENTAL DETERMINATIONS OF LINEAR VARIATION.

Plot the values given in the table as accurately as possible, marking each point by a minute dot surrounded by a small circle, or by a cross composed of a short vertical mark to indicate the exact value of the abscissa and a short horizontal line at the exact height of the ordinate.

Be sure that the page of the notebook rests in a perfectly flat position and stretch a fine black silk thread on it in such a position that it follows the general direction of the points. Move it a trifle toward the top or bottom of the page, also rotate it slightly, both clockwise and counter-clockwise. Attempt to get it into such a position that it lies among the points like a smoothed curve (§ 90), following their general trend but not necessarily passing exactly through any one of them. See that there are about as many points above the line as below it, but if the high points are more numerous toward one end of the thread and the low ones toward the other end rotate the thread enough to remedy the condition. When the thread is finally arranged in the most satisfactory position do not attempt to draw the line at once but *notice* where the thread cuts the x-axis and where it cuts the y-axis, or where it passes through some easily remembered intersections on the squared paper. From these numbers calculate the slope (gradient) of the thread, noticing whether its value is positive or negative. Write the equation of the line that is indicated by the thread, making y equal to a numerical value plus a certain number of times x; *i.e.* write the equation in the form $y=a+bx$ (§ 94).

109. Intercept Form of a Linear Equation.

The equation $x/m+y/n=1$ must be the equation of a straight line, since it is easily reducible to the form $y=a+bx$. Substitute zero for the value of x and notice that the corresponding value of y is n. Show likewise that when y is equal to zero x will be equal to m. In other words, the graph of $x/m+y/n=1$ passes through the points $(0, n)$ and $(m, 0)$ (compare § 94).

Draw the straight line $x/(-3)+y/2=1$ by ruling a line through $(-3, 0)$ and $(0, 2)$; then reduce the equation to the form $y=a+bx$ and see whether the coefficients a and b verify the y-intercept * and the gradient of the ruled line.

Since *m and n are the x-intercept and y-intercept of the line $x/m+y/n=1$* the equation of a line that cuts both axes can be written immediately without any calculation. Write the equation of your black thread determination in this form.

110. The Graph of $y=a+bx+cx^2$.

Just as the curve $y=a+bx$ can be considered as having its ordinate for each value of x built up of the ordinate a plus the ordinate bx (§ 105, Fig. 41), so the more elaborate equation $y=a+bx+cx^2$ can be considered as representing a curve which is made by piling up the parabola $y=cx^2$ upon the slanting line $y=a+bx$. Curiously enough this also represents a parabola in every case in which c is different from zero; the slant of the straight line does not cause the curve to be lop-sided.

On a single graphic diagram plot both $y=-0{\cdot}1x^2$ and $y=3+0{\cdot}5x$. To the ordinates of the latter add (or subtract, as the case may require) the ordinates of the former for each integral value of x, and draw as smoothly as possible the resultant curve of the equation $y=(3+0{\cdot}5x)+(0{\cdot}1x^2)$.

If a curve that is obtained from experimental measurements looks like a portion of a parabola it is possible to draw an approximate tangent, find its linear equation, and then determine a value of c that will make the equation $y=a+bx+cx^2$ fit the given curve. An algebraical procedure that accomplishes the same result is to measure the co-ordinates of some point on the curve (say $(2, 3)$) and substitute in the general equation (giving $3=a+b\times2+c\times2^2$);

* The points at which a locus cuts the x-axis and the y-axis are called its *x-intercept* and *y-intercept*, respectively.

repeating this with two more points gives three equations, from which the values of the unknown a, b, and c may be determined. For a graphical procedure it is usually more satisfactory to complete a free-hand parabola as far as its vertex, if that is not already present, and then continue as in the following example.

111. Law of Density-Variation for Water—an Example of a Parabolic Law.

The density of water in grams per c.c. at different temperatures is given in the table. If the density is called y and the temperature x, it is required to find the numerical values of the coefficients of the equation $y=a+bx+cx^2$ that will express the law of variation.

The first step is to make a careful graph. Use extended scales for both x-values and y-values so that the diagram will cover practically a whole page of your notebook, noticing that the y-values need not include zero but only extend from 0·995 near the bottom of the page through 0·996, 0·997, 0·998, and 0·999 to 1·000 near the top. The curve will be seen to have the appearance of an "arch-shaped" parabola, so that it is evident that its equation will be approximately $y=-mx^2$ if the origin is located at the vertex of the curve, namely at the point (4, 0·99997). With the origin in this position the ordinate for 4° will obviously be zero, for a temperature 2° higher than this the ordinate will be −0·00003 (*i.e.* 0·99997 −0·99994), for 4° higher it will be −0·00012, etc. These x-values and y-values have been given in the third and fourth columns, where for simplicity the negative signs and the decimal points have been omitted.

temp.	*dens.*	*x*	*y*	$\sqrt{y}$
0	0·99984			
2	0·99994			
4	0·99997	0	0	
6	0·99994	2	3	
8	0·99985	4	12	
10	0·99970	6	27	
12	0·99950	8	47	
14	0·99924	10	73	
16	0·99894	12	103	
18	0·99859	14	138	
20	0·99820	16	177	
22	0·99777	18	220	
24	0·99730	20	267	
26	0·99678	22	319	
28	0·99623	24	374	
30	0·99564	26	433	

RELATIONSHIP BETWEEN TEMPERATURE AND DENSITY OF WATER.

If the values in the third and fourth columns correspond to an equation of the form $y=kx^2$ the square root of y must be proportional to x itself. Using the slide rule, read off the square roots of the numbers in column y and enter them in the vacant column headed $\sqrt{y}$. Since the x's in the third column and the $\sqrt{y}$'s in the fifth one are proportional their relative magnitudes can be determined by the black-thread method.

Plot their values on another graphic diagram, unless special directions to the contrary are given by your instructor, and determine the slope with the black thread. Since this is a case of proportion the thread must of necessity pass through the origin (§ 107), even though it may appear to lie less evenly along the row of points than it otherwise would. (Measurements which must necessarily fulfil a certain conditional relationship are called *conditioned measurements* and will be considered later. In this case notice that the temperature and density of water are not in themselves conditioned measurements, but we are attempting to make them satisfy a "*condition*," namely, that they shall follow the law $y=a+bx+cx^2$.) The value obtained for the slope of the black thread will probably be in the neighbourhood of 0·83.

x	$0{\cdot}69x$	$y+3$
0	0	3
2	3	6
4	11	14
6	25	28
8	44	47
10	69	72
12	99	102
14	135	138
16	177	180
18	224	227
20	276	279
22	334	337
24	398	401
26	467	470

TABLE OF VALUES FOR $y=0{\cdot}69x^2$ AND FOR $0{\cdot}69x^2+3$.

If $\sqrt{y}=0{\cdot}83x$, then, it follows that $y=0{\cdot}69x^2$, y being expressed in hundred-thousandths and being measured downward from the level of the vertex of the parabola. Instead of measuring y-values downward from the original 99997 it will be found easier to increase them by 3, as in the third column of the next table, and then measure the results downward from the level 100000, which is probably one of the ruled lines of the plotting paper.

On the graphic diagram already used for plotting the density of water lay off the values of $y+3$ (in 0·00001's of a unit of density proper) downward from the line $y=1{\cdot}00000$, noticing particularly that $x=0$ is now at the vertex of the curve (at 4°, not at 0°). If the work has been carefully done it will be seen that this new curve of the equation $y=-0{\cdot}69x^2$, or more precisely, $100000y$ * $=-0{\cdot}69x^2$ forms a fairly good approximation to the unknown relationship of the empirical values of density and temperature.

The last step that remains to be taken is to reduce the equation $100000y=-0{\cdot}69x^2$ to the original axes of temperature and density. Since x is zero when t (temperature) is 4, and 2 when t is 6, etc., it is plain that $x=t-4$ everywhere; in the same way $y=d-0{\cdot}99997$. Substituting these values in the original equation $100000y=-069x^2$ gives

$$d=0{\cdot}99986+0{\cdot}0000522t-0{\cdot}0000069t^2$$

which is the required law expressing the relationship between density and temperature.

* The large coefficient is needed because the y-values do not properly run into the hundreds of units as would appear from the table, but are condensed within a small fractional range (0·99564 to 0·99997) if they are to represent densities correctly.

112. Typical Curves.

Any law of change (as far as finite values of the variables are concerned) can be expressed by an equation of the form $y=a+bx+cx^2+dx^3+ex^4+\ldots$ if enough terms are used, but it will easily be understood that the method soon becomes difficult to handle. Sometimes the appearance of the curve makes it possible to guess that its equation is of some particular form (Figs. 27–35), or the form may be deducible from theoretical considerations. For a curve that deviates only slightly from a straight line it often happens that the black-thread method will give a linear law that is a sufficiently good approximation for practical purposes. The parabola can usually be fitted fairly well to a curve that shows a single upward or downward sweep, and is easier to apply than the exponential curve, which is so often used instead.

113. Linear Relationship by Change of Variables.

In the previous section the values of y were not proportional to those of x, but a straight line was obtained by plotting x and $\sqrt{y}$ instead of x and y. A transformation of this sort can always be made when the type of equation has been picked out and its numerical constants are to be determined. Thus if $y=e^{ax}$ the transformed equation $\log y=ax \log e$ shows at once that x and $\log y$ are proportional; if $y=ax/(b+x)$ it will be found that y and y/x follow the straight line law; if $y=a/x$ either of the variables will be directly proportional to the reciprocal of the other, etc.

pressure	*volume*
760 mm. Hg	8·1 cm.³
830	7·2
926	6·6
1022	5·8
1125	5·3
1230	4·9
1340	4·5
1410	4·1
1520	4·0
1600	3·9

EXPERIMENTAL DETERMINATION OF pv-VARIATION OF A GAS.

The law of variation for a constant mass of gas is known to be of the form $v=k/p$.

The volume of a certain mass of air was found to vary, under changes of pressure to which it was subjected, according to the numbers in the adjoining table. Assuming that the pressure and volume are inversely proportional, represent the relation between them by an equation after plotting a smoothed curve on a graphic diagram with a condensed x-scale (pressures) and an expanded y-scale (volumes): Since y is inversely proportional to x a new variable, $1/y$, can be obtained, which will vary directly with x. By plotting x and $1/y$ and holding a black thread so as to pass through the origin find an equation connecting x and $1/y$, and reduce it to a simple equation between x and y. The result will probably be in the neighbourhood of $v=6000/p$. Plot this equation on the same graphic diagram as the smoothed curve and notice how closely the two loci correspond.

114. Use of Logarithmic Paper.

A very common type of variation is that in which one of the variables is proportional to some power of the other one. For example, the distance traversed by a falling body is proportional to the square of the time of fall; the time of rotation of a planet or satellite about a particular central body is proportional to the 3/2 power of its mean distance; friction of water flowing through a pipe varies (approximately) as the 1·8 power of the velocity. The determination, for a set of experimental data, of the proper value of the exponent is not easily accomplished by means of an ordinary graph, because the various curves $y=x^2$, $y=x^3$, $y=x^4$, etc., all have the same general shape (Fig. 42).* If logarithms are taken, however, of both sides of the equation $y=x^n$ the equivalent equation $\log y=n \log x$ is obtained, showing that $\log x$ and $\log y$ are proportional. Accordingly, if $\log x$ and $\log y$ are plotted on an ordinary graphic diagram a straight line through the origin will be obtained. Now, just as a slide rule is constructed by marking the numbers 1, 2, 3, 4, etc., at distances which are really log 1, log 2, log 3, log 4, etc., so plotting paper can be constructed with logarithmic scales like those of the slide rule along each axis; consequently plotting x and y according to the numbered scales will really be plotting points at actual distances of $\log x$ and $\log y$ from the origin; and if these logarithms are known to be proportional it will be evident that *the graph of* $y=x^n$ *will be a straight line through the origin.* Such paper is called *logarithmic paper* and can be obtained from dealers who handle drawing materials. The lower left-hand corner of the sheet is the origin and is marked 1, 1, meaning of course log 1, log 1, or 0, 0; and the scales extend both upward and to the right from 1 to 10 as in the *C* and *D* scales of the slide rule, or sometimes from 1 to 100 as in the *A* and *B* scales. The latter arrangement would allow a single unbroken line to represent a greater part of the graph.

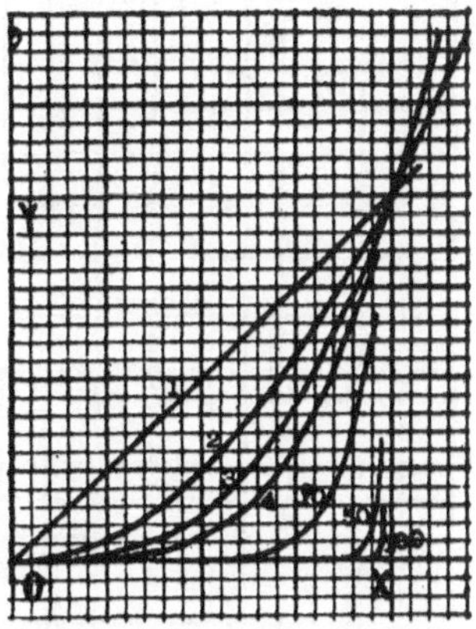

FIG. 42.—LOCI OF $y=x^n$.

The curves are drawn for $n=1, 2, 3, 4, 10, 50$, and 100. The letters X, Y, are placed at unit distance along their respective axes.

If a straight line is drawn on logarithmic paper without passing through the origin it must cut the y-axis at some point, such as k on the logarithmic y-scale. Then $(\log y)=(\log k)+n(\log x)$

* Experimental measurements are usually of positive quantities, so that the shape of these curves to the left of the y-axis or below the x-axis is not a determining factor.

(compare the ordinary equation of the straight line $y=a+bx$), or log $y=\log k+\log (x^n)$, or $\log y=\log (k\times x^n)$, or $y=kx^n$. In other words, *a straight line on logarithmic paper represents the equation* $y=ax^n$, *a* being the y-intercept as measured by the logarithmic scale, and n the true slope as measured by uniform scales. The diagram (Fig. 43)

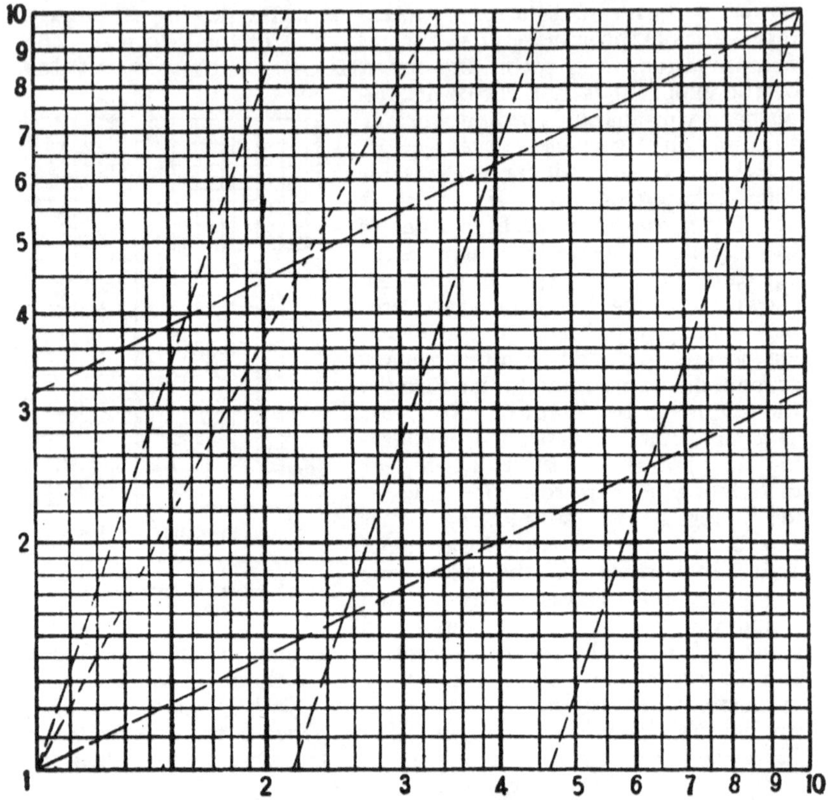

FIG. 43.—LOGARITHMIC PAPER.

The logarithmic scales cause the graph of any equation of the form $y=ax^n$ to appear like $y=a+nx$ on ordinary paper, *viz.*, as a straight line.

shows $y=\sqrt{x}$ and $y=x^3$ on logarithmic paper, the two lesser inclined parallel lines showing $y=\sqrt{x}$, and the three steeper parallel lines showing $y=x^3$. The remaining line shows $y=x^{1\cdot9}$. Any data that are suspected of following the law $y=ax^b$ may be plotted directly on this ruled diagram and their equation determined at once.

115. Semi-Logarithmic Paper.

Consider the equation of a straight line on paper that has a uniform scale along the x-axis and a logarithmic scale along the y-axis.

Instead of $y=a+bx$ or $\log y=\log a+b \log x$ its equation must

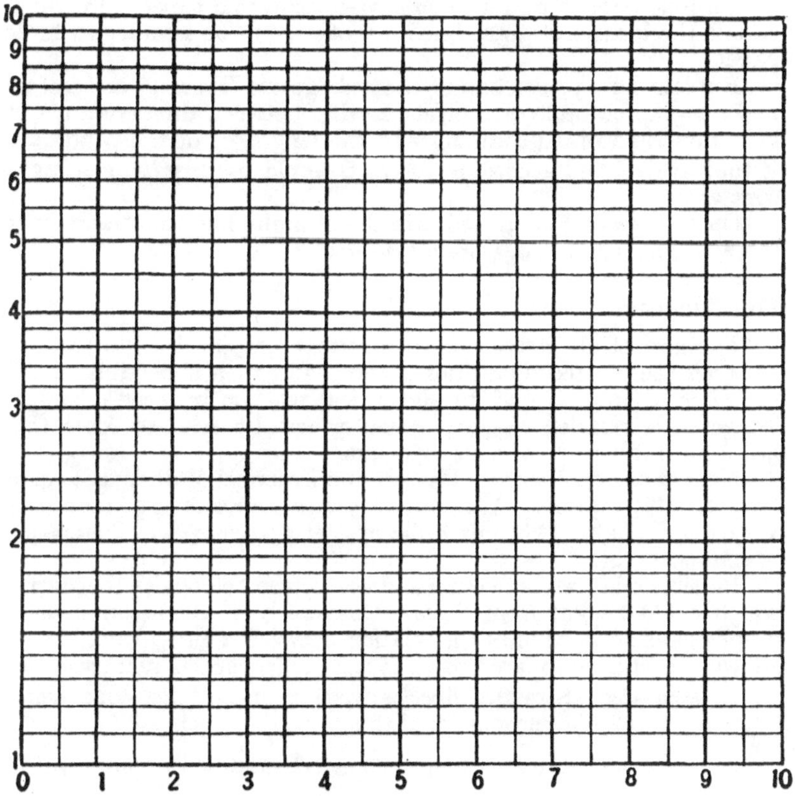

FIG. 44.—SEMI-LOGARITHMIC PAPER.

The logarithmic scale along the y-axis causes a straight line through the origin to represent proportionality between x and $\log y$, *i.e.*, $\log y=kx$. The exponential law of variation, $y=ae^{mx}$ is obviously reducible to the form $\log y=kx$.

now be $\log y=\log a+bx$. Clearing this of logarithms gives $y=a\times 10^{bx}$ or $y=a\times e^{bx}$ according to the base that is used.* Consequently, *a straight line on semi-logarithmic paper (Fig. 44) represents the "exponential" type of variation,* $y=ae^{mx}$.

* These two equations are of the same *form*; the first can be put into the form of the second merely by changing the *value* of b; for $10=e^{2\cdot 30}$, so 10^{bx} is identical with $e^{2\cdot 30bx}$.

Draw a straight line or hold a black thread on either Fig. 43 or Fig. 44 so as to represent the area of a circle (on the vertical scale) that corresponds to the radius (as indicated on the horizontal scale). The formula is $a=\pi r^2$, or $y=\pi x^2$.

Plot the locus of $y=2^x$ on squared paper for positive integral values of x up to 6 or 7. Plot the same equation up to $x=10$ on logarithmic paper,* and also on semi-logarithmic paper. In which case is the locus a straight line? Plot $y=2^{-x}$ on semi-logarithmic paper.

Explain how to use one of these diagrams (Figs. 43 and 44) to make a graphic table of the relationship between the period (p) of vibration time of a pendulum and the length (l) of the pendulum if they are related according to the formula $p=2\pi\sqrt{l/g}$, g being a constant.

On Fig. 43 or Fig. 44 indicate the straight line that represents the equation $xy=a$ (suggestion: $y=ax^{-1}$).

116. Summary.

The aim of all plotting is to illustrate graphically the relation between the plotted quantities x and y and then if possible to aid the experimenter to find an algebraical relation between x and y. To get this relation the experimenter usually tries to force this graph into the shape of a straight line.

Most books on Practical Physics and Practical Mathematics give help on this subject. Watson's *Practical Physics*, pp. 20–24, is very helpful, also Gibson's *Graphs*, Perry's *Practical Elementary Mathematics*, etc.

Possible relations between x and y are given below with the graphs which could be got from them. Conversely, if the recommended graphs gave straight lines the relation between x and y would be as shown. Ordinary squared paper is used unless specially mentioned,

1. $y=a+bx$. Straight line between x and y, the slope being proportional to b.
2. $y=\frac{a}{x}$. Curved line between x and y the slope being proportional to $1/x^2$. Straight line between y and $1/x$, and x and $1/y$.
3. $y=a+bx+cx^2$. Parabola between x and y (cf. § 111). Straight line between $\frac{dy}{dx}$ (*i.e.* the slope of the curve) and x.
4. $y=a+bx^2$. Straight line between y and x^2, also straight line between $\log(y-a)$, and x (a to be found by trial if proceeding from the x,y graph to the equation).

* If the specially ruled paper is not on hand the work can be done on thin paper laid over Fig. 43, and Fig. 44 of this book.

5. $y=\frac{a+bx}{x}$
$=\frac{a}{x}+b$
or $xy=a+bx$

Straight line between y and $\frac{1}{x}$, also straight line between xy and x.

6. $y=\frac{ax}{1+bx}$
or $\frac{y}{x}=a-by$
or $a\frac{x}{y}=1+bx$

Straight line between $\frac{y}{x}$ and y, also straight line between $\frac{x}{y}$ and x.

7. $y=\frac{ax^2}{1+bx^2}$ Straight line between $\frac{y}{x^2}$ and y, and between $\frac{x^2}{y}$ and x^2.

8. $y=ax^n$ Straight line between log y and log x. Slope of line gives n. n may be + or −.

9. $y=a(b+x)^n$ Straight line between log y and *log* $(x+b)$, b to be found by trial.*

10. $(y+a)=c(x+b)^n$ Straight line between log $(y+a)$ and $(x+b)$, a and b to be found by trial.

11. $y=a\,.\,10^{nx}$ Straight line between log y and x.

12. $y=ae^{nx}$ Straight line between log y and x.

13. $y=a+be^{nx}$ Straight line between log $(y-a)$ and x, a to be found by trial.

Equations 8 to 13 may, in many cases, be more simply obtained by graphing on logarithmic and semi-logarithmic paper.

14. $y=\frac{ax+b}{cx+d}$ See Gibson's *Graphs*, p. 96.

15. $\log y=a-\frac{b}{x}$ Straight line between log y and $1/x$.

16. $y=\frac{a}{(1+bx)^m}$ See Watson's *Practical Physics*, § 5.

The use of an interpolation table (see later) helps sometimes in the solution.

Exercises IX

1. With the black thread find the best straight-line approximation for the data or smoothed curve of § 113 over the range of the experiment.

---

* See Watson's *Practical Physics* (Longmans), § 5. If n is known plot $\sqrt[n]{y}$ against x. This will enable b to be found.

2. Name the kind of curve that would correspond to the equation obtained by solving

$$\begin{cases} 8{\cdot}1 = a + b(760) + c(760)^2, \\ 5{\cdot}3 = a + b(1125) + c(1125)^2, \\ 4{\cdot}0 = a + b(1520) + c(1520)^2, \end{cases}$$

for a, b, and c, and substituting the values so obtained in the equation $y = a + bx + cx^2$. For what purpose could the resultant equation be used ?

3. Explain how an equation of the form $y = e^{ax}$ could be determined for the smoothed curve of the last section.

4. The resistance R of a platinum thermometer is connected with the temperature t by the equation $R_t = R_0(1 + at + bt^2)$, where a and b are constants. The resistances at 0·0°, 100·0°, 443·8° C. are 257·4, 356·9, 673·0 units, respectively. Find a and b. What is the shape of the curve plotted between values of R and t ?

5. If e^{ax} is always identical with $(e^a)^x$ and if $2{\cdot}718^{2{\cdot}303}$ is equal to 10, turn to Fig. 32 and see how the curve $y = 10^x$ would run. (The reciprocal of 2·303 is 0·4343.) Decide how the curve $x = 10^y$ would compare with it. Have you drawn the latter curve previously ?

6. A line is indicated on Fig. 43, which passes through the points (1, 1) and 3, 8) of the logarithmic paper. Determine the equation which it represents.

7. The vanes of a Crookes' Radiometer are caused to revolve by the radiation from a red hot ball. Observed data are :—

Distance of vanes from ball, D	30	40	50	60	70	80	80	100 cms.
Revolution per minute, N . .	43·5	24·5	15·8	11·1	8·15	5·75	4·23	3·45

Find the relation between distance and $R.P.M.$

8. A lighted Pointolite lamp caused the vanes of a Crookes' radiometer to revolve. The relation between the distance of the radiometer from the source of radiation and the period of revolution is as follows :

D cms.	8·3	10·0	11·8	14·3	16·0	18·0	22·5	26·8 cms
Period	1·33	1·96	2·98	4·1	5·4	7·7	12·5	23·1 secs.

Find the equation between D and the $R.P.M.$

9. In an experiment to determine the relation between two variables the following results were obtained :

x pounds.	y inches.
10	−1·30
20	−0·91
30	−0·49
40	−0·03
50	0·34
60	0·81
70	1·16
80	1·65

Plot a graph showing the relation between x and y, use the black thread and express the relation in the form

$$y = mx + C$$

10. The following numbers were obtained in an experiment with a small crane:

E, lbs.	4·5	10·7	15·6	21·7	27·0	32·8	38·1	43·6
R, lbs.	0	20	40	60	80	100	120	140

Find E in terms of R.

11. The following results were obtained in an experiment with a machine:

Effort.		Resistance.
lbs.	oz.	lbs.
3	7	0
4	9	7
5	6	14
6	10	21
7	8	28

Plot the results on squared paper and find the equation expressing the relation between Effort and Resistance.

12. The following readings were taken in an experiment with a pulley block. E is the effort which just overcomes the resistance R.

E pounds.	R pounds.
1½	0
3¼	14
5	28
6¾	42
8½	56
10	70
11½	84
13¼	98
14¾	112

The effort was read to the nearest quarter-pound.
Find graphically and in *one* other way the relationship between E and R.
Express this in the form $E = mR + c$.

13. The following numbers were obtained with a machine:

Effort, E .	6	7	8½	10	11	13	16	17 pounds
Resistance, R .	22½	27½	37	46½	5·1	65½	81	89½ pounds

The results are known to follow a simple law $E = aR + b$.
Find, by the use of squared paper, the most probable results of a and b.
The velocity ratio being 6, find the efficiency of the machine when $E = 12$ pounds.

14. Find the relation between p and v in the following set of readings:

p	333	285	250	200	160	140	120
v	120	140	160	200	250	286	333

15. The following are corresponding readings of pressure (p) and volume

(v) with a given body. Find the relation between p and v. To do this try plotting: (1) p against v, (2) p against $\frac{1}{v}$, (3) $\log p$ against $\log v$.

p, cms. of mercury	81·5	86·7	92·9	98·7	118·6	125·9	132·8	141·2	148·6
v, c.c.	37·65	35·41	33·11	31·20	26·00	24·54	23·31	21·94	20·76

16. Find the relation between p and v in the following set of readings:

p	376	436	543	583
v	419	365	297	277

17. Find the relation between s and t in the following set of readings:

t, secs.	0·2	0·4	0·6	0·8	1·0	1·2
s, cms.	20	78	176	314	490	706

18. The Cambridge Instrument Co. give the following table for the E.M.F. generated by a platinum-platinum iridium couple at the following temperatures (the cold junction is at 0° C.):

Temperature	0	25	100	200	300	400	500	600	700
Millivolts .	0	0·31	1·12	2·47	3·91	5·41	6·97	8·57	10·21

Temperature	800	900	1000	1100	1200	1300	1400	1500	1600
Millivolts .	11·88	13·57	15·30	17·05	18·82	20·60	22·41	24·23	26·07

Find the equation between millivolts (e) and temperature (t° C.).

19. The following observations were made of the excess temperature of a cooling body at the times stated:

Time, t, in minutes.	Excess temperature, θ, in degrees.
0	40·5
4	27·0
8	18·0
12	12·0
16	8·0
20	5·3

Tangents to the cooling curve will give the rate of cooling.

Plot a graph between the rate of cooling and the excess temperature and find its equation. This gives us the Law of Cooling (known as Newton's Law)

20. The following readings of excess temperature were taken at two-minute intervals in an experiment on cooling:

46·2	38·5	32·3	27·1
44·8	37·5	31·4	26·2
43·6	36·4	30·5	25·4
42·1	35·4	29·4	24·8
40·8	34·3	28·6	23·9
39·7	33·4	27·9	23·3

Plot the cooling curve and confirm Newton's Law of Cooling.

Plot also a curve between the logarithm of the excess temperature and the time and show that the equation of the curve may be written

$$\log \theta = A - \lambda t$$

Find the values of A and λ.

21. The table gives the maximum vapour pressures of water at the temperatures stated.

°C.	cms. of mercury.	°C.	cms. of mercury.
0	0·46	120	148·9
20	1·75	140	270·9
40	5·51	160	463·3
60	14·92	180	751·4
80	35·51	200	1164·7
100	76·00	220	1738·0

Plot the vapour pressure curve of water, taking values at every 20° C. from 0° C. to 120° C. Find by drawing tangents the value of $\frac{dp}{d\theta}$ in mm. per degree at 60°, 80°, and 100° C.

22. Plot the logarithm of the vapour pressure of water against the reciprocal of the *absolute* temperature for temperatures from 0° to 100° C. Find the equation of the graph as accurately as you can.

Substitute the temperature 323° A. in your equation and find the difference between the calculated pressure and the pressure given in the tables for 50° C. (9·23 cms.)

23. Do the same from 100° C. to 200° C. Find also the differences between the values calculated from the graph and the correct values at the temperatures given in the above table.

24. The table shows corresponding values of x and y. Find the relation between x and y.

x	0	0·20	0·50	1·0	2·0	3·0	4·0	5·0	6·0	9·0	19·0
y	0	0·33	0·50	0·60	0·67	0·69	0·70	0·71	0·72	0·73	0·74

25. (*a*) The table contains corresponding values of x and y. Plot a curve between x and y. Now endeavour to get a straight line plot between some functions of x and y and deduce the equation between x and y.

x	1·0	2·0	3·0	5·0	8·0	10·0	20·0	70·0	120·0	220·0	470·0	970·0	1700·0
y	3·2	6·3	9·1	14·3	21·0	25·0	40·0	70·0	80·0	88·0	92·0	97·0	98·3

Plot first over the short range $x=1$ to $x=10$, and then over the whole range.

(*b*) Do the same with the following values of x and y:

x	3·0	6·0	9·0	15·0	30·0	60·0	90·0	150·0
y	9·5	18·5	25·9	39·3	63·2	86·5	95·0	99·3

Finally, compare the forms of the two curves and the forms of the two equations.

26. The coefficient of viscosity (η) of glycerine at 0°, 5°, 10°, 15°, 20°, 25°, 30° C. is 46·0, 31·5, 21·0, 13·5, 8·5, 5·5, 3·5 C.G.S. units. Find the relation between η and the temperature.

27. If p is the pressure in pounds per square inch and v the volume in cubic feet of one pound of saturated steam, p and v are related as stated in the table:

p	6·86	14·70	28·83	60·40	101·9	163·3	250·3
v	53·92	26·36	14·00	6·992	4·280	2·748	1·853

Find the relation between p and v. Use logarithmic paper.

28. The following results have been obtained with a U-3 electric vacuum valve. V denotes the voltage applied between anode and cathode and I the current sent through the valve.

V	13	20	32	41	50	58	66 volts
I	2½	5	10	15	20	25	30 milliamperes

Find the relation between I and V in the form $I=f(V)$. Use logarithmic paper.

29. It has been suggested that the vapour-pressure curve of water between 0° C. and 100° C. has the equation:

$$\log_{10} p = 7{\cdot}39992 - \frac{a}{\theta^b}$$

where a and b are constants, p is the pressure in cms., and θ is the absolute temperature on the C scale. Plot and find the value of b. (Hint: Plot $(7{\cdot}39992 - \log p)$ against θ on logarithmic paper.)

30. The saturation pressure of steam in pounds per square inch is as follows:

Temperature.	p	Temperature.	p
°C		°C	
0	0·0892	120	28·808
20	0·3399	140	52·482
40	1·0703	160	89·800
60	2·8873	180	145·59
80	6·8627	200	225·24
100	14·689	215	304·01

Rankine suggested the formula: $\log_{10} p = A - \frac{B}{T} - \frac{C}{T^2}$, where A, B, C are constants and T is the absolute temperature. Find A, B, C. (Hint: Write y for $\log p$ and x for $\frac{1}{T}$, thus getting $y = A - Bx - Cx^2$.)

31. The table shows corresponding values of x and y. Find the relation between them.

x	1	2	3	4	5	6	8	10	12
y	138	78	58	48	42	38	33	30	28

32. The following values of pressure and volume were obtained in a gas engine test. Find the relation between them.

p, pounds per sq. in.	150	130	125	102	88	76	66	58	45
v, cubic feet . . .	3·36	3·75	3·96	4·50	5·00	5·60	6·15	6·84	8·30

33. The following values of p and v (in arbitrary units) were measured in a gas engine test. Find the relations between p and v.

v	12·95	15·91	18·87	21·80	24·78	27·73
p, on expansion .	27·17	21·32	17·23	14·26	12·30	10·65
p, on compression	9·38	7·55	6·20	5·19	4·40	3·75

34. In an experiment on the friction of a cord over a post the values of N given below were just sufficient to overcome a value of $M=50$ gms. for the stated amounts of lap n. Find the relation between M, N, and n.

Lap n .	$\frac{3}{4}$	1	$1\frac{1}{4}$	$1\frac{1}{2}$	$1\frac{3}{4}$	2	$2\frac{1}{4}$	$2\frac{1}{2}$	$2\frac{3}{4}$
N, gms.	105	145	185	250	320	400	530	700	920

35. Find the relations between x and y.

(*a*)

x	0	1	2	3	4	5	6
y	18	32	50	72	98	128	162

(*b*)

x	0	1	2	3	4	5
y	16	92	240	484	848	1356

36. Find the relation between x and y.

x	y	x	y
0·0	12·0	9·0	2·1
1·0	10·1	10·0	2·0
2·0	8·4	11·0	2·1
3·0	6·9	12·0	2·4
4·0	5·6	13·0	2·9
5·0	4·5	14·0	3·6
6·0	3·6	15·0	4·5
7·0	2·9	16·0	5·6
8·0	2·4	17·0	6·9

37. The viscosity of water varies with the temperature as shown.

Temperature, t	0°	10°	20°	30°	40°	50°	60°	70°
Viscosity, η .	·01793	·01311	·01006	·00800	·00657	·00550	·00469	·00400

Find the relation between η and t. (Hint: Plot log η against log $(t+40.)$)

38. Raoult found the following values for the lowering of the freezing-point of water when 1 gm. of the substance stated was dissolved in 100 gms. of water:

	°C.		°C.
Methyl alcohol .	0·541	Acetone	0·294
Ethyl alcohol . .	0·376	Formic acid . . .	0·419
Glycerine . . .	0·186	Acetic acid . . .	0·317
Glucose . . .	0·107	Oxalic acid . . .	0·182
Cane sugar . .	0·054	Ammonia . . .	1·117

Find the relation between the molecular weight and the depression of the freezing-point.

39. Sir George Stokes calculated the velocities of fall of very minute drops of water through the air:

Radius, cm.	Velocity, cm./sec.
0·0005	0·3
0·001	1·3
0·005	32·0
0·010	126·0

Find the relation between radius and velocity.

I

40. The surface tension T at 20° C. of solutions of common salt and water of specific gravity, d, are as follows:

d (grms. per c.c.).	T (dynes per cms.).
1·0000	75·2
1·0360	77·6
1·1074	80·5
1·1932	85·8

Express T in terms of d.

41. Coils of wire could be wound on a circular hoop of 65 cms. diameter placed with its axis in the magnetic meridian and a magnetic needle at the centre was set in vibration. A constant current of 1·17 ampere was maintained in the wire from which the loops were formed.

Conditions on loop.	Time of 20 swings.
No current	124·8
One turn	115·7
Two turns	109·9
Three turns	104·2
One turn (current reversed)	134·6

Find the relation between the number of turns n and frequency of vibration N.

Find the value of the horizontal component of the earth's magnetic field. $\left(\text{Field at centre due to current and earth} = \frac{2\pi nC}{r} \pm H\right)$

42. A constant current of 1·0 ampere was sent through loops of wire wound on a hoop of diameter 32·6 cms. placed in the magnetic meridian; the number of loops could be varied by winding or unwinding the wire. A magnetic needle at the centre was deflected and the readings of its extremities taken.

Number of turns.	Reading of ends of needle.			
	As number of loops were on the increase.		As number of loops were on the decrease.	
	Near end.	Far end.	Near end.	Far end.
0	0°	0°	0°	0°
1	$12\frac{1}{2}$°	$12\frac{1}{2}$°	$12\frac{1}{2}$°	13°
2	$24\frac{1}{2}$°	$24\frac{1}{2}$°	25°	25°
3	$34\frac{1}{2}$°	34°	35°	33°
4	$42\frac{1}{2}$°	$42\frac{1}{2}$°	42°	43°
5	49°	49°	49°	49°

Find the relation between number of turns n and angle of deflection.

If $C = \frac{Hr}{2\pi n} \tan\theta$, find H.

43. Plot $y_1 = (\tan x - x)$ and $y_2 = (x - \sin x)$ for values of an angle x (expressed in circular measure) up to approximately $\frac{\pi}{6}$ radians. Find for what values of x the differences y_1 and y_2 amount to 5 per cent. of x.

44. Plot the curve $y = \log x$ from $x = 1$ to $x = 100$. Draw tangents at points whose abscissæ are 5, 10, 30, 60, 90 and from their slopes draw the first derived curve. Find now the equation of the derived curve.

45. In an experiment with long bars of brass and steel for the determination of the relative thermal conductivities of these two substances the following temperature readings (degrees C.) were taken at equidistant holes 15 cm. apart:

Brass	122	73	47	34	27	$23\frac{1}{2}$	22	21	20	20
Steel	107	51	31	24	$21\frac{1}{2}$	$20\frac{1}{2}$	20	20	20	20

Plot the temperature gradient curves and from three pairs of corresponding points deduce three values of the relative conductivity of the two alloys. Find also the temperature gradients at the second holes on both bars.

(Corresponding points are points along the two bars which have the same temperatures and conductivities are proportional to the squares of the distances between corresponding points. See a text-book on Heat.)

46. The table gives the relation between the coefficient of viscosity of air and the temperature.

$t°$ C. .	−60	−40	−20	0	20	40	60	80	100	120	140	160	180	200
$\eta\times10^6$	141	151	162	172	182	192	202	211	221−	229+	238	247	256+	265

Let θ = the absolute temperature. Find the relation between η and θ. If one equation will not extend over the whole range divide into two ranges.

47. In an experiment on the grinding and sifting of sand, Martin obtained the following results:

Diam. of particles (d), ins.	Percentage number of particles (N).
$\cdot1\times10^{-4}$	58·75
2×10^{-4}	23·38
3×10^{-4}	9·75
4×10^{-4}	3·97
5×10^{-4}	2·17
6×10^{-4}	1·44
7×10^{-4}	0·54

Find the law connecting N and d.

48. A sand is screened. The table gives the fraction of the weight of sand which passes through the given screens.

Screen openings in ins.	Fraction of weight.
x	W
0·185	0·97
0·130	0·88
0·094	0·78
0·065	0·69
0·046	0·59
0·033	0·47
0·0235	0·345
0·0165	0·250
0·0118	0·195
0·0082	0·142
0·0058	0·101
0·0041	0·077
0·0029	0·060

Find the relation between W and x.

49. A tank is being emptied in such a way that half the water runs out in 5 minutes. Plot curves between (1) the volume of water (V) in the tank and the time (x) in minutes, (2) the logarithm of the volume and the time. If the original volume is 128 cubic feet, find the equation between volume and time.

50. The rotation (ϕ) of plane polarized light by a slab of quartz 1 mm. thick has been measured for light of different wave-lengths (λ).

ϕ	17° 24′	17° 30′	21° 43′	27° 46′	32° 40′	42° 20′
λ	6867	6563	5893	5270	4861	4308

Find the relation between ϕ and λ.

51. The activity of thorium emanation was measured at different instants with the following results:

Time .	0	1	2	3	4	5 minutes
Activity	50	22½	10	4½	2	1 units

Plot the curve of decay and find its equation, also find the time for the activity to decay to half value.

52. The following results were obtained in a curve on magnetic repulsion:

Polar distance $=d$	29	25	23	21	18	17	15½	14	13
Force $=F$ . .	3	4	5	6	8	9	10½	13	15

Find the relation between F and d.

53. Shaxby and Evans measured the pressure at the bottom of a column of small lead shot for different heights of columns.

Height, cms.	1·8	3·2	5·2	8·1	11·0	14·4	18·0	26·0	41·2	46·1	53·5
Pressure $\frac{\text{gm.}}{\text{cm.}^2}$	13·7	20·8	24·5	28·6	31·2	33·4	35·3	36·4	36·7	37·1	37·1

Plot a curve between p as ordinate and h as abscissa.
Show that the relation between h and P is of the form

$$p=p_m(1-C\epsilon^{-kh})$$

where p_m = the maximum pressure and C and k are constants. Find C and k.

54. Porter measured at different depths (y) the numerical concentration of gamboge particles (n per c.c.) in a gamboge suspension in distilled water. Results:

y, cms.	0·023	0·033	0·043	0·063	0·083	0·103	0·123	0·143
$n\times10^{-6}$	0·95	2·14	3·01	7·14	8·69	9·52	9·76	9·76

Plot the curve between y (abscissa) and n (ordinate). Porter suggested the equation

$$\log_e\frac{n}{1-bn}+\frac{1}{1-bn}=Ky+A$$

where b, K, and A are constants.

$$\text{(Note: } b=\frac{1}{n_\infty}=\frac{1}{9{\cdot}76\times10^{+6}}=10{\cdot}2\times10^{-8}\text{)}$$

Find k.

55. The emission of radiation of wave-length λ from a black body at temperature 2000° absolute is given by the equation

$$E_\lambda = \frac{3{\cdot}72 \times 10^{-5}}{\lambda^5} \cdot \frac{1}{\epsilon^{\frac{1{\cdot}44}{2000\lambda}} - 1} \text{ ergs per sq. cm. per sec.}$$

For what value of λ is E_λ a maximum? (Start at $\lambda = 0{\cdot}00001$ and work by steps up to $\lambda = 0{\cdot}0002$. Plot E_λ against λ to find the maximum.) Find the maximum value of E_λ.

56. An example of the speed of a chemical reaction.

When cane sugar in aqueous solution is treated with a little acid it inverts, *i.e.* changes into a mixture of dextrose and lævulose.

$$\underset{\text{Cane sugar.}}{C_{12}H_{22}O_{11}} + H_2O = \underset{\text{Dextrose.}}{C_6H_{12}O_6} + \underset{\text{Lævulose.}}{C_6H_{12}O_6}$$

One experiment yielded the following results:

Time in minutes.	Amount of cane sugar present.
0	130·9
45	113·9
90	98·9
150	81·4
210	67·4
285	53·9

If the equation between the amount q and the time t is

$$\log_{10}\frac{a}{q} = kt$$

find k. (Hint: First plot q against t, then $\log q$ against t.)

57. The hydrolysis of methyl acetate. The methyl acetate decomposes into methyl alcohol and acetic acid, which is determined by titration with barium hydroxide solution ($CH_3.COOCH_3 + H_2O = CH_3.COOH + CH_3OH$).

The first line of the table gives t, the time in minutes from the moment of mixing the acetate with hydrochloric acid (1 c.c. methyl acetate + 10 c.c. normal hydrochloric acid + 15 c.c. water).

The second line gives x, the number of c.c. of baryta solution required (in excess of that required to neutralize the hydrochloric acid).

t	14	34	59	89	119	159	199	239	299	399	539	∞
x	0·92	2·14	3·52	4·91	6·15	7·59	8·82	9·77	10·88	12·13	13·09	14·11

Find the relation between t and x.

(Hint: calling the last value of x, A, we see that $A - x$ at any time is proportional to the amount of methyl acetate present at that time.)

58. Water and benzene are mutually insoluble. They are put into a test tube and some phenol (carbolic acid) added. The contents are shaken. The phenol dissolves in both water and benzene. The solutions are analysed. Let c_1, c_2, denote the concentrations of phenol in the water and benzene layers.

c_1	0·050	0·123	0·200	0·327	0·500	0·750
c_2	0·100	0·159	0·200	0·253	0·316	0·390

Find the relation between c_1 and c_2.

CHAPTER X

INTERPOLATION AND EXTRAPOLATION

117. Definitions.

The process of drawing the locus of an equation by plotting a few isolated points and filling up the intermediate positions with a smooth curved line involves the tacit assumption that the values of y for the intervening values of x would have been found to vary in this predeterminate manner if they had been calculated. That is, the value of a function that corresponds to a certain magnitude of its independent variable need not be obtained by calculation in all cases, but may often be determined by comparing it with the values which the function is known to have when the variable is larger and smaller than in the particular case that is under investigation. When a few known values are used for the purpose of determining an intermediate unknown value the latter is said to be found by *interpolation*. For example, if the population of a city were known for each of the years 1881, 1891, 1901, and 1921, one could estimate fairly accurately what the population amounted to in the year 1895, even if the given values should not follow any known law or any recognizable type of curve.

The process of using a certain range of values for determining a value that lies outside of that range is called *extrapolation* (Latin, *extra*, outside ; *inter*, between ; *polire*, to make smooth). Thus, from the data mentioned above it would be possible to make some kind of an estimate of the population for the year 1875, or for 1930, or perhaps even for 1940.

Turn to your diagram of the daily variations in body temperature (§ 89) and determine the normal temperature of the human body at 9.30 a.m. It should be about 37°·07, ·08, or ·09. What was the temperature of this individual at 9.30 a.m. on the day of the experiment. Ans. : probably about 37°·15 or 37°·16.

Interpolation is a process that is trustworthy only when the data are sufficiently numerous and are given at sufficiently close intervals, and when their variation is not too irregular. It would be impossible to interpolate the values $y=x^3-x$ (Fig. 25) from the three points $(-1, 0)$, $(0, 0)$, $(+1, 0)$; or to fill in a free-hand parabola for $y=x^2-2x-3$ if the only data were the points $(-1, 0)$ and $(+2, -3)$.

118. The Principle of Proportionate Changes.

It is always necessary to make an assumption of some kind when a process of interpolation is used. In obtaining logarithms from a table by interpolation it is assumed, for example, that the logarithm of 2718 is 8 tenths of the way from log 2710 to log 2720 (§ 59), or in general that any change in a logarithm is proportional to the change in the corresponding natural number. This is an instance of the *linear law* (§ 107), and may be expressed as $\Delta(\log x)=k\Delta x$. It is true only for *small* differences (log 300 − log 200 is not equal to log 200 − log 100), because the curve of $y=\log x$ is nowhere nearly a straight line unless a very small stretch of it is considered by itself.

Plot a graphic diagram of $y=\log x$ from $x=0$ to $x=10$, using a large scale on the x-axis (10 □'s = 1 unit) and a much larger one on the y-axis (10 □'s = 0·1 unit). Do *not* use any x-values except 0·1, 0·2, 0·4, 0·6, 1·0, 2·0, 3·0,.. . . 10·0. Draw a free-hand curve as smoothly as possible through the corresponding points. *Measure* the y-values for $x=3·5$, 2·5, 1·4, and 0·8; then verify each by referring to the tables. Notice that the points on the curve for which $x=1, 2, 3$, do not lie in a straight line. Notice that the short

FIG. 45.—ELEMENTARY STRAIGHTNESS.

When a circle that is drawn around any point on a *smooth* curve is made smaller and smaller it is cut more and more nearly into exact halves by the curve itself.

stretch of curve that includes the points $x=2710$, 2718, 2720 has no perceptible curvature. If the points are specified with sufficient accuracy, however, it will be found that no three of them lie in the same straight line; thus four-place logarithms may be safely interpolated if the values for every three-figure natural number are given, but five-place accuracy for the logarithms necessitates four-figure values for the natural numbers in the first part of a logarithm table, that corresponds to the more sharply curved part of the graph.*

The fact that a curve is "smooth" means that a very short stretch of any part of it deviates very little from a straight line. In other words, if a small circle is described around any point of such a curve its circumference will be cut by the curve at two points whose angular distance approaches 180° as the circle is made smaller and smaller (Fig. 45). This property of a curve is known as *elementary straightness*, the term "element" being used in the sense of "a very small portion," and is characteristic of the curves of all

* Reference to Kaye and Laby's tables, shows that although the same set of proportional parts may be used for any one horizontal line in a four-figure logarithm table for numbers greater than 20, yet for numbers less than this it is advisable to use one set of proportional parts for the first half of a line and another set for the second half.

equations in which y can be expressed as a rational function of x. Some " transcendental " equations * lack elementary straightness at a few points, but in general the process of linear interpolation is applicable to any tabular values that are given at sufficiently small intervals.

119. Examples of Linear Interpolation.

Consult the table in § 111 and state the density of water at 21° C.

Water boils at 100° C. when the barometric pressure is equal to that of a 760-mm. column of mercury; to make it boil at 101° C. the pressure must be raised to 788 mm. What should an accurate thermometer register in boiling water when the barometer stands at 775?

Plot a few points for the equation $y=x^2-2$ and fill out a free-hand curve. When x is 1 y is -1; when x is 2 y is $+2$. Accordingly, if the locus were not curved it would intersect the x-axis at $x=1\frac{1}{3}$. Substituting $x=1{\cdot}3$ gives $y=-0{\cdot}31$; substituting $x=1{\cdot}4$ gives $-0{\cdot}04$; substituting $x=1{\cdot}5$ gives $+0{\cdot}25$. The stretch of the curve that lies between 1·4 and 1·5 (Fig. 46) is so nearly straight that the point where it intersects the x-axis can be found quite accurately by the law of proportionality: $\Delta y=0{\cdot}29$ for $\Delta x=0{\cdot}1$, so that the Δx required to be added to 1·4 to give a Δy of 0·4 equals 0·04/0·29 of 0·1, or 0·014. Substituting $x=1{\cdot}414$ gives $y=-0{\cdot}000604$; substituting $x=1{\cdot}415$ will be found to give $y=+0{\cdot}002225$. Another application of the principle of proportionality gives $x=1{\cdot}414+0{\cdot}000214$ or 1·414214, and thus the calculation of the value of $\sqrt{2}$ proceeds with continually increasing rapidity. This, of course, is a determination of the value of one root of the equation $0=x^2-2$, and is known as the method of *false position*, or of *double position*. It can be used for any equation whatever containing one unknown quantity by arranging it in the form $\phi(x)=0$ † and drawing a graphic diagram of $y=\phi(x)$.

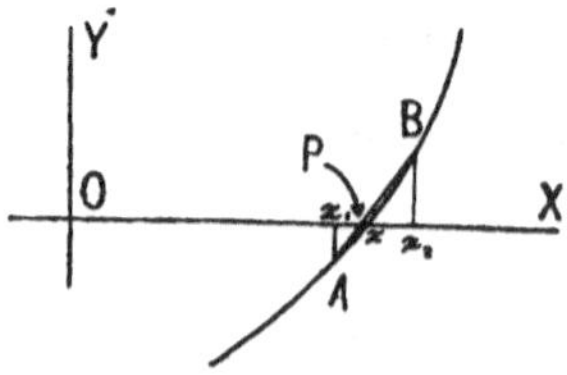

FIG. 46.—SOLUTION BY " DOUBLE POSITION."

Any equation can be solved approximately by aid of a graph. Then by using two x-values that are near the required root, x, and interpolating a y-value along a straight line it is possible to find as close an approximation, P, as is desired.

A delicate beam balance has a pointer which swings to a position of rest at 8·0 on an arbitrary scale when an unknown mass is balanced

* A *transcendental* equation is one that involves non-algebraical functions; for example, $y=x \log x$. The curve of this equation cuts the elementary circle around (0, 0) at only *one* point, *i.e.* comes to an abrupt end at the origin.

† The expression $\phi(x)$ means " any function of x," and may be used as a general form to denote $3x^2$, $\log x$, $2a^x$, or any other function, just as the general symbol m may be used to stand for the number 2, or 100, or any number whatever. As alternative notations $f(\ )$, and $F(\ )$ are often used.

against standard weights of 14·837 grams, but stops at 10·5 when the weights are changed to 14·836. What weights would be required to bring the pointer to the central position, which is at 10·0 on the scale? Ans.: $14{\cdot}837-\frac{1}{5}(0{\cdot}001)$; or 14·8362.

No. of messages	Charge
600	£10
800	12
1000	14
1200	16

EXAMPLE OF A "READINESS-TO-SERVE" CHARGE INCORPORATED WITH A RATE CHARGE.

A telephone company charges for measured service at the yearly rates given in the table. It will be plain that an increase of 200 messages means an additional cost of two pounds, making the rate for an increased number of messages equal to one pound per hundred messages. At this rate 600 messages should cost £6; accordingly, it is evident that the rate for any number of messages is made up of two parts, a flat charge of £4 plus a message rate of £1 per hundred. If the number of messages is represented by x and the total cost by y pounds, then the equation connecting them will be $y=4+\frac{1}{100}x$, showing straight-line variation without proportionality.

120. Graphic Interpolation.

In general, tabular values do not follow the straight-line type of variation, and rather complicated formulæ may be required for purposes of interpolation (see § 125). In case a graph can be drawn, however, there is usually no difficulty in constructing a smooth curve (or "smoothed" curve, as occasion may require) and obtaining any intermediate values simply by measuring them on the graph. The process is sometimes uncertain or erroneous if the given values are not close enough together or if their variation is too irregular. It will have been noticed that the problem of finding intermediate values is closely allied to the problem of finding a law of variation or of finding the equation of a given curve. In case a law or equation is known, unless it is a complicated one it will usually be found easier to substitute and calculate values than to interpolate them.

121. Graphic Tables.

If the variation of two quantities is known to follow the linear law it is often convenient to make a graphic table by plotting any two points and ruling a straight line through them.

Lay off a scale of values from 0 to 20 along the x-axis and label it "inches." Lay off a scale from 0 to 50 along the y-axis and label it "centimetres." Rule a straight line through the two points (0, 0) and (13, 33). Explain how a diagram of this sort can be utilized.

The temperatures $-40°$ F.$=-40°$ C., $32°$ F.$=0°$ C., $212°$ F. $=100°$ C. Plot a graph laying off the Fahrenheit readings and the corresponding Centigrade readings. Rule a straight line through the points. Read off the C. temperature corresponding to 68° F., the F. temperature corresponding to 50° C., and so on. Check by the formula $F° = \frac{9}{5} C° + 32°$.

According to Hooke's Law, the difference in length (stretching) of a spring is proportional to the difference in force applied to it. If a spring that is hung in front of a scale has a length of 12·00 cm. when no weight is attached to it, and becomes 14·85 cm. long when a weight of 1 gm. is hung on it, construct a graphic table which will enable you to reduce its indicated centimetres to grams of weight (see § 93).

122. Interpolation along a Curve.

If the change in two variables is not in accordance with a linear law it is possible to use certain interpolation formulæ for obtaining intermediate values (see § 125), but it is usually much easier to make use of graphic methods. The known data are plotted as a series of points, and these are either connected by a smooth curve or are investigated with a view to discovering an equation that will adequately represent them. If they appear to lie along a curve that has a vertical or a horizontal asymptote (Fig. 28) the hyperbola $xy=k$ may be tried, with a suitable choice of temporary axes and scales. If the curve has a single upward or downward sweep the exponential curve $y=e^{ax}$ (Fig. 31) is frequently used, but the parabola $y=ax^2$ (Fig. 27) is generally easier to handle and can usually be fitted to the given points just as satisfactorily. It may be turned so as to have its axis horizontal, if this position seems more suitable, by interchanging the variables and writing the equation $ax=y^2$.

In trying to fit a parabola to the part of the curve of $y=\log x$ (Fig. 32) that lies between $x=5$ and $x=15$ would you prefer to have its axis horizontal or vertical? If vertical, would its vertex be directed upward or downward? If horizontal, would its vertex be directed to the left or to the right? Plot a logarithmic curve rapidly, on a small scale, if there is any difficulty in answering the questions; compare it with the curves for $y=x^2$, $y=-x^2$, $x=y^2$, and $x=-y^2$.

123. Insufficiency of Data.

When certain tabular values are given and others are to be obtained by interpolation it must always be remembered that the known values are the only actual data and that nothing else can be obtained without making some kind of an assumption (§ 118). If it is assumed that the points all lie along a smooth curve there is always a possibility that the assumption is incorrect. It may even happen that the given points appear to be irreconcilable with a smooth curve or with a uniform law, as in the following case: Electricity is sold by the kilowatt-hour (abbreviated KWH.)

and four consumers pay the same rate to one company. The first is charged \$1·08 for 9 KWH.; the second, \$1·47 for 21 KWH.; the third, \$0·99 for 11 KWH.; and the fourth, \$1·26 for 18 KWH. Find the rate.

Plot the four points, using 1 square along the x-axis for each KWH. and 1 square along the y-axis for each \$0·10. It will be seen that it is impossible to decide where a smooth curve should run. This is a case of what is called a "step meter rate" and gives a broken line, not a smooth curve. The rate is "12c. per KWH. if less than 10 KWH. are used; 9c. per KWH. if the consumption is between 10 and 15 KWH.; 7c./KWH. if over 15." Draw the locus for this rate on the same graph and notice that it passes through the four points. The objectionable feature of sometimes charging less when the use of the current is greater is so apparent that a rate of this kind is not often used. The following is a "block meter rate," which is not so objectionable and gives approximately the same income to the company: "10c. each for the first 10 KWH. used, 7c. each for the next 5, 3c. each for all after the 15th." Plot this rate on the same diagram; also the rate, "8c. per KWH., with a minimum charge of 50c." Find a smooth curve which will come fairly close to all three of these rates.* (Suggestion: consider the equation $y=ax-bx^2$ with suitable values for a and b.)

124. Extrapolation.

The principles of extrapolation are like those of interpolation, but the former process is naturally more uncertain than the latter and can be trusted to give good results only when the extrapolated point is relatively near the points that correspond to the known data. As an example of extrapolation, take the census figures for England and Wales for the years 1861–1901 (see table) and extrapolate to find the population in 1911.

Year	*Population in millions*
1861	20·07
1871	22·71
1881	25·97
1891	29·00
1901	32·53

Draw the curve AB (Fig. 47), and produce it in the way you think it would run (along dotted line BD, say), and note where it cuts the 1911 ordinate. Read the population.

* In general, a "broken line" cannot be represented, by itself, without using an equation containing an infinite series. This is the case with the last of the above rates, which is represented by the straight line $y=50$ from $x=0$ to $x=6·25$ only, and the straight line $y=8x$ for the part further to the right only.

If we are not restricted to *finite* stretches of lines it is always possible to find an equation for *both* of two loci each of which has its own ascertainable equation, by putting the individual equations in the form $\phi(x, y)=0$ and multiplying them together. Thus, the two lines $y=50$ and $y=8x$ are the locus of the single equation $(y-50)\times(y-8x)=0$, or $400x-8xy-50y\times y^2=0$, as the student can readily show by plotting a few points or by algebraical treatment.

The actual population by the 1911 census was 36·07 millions. If your point *D* is not right re-plot it and now continue *BD* in the

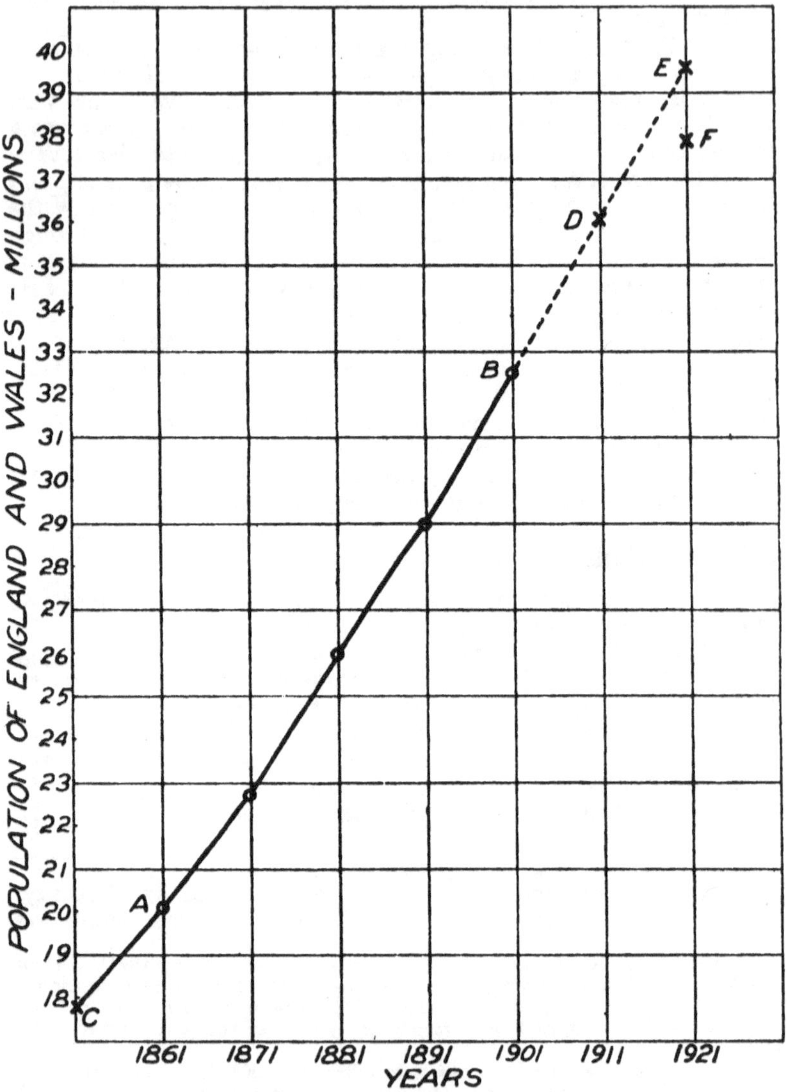

FIG. 47.

same way (along *DE*, say) to give the population in 1921. The population corresponding to *E* will be about 39·6 millions. As a matter of fact, the census of 1921 gave only 37·88 millions (F, in Fig. 47). This shows the danger of extrapolation. What happened between 1911 and 1921 to change the trend of the curve? Could you extrapolate to find the population in 1931?

Extrapolate to find the population in 1851 (extend BA to C, Fig. 47).

The following example shows that extrapolation may be a very definite and decisive process if the given values follow a consistent law of change and can be carried close to the required value: Find the instantaneous velocity of a body at a certain point of time if it is known that immediately after that instant it travels 10 cm. in the first second, 7·0711 cm. in the first half second, etc., as given in the next table.

Time (*sec.*)	*Space* (*cm.*)	*Velocity* (*cm./sec.*)
1	10·000	10·000
0·5	7·0711	14·142
0·3	4·5399	15·133
0·2	3·0902	15·451
0·1	1·5643	15·643
0·05	0·7846	15·692

AVERAGE VELOCITY OF A PISTON ROD THAT MOVES WITH SIMPLE HARMONIC MOTION.

Its total path is assumed to measure 20 cm.; its period, 4 seconds; the required velocity is that at the central point of its motion.

Plot the tabular values with a scale of time along the x-axis and a scale of average velocity along the y-axis and extrapolate graphically to find the velocity at the beginning of the time.*

125. An Interpolation Formula.

If the dependent variable y does not vary linearly with the independent variable x and we wish to get the value of y for some value of x intermediate between the tabulated values of x we cannot use proportional parts. Of course, if we have an equation between x and y or a smooth graph we can always get what we want; but sometimes if our values of x contain many significant figures our graph may not be accurate enough for the purpose. To solve such problems interpolation formulæ have been invented. We take up here the simplest case. It is developed as follows: The accompanying table shows in the first column the numbers 0, 1, 2, 3, 4, 5, 6, and in the second the cubes of these numbers. The two columns may be regarded as corresponding values of x and y where $y=x^3$. The third column gives the differences between successive values of

* In *no time* a moving object will of course traverse no space if its velocity is not infinite, and there is strictly speaking no meaning for such a phrase as "instantaneous velocity." It is convenient, however, to consider that a body which speeds up from a condition of rest to a definite velocity must have passed through all intermediate velocities successively, and as its velocity must have been continually changing the concept of a definite *velocity at a certain point* of space or time becomes almost a necessity. For purposes of rigorously logical deduction this concept is defined as a limiting value in the way indicated above. The "initial" velocity in the case considered = 15·7080 cm./sec.

y. This column is headed Δ^1. The fourth column gives differences between successive values of Δ^1. It is headed Δ^2. In the same way we proceed to other columns and the numbers in the columns headed Δ^1, Δ^2, Δ^3, . . . are called first differences, second differences, third differences, and so on.

x	y	Δ^1	Δ^2	Δ^3	Δ^4
0	0	1	6	6	0
1	1	7	12	6	0
2	8	19	18	6	0
3	27	37	24	6	
4	64	61	30		
5	125	91			
6	216				

In this case the third differences happen to be constant, and consequently the fourth differences are zero.

Having this table we could get the cubes of numbers higher than 6 by working backwards from the last difference column, filling in the columns of differences by adding as we go back. Thus, 6 and 30 give 36, 36 and 91 give 127, 127 and 216 give 343. Which is the cube of 7? And so on.*

It is often useful to draw up a difference table from corresponding values of x and y, obtained in experimental work when x has successive equal increments, for any marked irregularity in the differences will indicate experimental or other errors occurring in the values of y.

Take a more general case. Let x be an independent variable and y a dependent variable; the relationship between x and y, *i.e.* $y=f(x)$, is supposed capable of representation by a smooth curve. The table indicates successive values of x increasing each time by h. Note how the differences are labelled.

		Δ^1	Δ^2	Δ^3	Δ^4	Δ^5
x	y_x	Δ^1_0	Δ^2_0	Δ^3_0	Δ^4_0	. . .
$x+h$	y_{x+h}	Δ^1_1	Δ^2_1	Δ^3_1	. . .	
$x+2h$	y_{x+2h}	Δ^1_2	Δ^2_2	. . .		
$x+3h$	y_{x+3h}	Δ^1_3	. . .			
$x+4h$	y_{x+4h}	. . .				
. . .	. . .					

Then we have

$$y_{x+h}=y_x+\Delta^1_0$$

$$y_{x+2h}=y_{x+h}+\Delta^1_1$$

$$=y_{x+h}+\Delta^1_0+\Delta^2_0$$

$$=y_x+2\Delta^1_0+\Delta^2_0 \quad \text{. . . . . (1)}$$

* This is the basic principle of Babbage's calculating machine.

$$
\begin{aligned}
y_{x+3h} &= y_{x+2h} + \Delta^1{}_2 \\
&= y_{x+2h} + \Delta^1{}_1 + \Delta^2{}_1 \\
&= y_{x+2h} + (\Delta^1{}_0 + \Delta^2{}_0) + (\Delta^2{}_0 + \Delta^3{}_0) \\
&= y_x + 3\Delta^1{}_0 + 3\Delta^2{}_0 + \Delta^3{}_0 \quad . \quad . \quad . \quad . \quad . \quad . \quad . \quad (2)
\end{aligned}
$$

Similarly

$$y_{x+4h} = y_x + 4\Delta^1{}_0 + 6\Delta^2{}_0 + 4\Delta^3{}_0 + \Delta^4{}_0 \quad . \quad . \quad . \quad . \quad . \quad (3)$$

Note that the successive numerical coefficients in (1), (2), and (3) are the binomial coefficients which occur in the expansions of $(1+a)^2$, $(1+a)^3$, $(1+a)^4$, and that by induction we may write:

$$y_{x+nh} = y_x + n\Delta^1{}_0 + \frac{n(n-1)}{1 \cdot 2}\Delta^2{}_0 + \frac{n(n-1)(n-2)}{1 \cdot 2 \cdot 3}\Delta^3{}_0 \ldots$$

If we take our first two terms x and y as zero terms in our series we may amend the last formula to

$$y_{nh} = y_0 + n\Delta^1{}_0 + \frac{n(n-1)}{1 \cdot 2}\Delta^2{}_0 + \frac{n(n-1)(n-2)}{1 \cdot 2 \cdot 3}\Delta^3{}_0 \ldots$$

Now let n be fractional so that our new y_x is part way between y_0 and y_h. Writing our new value of x equal to nh, we get

$$
\begin{aligned}
y_x &= y_0 + \frac{x}{h}\Delta^1{}_0 + \frac{\frac{x}{h}\left(\frac{x}{h}-1\right)}{1 \cdot 2}\Delta^2{}_0 + \frac{\frac{x}{h}\left(\frac{x}{h}-1\right)\left(\frac{x}{h}-2\right)}{1 \cdot 2 \cdot 3}\Delta^3{}_0 \ldots \\
&= y_0 + \frac{x}{h}\Delta^1{}_0 + \frac{x(x-h)}{1 \cdot 2}\,\frac{\Delta^2{}_0}{h^2} + \frac{x(x-h)(x-2h)}{1 \cdot 2 \cdot 3}\,\frac{\Delta^3{}_0}{h^3} \ldots
\end{aligned}
$$

which is the Gregory-Newton interpolation formula.

Two examples will illustrate its use.

Example 1.—Find the square root of 26·5 given the square roots of 26, 27, 28, 29.

Arrange the work thus:

	y	Δ^1	Δ^2	Δ^3
26	5·0990			
27	5·1962	972	−19	
28	5·2915	953	−16	3
29	5·3852	937		

It will be sufficient to stop at the Δ^3 column and it is not necessary to use the decimal point in the difference columns.

Here the argument increases regularly by 1 and we wish to proceed just halfway from 26 to 27. Therefore

$$x/h = \tfrac{1}{2}$$

and

$$y_{26\cdot5}=50990+\tfrac{1}{2}0\cdot972+\frac{\frac{1}{2}(\frac{1}{2}-1)}{1\,.\,2}(-19)+\frac{\frac{1}{2}(\frac{1}{2}-1)(\frac{1}{2}-2)}{1\,.\,2\,.\,3}(3)+\ \ldots$$
$$=50990+486+2+\text{negligible terms}$$
$$=51478$$

Therefore $\sqrt{26\cdot5}=5\cdot1478$. Simple linear interpolation between $\sqrt{26}$ and $\sqrt{27}$ gives 5·1476.

Example 2.—What is the value of tan 68° 20′ given the values of tan 68°, 69° 70°, 71°, 72°, respectively?

Here the argument increases by 60′ therefore $h=60'$. Putting $x=20'$ we get $x/h=\frac{1}{3}$.

Arrange the work thus:

	y	Δ^1	Δ^2	Δ^3	Δ^4
68	2·4751				
		1300			
69	2·6051		124		
		1424		19	
70	2·7475		143		6
		1567		25	
71	2·9042		178		
		1735			
72	3·0777				

The result

$$=24751+\tfrac{1}{3}\times1300+\frac{\frac{1}{3}(\frac{1}{3}-1)}{1\,.\,2}(124)+\frac{\frac{1}{3}(\frac{1}{3}-1)(\frac{1}{3}-2)}{1\,.\,2\,.\,3}(19)+\ \ldots$$
$$=24751+433\cdot3-13\cdot8+1\cdot2$$
$$=25172$$

Thus tan 68° 20′=2·5172.

Simple proportional parts between tan 68 and tan 69 gives 2·5184.

We have taken only the simplest case of an interpolation formula. Cases where the argument does not proceed by equal steps are dealt with by Lagrange's interpolation formula. If the function is periodic we use Gauss' formula.

Also in some cases instead of using the upper differences as in Newton's formula it is more accurate to arrange the numbers so that the differences near the middle row of the table may be used. Stirling's formula has been developed for this case.*

Exercises X

1. Write a definition of what you understand by the terms *interpolation* and *extrapolation*.

2. Turn to the table of values for Chauvenet's criterion (Ch. XVII) and plot l as a function of n, using enough points to determine whether the values

* For a fuller treatment see Mellor's "Higher Mathematics for Students of Chemistry and Physics," and Whittaker and Robinson's "A Short Course in Interpolation."

of l for integral values of n that are not given in the table (*e.g.* $n=31$, $n=70$) can be satisfactorily obtained by graphic interpolation.

3. Turn to your graphic diagram of $y=\log x$ and draw a straight line from the point (2, log 2) to the point (3, log 3), thus making a *chord* of the curve. Mark the middle point of the chord. Is the ordinate of this point equal to the average of log 2 and log 3 ?

4. What operation can be performed on any two numbers by averaging their logarithms ? Label your diagram so as to show clearly the distance that corresponds to $\log\frac{2+3}{2}$ and the distance that corresponds to $\log[(2\times 3)^{1/2}]$. Which of the δ-formulæ for approximate calculation with small magnitudes does this diagram illustrate ? In what way does it show the *equality* expressed by the formula, and in what way does it show that this equality is not exact but only approximate ?

5. In a Stephenson Link Motion the following measurements were made :

Angle of crank . .	−45°	−15°	15°	45°	75°	105°
Displacement of value	−0·25″	0·98″	1·80″	2·24″	2·05″	1·32″

Determine (*a*) the displacement of the valve when the crank angle is zero.
(*b*) The angle of the crank when the displacement is zero.
(*c*) The maximum displacement of the valve.

6. The table gives the value of the sun's declination at the stated dates.

Jan. 1	Jan. 31	Mar. 2	Apl. 1	May 1	May 31	June 30
−23·0°	−17·5	−7·4	4·4	14·9	21·8	23·2
July 30	**Aug. 29**	**Sept. 28**	**Oct. 28**	**Nov. 27**	**Dec. 27**	
18·6	9·5	−1·8	−13·0	−21·0	−23·4	

Add your latitude to these figures and get the meridian altitudes of the sun at these dates. Plot a curve between altitude and date and deduce the meridian altitudes at Feb. 15, Mar. 21, June 21, Aug. 14.

7. A spectrometer was calibrated by means of known spectrum lines.

Element.	Wave-length.	Micrometer reading.
Ca	3933·5	3·729
Hε	3970·2	4·270
Hδ	4102·8	6·160
Fe	4233·3	7·798
Hγ	4340·7	8·993
Mg	4481·3	10·387
Hβ	4861·5	13·445

Plot the calibration curve and find the wave-lengths corresponding to micrometer readings 5·40, 6·97, 8·22, 9·55, 11·39.

8. The following are the corrections to be applied at every 5 c.c. mark on a burette to give the true volume to the 50 c.c. mark. Plot the correction curve.

Mark .	0	5	10	15	20	25	30	35	40	45	50
Correction	+·114	+·107	+·154	+·132	+·132	+·059	+·141	−056	+·037	+·002	0

Find the correction to be applied at 12·5 c.c., at 42·5 c.c.

9. Plot the correction curve for the given thermometer.

0°	10°	20°	30°	40°	50°	60°	70°	80°	90°	100°
−0·20	−0·12	−0·04	−0·05	−0·14	−0·29	−0·35	−0·41	−0·48	−0·52	−0·60

Find the corrections at 35°, 75°.

10. The table gives the relationship between the height of the barometer, the boiling point of water and the altitude.

Barometer, inches.	Boiling point, °F	Altitude, feet.
32·3	216	−2,100
31·1	214	−1,000
29·9	212	0
28·7	210	1,200
27·0	207	2,800
25·4	204	4,500
23·5	200	6,700
20·7	194	10,000

Plot the curves and interpolate to find the missing data for barometric heights of 28·2 and 25·0 inches, for boiling points of 208°, 198° F., and for altitudes of −1,600 and 3,400 feet.

11. In finding the root of an equation by the method of double position would it be satisfactory to extrapolate instead of interpolating? Explain why.

12. Solve graphically the equations.

(a) $\begin{cases} 2{\cdot}63x+3{\cdot}12y=12. \\ 2{\cdot}14x-2{\cdot}36y=5. \end{cases}$

(b) $4x^2-8x-7=0$.

(c) $2x^3-7x+3=0$.

(d) $25x^3+18x-5=0$.

13. Given that $y=1-4{\cdot}818x+7{\cdot}514x^3$, calculate the values of y for values of x ranging from −1·0 by steps of 0·1 up to 1·0. Plot the curve between y and x, and find where $y=0$. Drawing these parts of the curve on an enlarged scale find to five significant figures the roots of the equation

$$1-4{\cdot}818x+7{\cdot}514x^3=0$$

14. A hollow box 6 inches square in cross-section contains 2 inches of water. What size must a cubical block be if, when it is placed resting on the bottom of the box, the top of the block comes exactly level with the surface of the water? Use squared paper to get the solution. Is there more than one solution?

15. Plot $y=\tan\theta-2{\cdot}75\theta$. Find the value of θ which makes $y=0$, and by replotting on an enlarged scale solve the equation $\tan\theta=2{\cdot}75\theta$ to a seven-figure accuracy.

16. When interpolating vapour pressures it is usually taken that the increase of the logarithm of the vapour pressure is proportional to the increase of temperature.

(a) The maximum vapour pressures of water at 10° C. and 20° C. are 9·2 and 17·5 mm. respectively. Find the m.v.p. at 15° C.

(b) The maximum vapour pressures of water at 95° C. and 100° C. are 633·9 and 760 mm. respectively. Find the maximum vapour pressure at 96, 97, 98, 99, 101°, respectively.

(*c*) The maximum vapour pressures of water at 100° C. and 110° C. are 760·0 and 1074·5 mm. of mercury, respectively. Find the pressures at 102°, 104°, 106°, 108° C.

17. The density of a common salt solution is related to the concentration as follows:

C Gms. of salt per kilogram of water.	*D* Density at 20° (gm per c.c.).
0·0	0·998
5·1	1·002
10·8	1·005
20·1	1·013
46·9	1·033
105·0	1·068
209·0	1·127
240·0	1·192

(*a*) Plot a curve between Concentration (*C*) and Density (*D*).

(*b*) Find the concentrations corresponding to densities of 1·030, 1·050, 1·150 gm. per c.c.

(*c*) Find the densities corresponding to concentrations 50, 100, 150, 250 gms. of salt per kilogm. of water.

(*d*) Draw a tangent to the curve representing the dilute solutions and find the relation between concentration and density in the form $C=k(D-0{\cdot}998)$.

(*e*) Find the densities of the solutions containing respectively 2·0 and 5·0 gm.-molecules of salt per kilogram of water. (Molecular weight of common salt=58·5.)

18. The specific gravity of a sugar solution varies with the concentration as shown in the following table:

Concentration in gram-molecules per kilogram of water.	Specific gravity at 20° C.
0·1	1·0129
0·2	1·0253
0·4	1·0485
0·6	1·0699
0·8	1·0896
1·0	1·1079

(*a*) Plot a curve between specific gravity and concentration.

(*b*) Find the specific gravities of solutions containing 170 and 300 gms. of sugar to the litre of water. (Molecular weight of sugar=342.)

19. The table gives the number of grams of anhydrous potassium nitrate which when dissolved in 100 grams of water at the temperature stated gives a saturated solution.

Degrees C.	0	10	20	40	60	80	100
Grams .	13·3	20·9	31·6	63·9	109·9	169·0	246·0

What quantities are required for saturated solutions at 15°, 50°, 90°?

20. The solubility of copper sulphate in water varies with temperature as follows. s denotes gms. of anhydrous salt in 100 gm. water.

Temperature (°C.)	10	20	60	100
s	17·4	20·7	40·0	75·0

Find the value of s at 40°, 80°.

21. The solubility of sodium sulphate in water varies with the temperature as shown. s denotes the gms. of anhydrous salt per 100 gm. water.

Temperature (°C.)	0	10	20	40	80	100
s	5·0	9·0	19·4	49	44	42

Find the value of s at 15°, 60°.

22. The freezing point ($-D$°C) of a cane-sugar solution is related to the concentration (n gms. of sugar per litre of water) as shown.

n	190	570	950	1330
D	1·15	3·62	6·68	10·60

Find the freezing points of solutions containing 380, 760, 1,140 gm. of sugar per litre.

23. The population of Toronto has increased as follows. Numbers in hundreds of thousands.

1834	9	1884	105
1844	18	1894	168
1854	38	1904	226
1864	45	1914	470
1874	68	1923	539

Plot (1) population against year;
(2) logarithm of population against year.
In 1914 it was estimated that the population in 1921 would be over a million. Give one reason for this expectation not being fulfilled.
Forecast the population for 1929, 1934.
When will it be a million?

24. The number of telephones in use in Toronto has grown as follows

Date (Y)	Number (N)
1881	400
1885	1,200
1890	3,400
1895	4,900
1900	7,100
1905	15,300
1910	32,300
1915	58,200
1920	88,280

Find the equation between N and $(Y-1881)$.
If the rate of growth is maintained find the value of N in (a) 1922, (b) 1925.

25. The table gives the population of England and Wales.

Y—year .	1801	1811	1821	1831	1841	1851	1861	1871	1881	1891	1901	1911	1921
P—millions	8·89	10·16	12·00	13·90	15·91	17·93	20·07	22·71	25·97	29·00	32·53	36·07	37·88

Plot the growth curve. Estimate the population in (*a*) 1846, (*b*) 1931.
(*c*) What is the probable rate of increase per 1000 in 1901?
(*d*) Find a mathematical expression between *P* and *Y* for the period 1831–1901.

26. The table gives the population of Canada and the numbers of British Descent.

Y—year . .	1851	1861	1871	1881	1891	1901	1911	1921
P—millions .	2·38	3·16	3·69	4·32	4·83	5·37	7·21	8·79
B.D.—millions	—	—	2·11	2·55	—	3·06	3·90	4·87

Plot the growth curves. Estimate the populations in 1931. Find a mathematical expression between *P* and *Y*.

27. Exercises in the use of the Gregory-Newton interpolation formula.
(*a*) The logarithms of 12, 13, 14, 15, and 16 are 1·0792, 1·1139, 1·1461, 1·1761, and 1·2041, respectively. Find the logarithm of 12·4.
(*b*) The cube roots of 21, 22, 23, 24, and 25 are: 2·759, 2·802, 2·844, 2·884, and 2·924, respectively. Find the cube root of 21·7.
(*c*) The maximum vapour pressures of water at 40°, 42°, 44°, 46°, and 48° C. are 55·13, 61·30, 68·05, 75·43 and 83·50 mm. respectively. Find the m.v.p. at 41° C.
(*d*) Find the m.v.p. at 41° C. by interpolating the logarithms of the vapour pressures at 40° and 42° C.

28. Show that if difference columns are being formed from the *n*th powers of the natural numbers the *n*th differences are constant.

29. A study of the records of twenty-two well-established North American families yields the following table (Crum., *Amer. Stat. Assn.*, xiv. 1914–1915):—

Marriage periods.	Average age of bride at marriage.	Average number of children per wife.
	Years	
Before 1700	21·4	7·37
1700–1749	21·7	6·83
1750–1799	22·0	6·43
1800–1849	22·3	4·94
1850–1869	22·9	3·47
1870–1879	23·1	2·77

Plot the curves and extrapolate to complete the table for periods
1880–1900
1920–1930.
Is extrapolation justified?

CHAPTER XI

CO-ORDINATES IN THREE DIMENSIONS

126. Co-ordinates of a Point in Space.

Just as the position of a point in a plane (*i.e.* in two-dimensional space) can be represented by two co-ordinates, an x-value and a y-value, giving its distance from two mutually perpendicular axes, so the location of a point in unrestricted, three-dimensional space can be fixed by three co-ordinates, a set of three numerical values

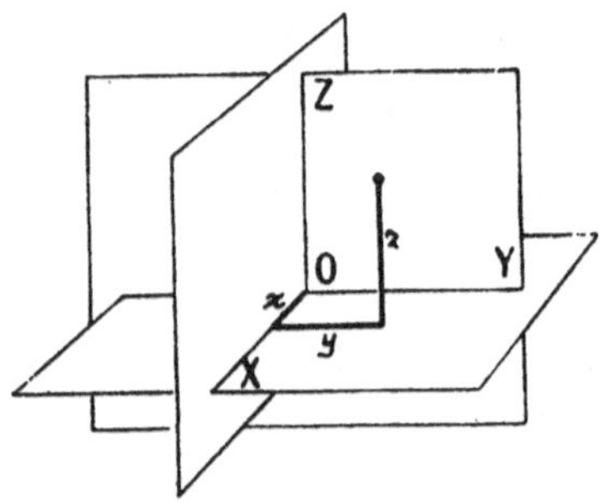

FIG. 48.—CO-ORDINATE PLANES.

These planes of reference for three-dimensional space correspond to the base lines of reference, or co-ordinate *axes*, that are used for two-dimensional space.

(x, y, z) which indicate its distance from each of three planes that intersect each other at right angles. Thus, the point in Fig. 48 is at a distance of x units along the x-axis from the plane YZ of the other two axes, and at a distance of y units, parallel to the y-axis, from the plane XZ, and at a distance of z, parallel to the z-axis, from the plane XY.

127. Convention in Regard to Signs.

The XY plane may be thought of as being represented by the same sheet of paper as that on which the flat graphs of x and y have previously been traced (Fig. 49). Then any point whatever must be located a certain distance above or below some definite position in the plane of the paper. The x-value and y-value for this *position* will be the x and y of the *point in space*, and the distance from the point to the plane will be the z of the point. It is customary among physicists and astronomers to consider a distance above the plane of the paper as a positive value of z, and distance below it as negative, according to the arrangement of axes shown in Figs. 48, 49, and 50*b*. The convention among pure mathematicians is just

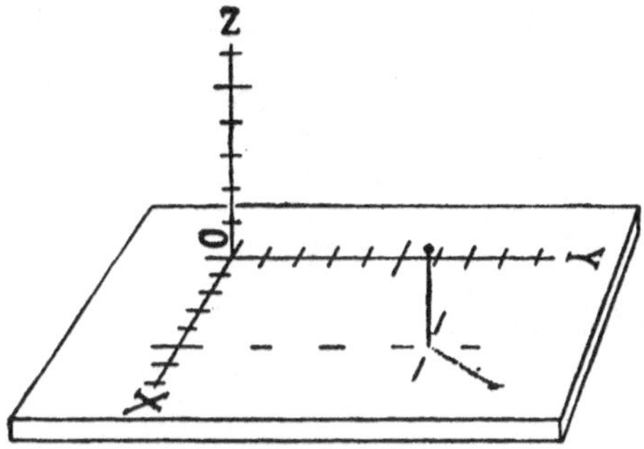

FIG. 49.—POSITIVE DIRECTIONS OF AXES.

Z is considered positive when measured upward from the normal plane of the x-axis and y-axis. The pin shown in this diagram would have the location of its head represented approximately by the position (5, 7, 3), the numbers in parentheses being used for x, y, and z, in order.

the opposite, viz. distances above the paper are called negative and those below are positive (Fig. 50*a*). The distinction is immaterial in the greater part of the study of pure mathematics, but it is important in many branches of applied science; for example, the equations of the curve of a right-hand screw thread in one system will represent a reversed or left-hand thread if the other system is used instead. Accordingly, there is a tendency among pure mathematicians to use the "physical" arrangement for the sake of uniformity, and it is the only one that will be used in this book.

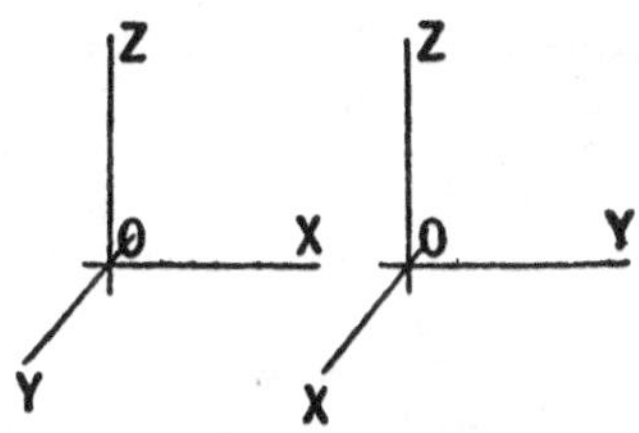

FIG. 50*a*. FIG. 50*b*.

CONVENTIONS AS TO SIGNS.

Fig. 50*a* corresponds to the arrangement that is understood in pure mathematics; Fig. 50*b* is that of applied mathematics.

That these two arrangements exhaust the possibilities may be seen by taking any arbitrary arrangement of axes at right angles to each other and rotating them as a rigid figure. They can always be made to coincide with either Fig. 50*a* or Fig. 50*b*. The two are essentially different because neither one can be rotated in three-dimensional space so as to coincide with the other.*

* Notice that a plane xy-diagram drawn with $+Y$ upward but with $+X$ to the left cannot be rotated in its own two-dimensional space so as to coincide with the conventional arrangement. It can be turned through the third dimension, however (turning the upper surface of the paper downward), and

In each of the following exercises the positive directions of the axes are indicated. Consider each in turn and state whether it is like Fig. 50*b* (" right ") or the opposite (" wrong "). To do this put the thumb, forefinger, and middle finger of the left hand at right angles and label them *Z*, *Y*, and *X* respectively.

1. *X* points northward; *Y*, upward; *Z*, westward. (Ans.: wrong.)
2. *X*, west; *Y*, up; *Z*, north. (Ans.: right.)
3. *X*, down; *Y*, east; *Z*, south.
4. *X*, down; *Y*, north; *Z*, east.
5. *X*, down; *Y*, west; *Z*, north.
6. *X*, down; *Y*, south; *Z*, east.
7. *X*, east; *Y*, north; *Z*, up.
8. *X*, west; *Y*, north; *Z*, up.
9. *X*, west; *Y*, north; *Z*, down.
10. *X*, up; *Y*, east; *Z*, north.

128. Loci of Simple Three-Dimensional Equations.

In studying geometrical relationships in a plane we have seen (§ 94) that the equation $y=a$ represents all the points that are a units above the x-axis, that is, a straight line parallel to the x-axis and situated at a distance a above it. In the same way, the equation $z=m$ must denote all points located m units above the xy-plane, *e.g.* the points $(2, 3, m)$, $(0, 0, m)$, $(0, 10, m)$, etc., since each of these groups of values ($x=2$, $y=3$, $z=m$, for example) will satisfy the equation. Similarly, $x=m$ or $y=m$ will denote a particular plane parallel to the plane YOZ or to XOZ, and so perpendicular to the x-axis or to the y-axis, respectively.

Consider next an equation that contains only two of the three variables. On a plane surface $y=x^2$ is a curve, a parabola. In space, any point above or below any point of the curve will satisfy this equation, for it has the same x and y, but a different z. Accordingly, the equation represents the surface that is made up of all the vertical lines that can be passed through points that lie on the plane curve.

If the equation contains all three variables, any arbitrary value may be assigned to x, any value whatever to y, and the value of z will then be determinate. That is, the points of the locus will be situated at varying distances above *all* the points in the xy-plane, and so will comprise a surface.

In general, then, *a single equation denotes a surface.* Since two surfaces intersect along some line *two simultaneous equations will*

made to coincide. Similarly, a solid model of Fig. 50*a* would need to be turned through a fourth dimension of space before it could be made to coincide with a solid model of Fig. 50*b*. Since we have no appreciation of a fourth dimension the two figures are as essentially different to us as a capital L and a Greek capital Γ would seem to a being whose sense perceptions were limited to space of two dimensions.

denote a curve in space. It has been seen (§ 100) that $x^2+y^2=5^2$ must be the equation of a circle; in the same way $x^2+y^2+z^2=5^2$ is the equation of a spherical surface that is everywhere 5 units distant from the origin. The simultaneous equations

$$\begin{cases} x^2+y^2+z^2=25 \\ z=3 \end{cases}$$

will have for their locus the points which satisfy the first equation and at the same time satisfy the second; *i.e.* each of the points that is located on the spherical surface $x^2+y^2+z^2=25$ and is at the same time in the horizontal plane $z=3$. Such points must lie on the intersection of the plane and the spherical surface; and this intersection is known to be a curve, namely, a circle. For other values of z the circular intersection would be larger or smaller; for the (tangent) plane $z=5$ the intersection would shrink to a point, for $z=6$ there would be no intersection.

129. Contour Lines.

If a were given all possible values in the equation $z=a$ the resultant loci would be horizontal planes at all levels, and they would intersect the sphere $x^2+y^2+z^2=5^2$ in all the horizontal circles that could be drawn on its surface. All of these intersections, conse-

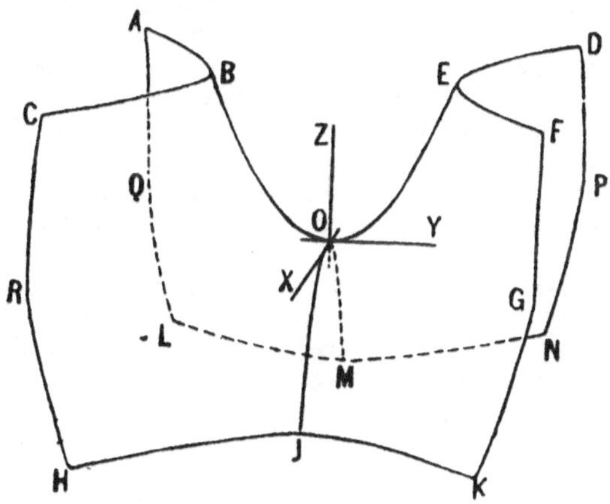

FIG. 51.—THE HYPERBOLIC PARABOLOID $x^2-y^2+z=0$.
A surface of this general character is sometimes spoken of as a "saddle-back."

quently, *would be* the spherical surface; a smaller number of them would form a sort of skeleton, from which the shape of the surface could be inferred if they were sufficiently numerous. An extremely useful way of representing a surface, especially an *irregular* surface, on paper is by outlining the intersections which would be formed by horizontal planes at various levels. The diagram (Fig. 51) shows a portion of the surface $x^2-y^2+z=0$, called a *hyperbolic paraboloid.* When x is zero the equation reduces to $z=y^2$; *i.e.* the intersection of the curve with the YZ-plane is the parabola $z=y^2$, BOE in the diagram. When $y=0$ the section MOJ is a parabola $z=-x^2$. At any level where $z=a$ the equation of the surface becomes $y^2-x^2=a$, a hyperbola such as ABC and DEF or for $z=-a$, a hyperbola such as HJK and LMN; for $a=0$ either of these degenerates into the two straight lines that are common asymptotes, $y=\pm x$. A series of horizontal sections of the surface are shown in Fig. 52, as they would appear if they were all viewed from above, the algebraical signs showing whether each curve is above or below the xy-plane and the smaller numbers denoting the lower levels while the higher numbers indicate the upper ones. If the numbered lines are imagined to be raised 1, 2, 3, 4, 5, 6, 7, 8, and 9 cms. respectively above the surface of the paper it will be evident that a good idea of the shape of the curved surface which they outline can be obtained without the necessity of consulting a perspective drawing like Fig. 51. Such a surface as this one (which is convex upward along the x-axis and concave upward along the y-axis) is commonly called a *saddle-back* and represents roughly the shape of the surface of the earth in a *mountain pass.*

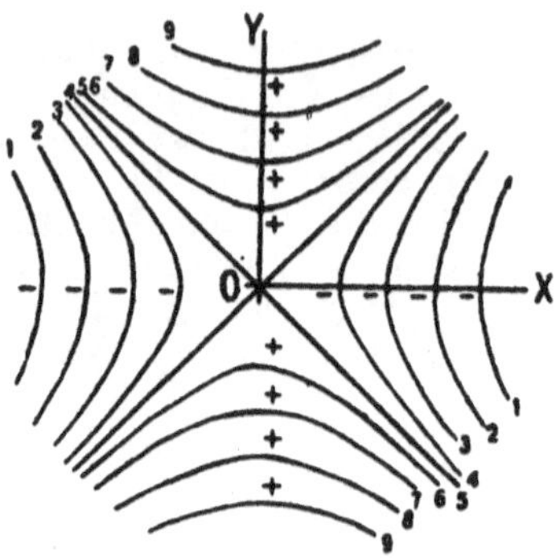

FIG. 52.—CONTOUR LINES OF A SURFACE.

Horizontal sections at different levels of the curved surface of Fig. 51.

The irregular surface of the earth is sometimes represented on maps by horizontal section lines (*contour lines*) which make it easy to find the location, height, slope, etc., of any hill, valley, or other surface character by inspection of the map.

130. Use of Contour Maps.

The lines of horizontal section usually correspond to heights taken at equidistant intervals, for example, at 20, 40, 60, 80, . . . feet above mean sea-level, and are called *contour lines.* Students usually find it most convenient to think of them as representing the new *shore lines* that would be formed if the sea-level were to rise 20 feet, 40 feet, etc., above its original level. The uniform interval, 20 feet in this example, is called the *contour interval,* and the level of reference, usually mean sea-level, is called the *datum plane.* The diagrams (Fig. 53) show a vertical section of a hill 80 feet high and a horizontal plan of its contour lines.

Examine a contour map and notice whether the datum plane and the contour interval are stated in the margin.

Find a hill or other elevated area on the map, and notice the arrangement of the contour lines. What is the difference, as indicated on the map, between a high hill and a low hill? What is the difference between a steep hillside and a more gradual slope?

Find a brook that runs down a hill. In what general *direction* do the contour lines cross the brook? Why? What is the general

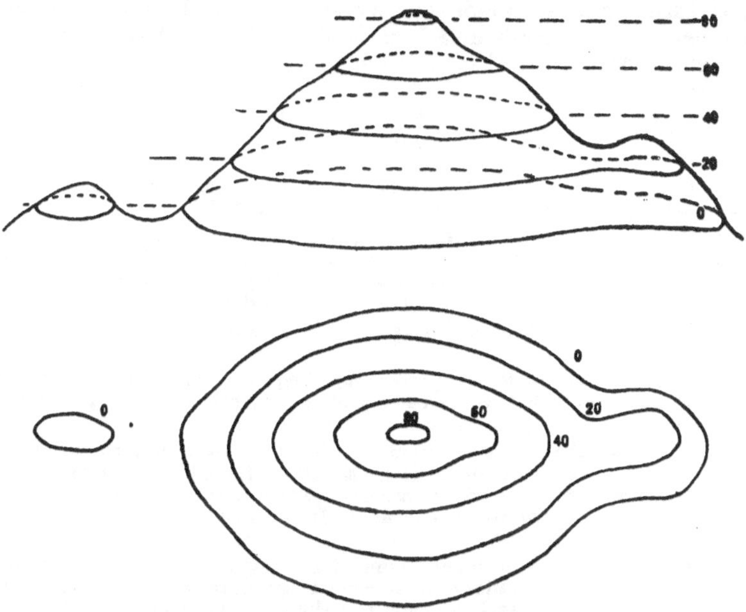

FIG. 53.—CONTOUR LINES OF A HILL.

The upper figure is a side view of a hill 80 feet high, showing in perspective the outlines that would be produced if it could be cut into 20-foot slices or the new shore lines that would be produced if the surrounding country could be flooded to depths of 20, 40, 60, and 80 feet.

The lower figure is a set of *contour lines* that represent a top view of the levels shown above. The usefulness of a map is greatly increased by having contour lines drawn or printed on it in some distinctive colour. Such maps usually have the contours in green and brown, rivers and lakes in blue, roads, buildings, boundaries, etc., in black.

The line 20 is the locus of all points where the surface is 20 ft. above the plane of reference. Notice that the ground is always higher on one side of a contour line and lower on the other.

shape of the contour lines where there is a watercourse? Where a hill sends out a projecting buttress or ridge what is the general contour form? How is a plateau formation indicated by contour lines?

Why is it that contour lines where they run across a roadway are never as near together as they often are in other localities? Find a road (on a topographical map) that appears to have been purposely so constructed as to cut the contour lines at considerable intervals of distance.

131. Practical Plane and Solid Geometry and Graphics.

The ordinary school geometries rarely give enough practical geometry to train the student into a quick and accurate use of drawing instruments. It is not possible here to give a treatment of the practical descriptive geometry of rectilinear figures, polygons, circles, ellipses, parabolas, and hyperbolas and their areas, of the solid geometry, plans and elevations of points, lines, planes, solid figures and their sections, but a few examples are given for practice. References may be made to Good's "Practical Plane and Solid Geometry," Harrison's and Baxandall's "Practical Geometry and Graphics" for practical instruction in these subjects.

EXERCISES XI

1. If $x/a+y/b+z/c=1$ is known to represent a plane surface what can you *prove* about the position of this plane (compare § 109)?

2. Write the equation of the cone (curved surface) produced by rotating the straight line $z=3x$ around the z-axis (suggestion: rotation changes each point that was located at a distance of x to the right of the z-axis into a horizontal circle whose radius is equal to that x-value).

3. What kind of a locus corresponds in general to three simultaneous equations?

4. Can two contour lines representing different levels ever intersect each other on a map? What would be true of the earth's surface at the point of intersection?

5. Show that the curves $RCAQ$ and $GFDP$ made by cutting the hyperbolic paraboloid (Fig. 51) at any distance to the left or the right of the origin by the plane $y=a$ are "arch-shaped" parabolas. Show that the curves $RHKG$ and $QLNP$ made by limiting the surface at the front or back by the plane $x=a$ are "festoon-shaped" parabolas.

6. The x, y, z co-ordinates of a point A are 7·0, 5·4, 4·8 cms., respectively. Find graphically the distance of A from the origin O, the angle of inclination of OA to the plane XOY and the length of the projections of OA on the three planes XOY, YOZ, ZOX.

7. The co-ordinates of the ends A and B of a straight line are for A $x=1$, $y=0$, $z=3$, for B $x=7$, $y=4$, $z=0$ inches. Find graphically the true length of AB and the inclination of AB to the planes XOY, XOZ.

8. A line directed straight down a slope measures 300 feet in plan and there is a difference of 100 feet in the elevations of its two ends. Find graphically the angular slope of this line.

9. The lines of steepest ascent up the plane face of a hill rise eastward. The slope is 1 (vertical) in 3 (horizontal). Find graphically:
(*a*) The inclination of the hill in degrees.
(*b*) The slope and inclination of a N.E. path.
(*c*) The slope and inclination of a N. path.
(*d*) The direction of an upward path which rises 1 in 7 (horizontal).
(*e*) The length of path (*d*) for a vertical rise of 160 feet.

10. The *dip* of stratified beds is the inclination of the strata to the horizontal. The *strike* of a bed is the direction of the intersection of the bed with a horizontal plane. If the bed crops out at a level surface the strike is the direction of the *outcrop*. A quarry was driven northward into a hillside. The floor of the quarry is horizontal and it has two vertical faces, the direction

of one being N. 45° W. and of the other N. 30° E. On the eastern face the apparent dip of the beds is 30°; on the western face 20°; both towards an observer entering the quarry. Find graphically the true dip; also the direction of the strike. Make a cardboard model of the quarry.

11. There is a similar quarry. Along a vertical face which runs S. 30° E. the apparent dip is 14°. Along a vertical face which runs S. 45° W. the apparent dip is 11½°. Find graphically the true dip and the strike.

12. The mineral tetrahedrite ($4Cu_2S,Sb_2S_3$) crystallizes in regular tetrahedra. Make a cardboard model of the crystal.

Such a figure stands on the ground with one edge perpendicular to the vertical plane. Draw the plan and elevation. Make the edge 2 inches long. The figure is cut by a vertical plane inclined to the ground line at 45°, cutting the edge mentioned above 1½ inches back from the front corner and passing in front of the upper vertex.

Draw the plan, elevation, and true shape of section.

Find the height of the model and the inclination of two adjacent faces to each other.

13. An octahedral crystal of alum ($Al_2SO_4.K_2SO_4.24H_2O$) has an edge exactly 2 inches in length. Make a cardboard model of the crystal.

Draw the plan and elevation of the crystal when

(*a*) One axis is vertical and two horizontal edges are perpendicular to the vertical plane;

(*b*) One axis is vertical and four horizontal edges are inclined at 45° to the vertical plane.

(*c*) The crystal is resting with one face on the horizontal plane and one edge of that face is perpendicular to the vertical plane.

What is the length of the crystal? What is the angle between two adjacent faces?

14. A rhombohedral crystal of calcspar ($CaCO_3$) has an edge exactly 2 inches long. The angles of the rhombus are 78° 5′ and 101° 55′. Make a model of the crystal. You must not let 3 acute angles meet at a corner. Draw its plan and elevation when

(*a*) Crystal is resting on the ground with four edges perpendicular to the vertical plane.

(*b*) Crystal is resting on the ground with four edges parallel to the vertical plane.

15. A quartz crystal consists of a regular hexagonal prism 1½ inch in side and 3 inches long. It is terminated at one end by a hexagonal pyramid, the angle of slope of a pyramidal face being 51° 47′, and at the other end by a basal plane. Make a model. This crystal stands upright. Draw its plan and elevation when

(1) A prismatic face is perpendicular to the vertical plane;

(2) A prismatic face is parallel to the vertical plane.

(*a*) Find the angle between alternate pyramidal faces.

(*b*) Find the angles of the triangles of the pyramidal faces.

In position (2) the crystal is cut by a plane perpendicular to the vertical plane and inclined at 50° to the horizontal plane. This cutting plane intersects the long axis of the crystal ½ inch from the top. Draw the plan and elevation of the section, also its true shape.

16. Sketch the shape of a pressure-volume-temperature surface for a perfect gas.

CHAPTER XII

ACCURACY

132. Significant Figures.

Significant figures have already been defined as all of those that compose a number except one or more ciphers at the extreme left or right which may be necessary to express the order of magnitude of the number (*i.e.* to locate the proper position of the decimal point) but are not needed in any other way for indicating its value. For example, 0·003803, 3·803, and 3803000 have four significant figures each; provided that the last one is a "round number" (*i.e.* is meant to be accurate only to a whole number of thousands). The ambiguity in the last case can best be avoided by writing $3{\cdot}803 \times 10^6$; and the same three numerical values could all be written with seven-figure accuracy by expressing them as $3{\cdot}803000 \times 10^{-3}$, 3·803000, and $3{\cdot}803000 \times 10^6$, respectively.

It is usually understood, in any kind of careful scientific work, that a number is never written with too many significant figures, which would appear to give it an unwarranted degree of accuracy; nor with too few significant figures, which would mean a neglect of the full extent of the accuracy that had been obtained. In other words, figures are known to be correct as far as they are stated and are unknown for all notational "places" beyond. For example, if 1 yard is found to be 83·8 cm., the best value of 1 foot that can be calculated from this determination is 1 ft. = 27·9 cm. and the ordinary arithmetical result, 1 ft. = 27·93333 cm., is not only unjustified, but in this case is positively wrong although the statement that 1 yard = 83·8 cms. is right.

Which four of the following values for e are correct and which six are incorrect: 2·718828182; 2·718828183; 2·718281; 2·7182; 2·7183; 2·7180; 2·7; 2·70; 2·71; 2·72?

133. Infinite Accuracy.

A number like the ratio of the diagonal of a square to its side or the ratio of circumference to diameter for any circle, which depends only upon theoretical considerations, can be calculated with any desired degree of accuracy, but no number whose value depends upon the measurement of a material thing can be stated with more than a definite degree of accuracy, on account of the limitations of the method of measurement. If an imaginary circle is made large enough to enclose the Milky Way its circumference, measured in terms of its radius, must be equal to $2 \times 3 \cdot 14159265358979323846264338327950288419716 94$, but if it were possible to measure a real line of corresponding dimensions the best microscope in existence would not enable us to decide whether the value of π should be as small as

$$3 \cdot 14159265358979323846264338 3279$$

or as large as 3·14159265358979323846264338 3280. The numerical value of π has been calculated as far as 707 decimal places, but this of course is not even an approach to perfect (*i.e.* infinite) accuracy.

134. Relative Errors.

If the thickness of a lead pencil is measured by holding it in front of a scale the separate measurements made by a class of students will be likely to vary as much as 0·03 cm. This is nearly 4 per cent. of the distance measured. If the length of a 6-foot table-top is determined with an ordinary steel tape measure the results that are obtained will be apt to vary 2 or 3 mm. This is about $\frac{1}{8}$ per cent. of the distance measured. In the latter case the error is about ten times as great, considered as an isolated length, as it is in the former. Considered in relation to the thing measured, however, 0·16 per cent. is much smaller than 4 per cent. Since the measurement of the table is obviously a more accurate process than that of the lead pencil it will be evident that accuracy is a matter of relative size, not of absolute size. To state that a measurement is uncertain by 4 per cent. means something definite, but the statement that a measurement is correct " within 0·03 cm." tells us nothing about its accuracy, if nothing is stated about the length measured.

Measure a lead pencil and a table in the manner described. The other members of the class should also measure the same objects with the same ruler and the same tape. Report the measurements to the instructor to be tabulated. Then determine their maximum discrepancy, both in centimetres and as two percentages.

By using the most refined methods it is possible to measure a distance of several miles with remarkable accuracy. If an accurate steel tape is used, which is stretched by a measured force when it is at a carefully determined temperature, it is possible in the course of a few weeks to measure a base line for surveying purposes with an error of about one unit out of 500000. Make a rough mental calculation (assume 5000 feet = 1 mile) of what such an error would amount to in measuring a distance of 10 miles. Is it larger or smaller than the 0·03 cm. of the lead-pencil measurement ?

Does the accuracy with which a number is stated depend upon the number of decimal places to which it is carried out, or upon the number of significant figures which it contains ?

135. Uncertain Figures.

The figure that follows the last trustworthy figure of a measurement may be uncertain to the extent of two or three units, and yet its approximate value may be definite enough to make one hesitate to discard it. This would probably be the case if the student attempted to estimate *hundredths* when using a scale of whole centimetres without millimetre graduations. In such cases it is better to retain the doubtful figure, keeping in mind however the fact that the result obtained from it in any calculation will be liable to have one uncertain figure also.

Another case in which it is sometimes desirable to keep a single uncertain figure is when a higher degree of accuracy is made possible by retaining it during the process of a calculation, although it is eventually discarded in stating the final result. This is illustrated in the treatment of the "figure last cancelled" in the processes of abridged multiplication and division.

136. Superfluous Accuracy.

An appreciation of the degree of accuracy that is required in particular cases often makes it possible to avoid needless trouble in measuring or calculating. If the dimensions of a rectangular block cannot be measured with more than three-figure accuracy its density can be obtained by weighing it merely with three-figure accuracy, and the determination of the fourth figure of its mass or weight will not enable a much better calculation of its density to be made.

Obtain the dimensions of a rectangular wooden or metal block as accurately as possible with a scale of centimetres and millimetres, estimating tenths of a millimetre. Calculate its volume, using the proper number of significant figures, and then find its density by weighing it and dividing mass by volume.

137. Finer Degrees of Accuracy.

A measurement may happen to be known with an accuracy greater than can be expressed by three significant figures and yet not with four-figure accuracy; that is, the step from any degree of accuracy to a tenfold greater degree may be too great to be suitable in all cases. For example, a period of time equal to $125\frac{3}{8}$ seconds may have been measured more accurately than to the nearest whole number of seconds and yet may not have been determined closely enough to enable tenths of a second to be stated. In such cases common fractions may be used instead of decimals; and a significant zero can be employed in such a form as "$125\frac{0}{8}$ seconds" without any lack of intelligibility, meaning of course "between 124·9 and 125·1 seconds," or "nearer to 125 seconds than to $124\frac{4}{8}$ or to $125\frac{1}{8}$."* It is usually more satisfactory, however, to state a numerical

* The latter is more likely to be true if a stop-watch is used for taking the time. The ordinary stop-watch stops at fifths of a second.

value for a measurement and then affix a statement of its uncertainty in the form of a percentage or otherwise.

What would you understand to be the largest possible error in a measurement stated to be 125 seconds? Ans.: 0·5 sec., or 0·4 per cent.

What would you understand to be the largest possible error in a measurement stated to be 125·0 seconds? What would it be for $125\frac{0}{8}$ sec.? Express the answers both in seconds and in percentages.

Consider carefully the stated measurement "$22\frac{1}{3}$ cubic centimetres." If it had to be written as a decimal for the sake of averaging it with a set of other measurements, would you prefer to write it 22·3 or 22·33, or would you round it off to 22? Explain why.

138. Possible Error of a Measurement.

Instead of stating that a measurement may have an error in a certain decimal place or significant figure although the figures that precede it are correct it is usually better to state the possible error in the form of a ratio or a percentage. Since the figures of a numerical statement are intended to be significant and correct as far as they go it is evident that the statement cannot have an error larger than half of a single unit in the last decimal place, or five units in the place that would follow the last one that is written. Accordingly, a statement, like that of § 51, that a length which is written 174·2 certainly cannot have an error of more than 1 out of 1742, is on the safe side but could be improved by making it read "not more than ½ out of 1742, or 1 out of 3484, or 0·0003, or 0·03 per cent."

Turn to your experimental determinations of sines (§ 36), find the greatest possible error of each as half of a single unit in the last written decimal place, and reduce this possible error either to a decimal fraction (*e.g.* 1 out of 25 is the same as 0·04) or to a percentage (1 out of 25=4 per cent.), according to which form appeals to you as being the more expressive. Then mark each error "small," "large," etc., as given in the accompanying table of errors.

SIZE OF ERRORS.

0·03 to 1 per cent.	"*very small*"
0·1 to 0·3 per cent.	"*small*"
0·3 to 1 per cent.	"*moderate*"
1 to 3 per cent.	"*large*"
3 to 10 per cent.	"*very large*"

Notice that a gradation which is as coarse as that expressed by enumerating significant figures is unsatisfactory in certain cases. The number 0·9624 has no more significant figures than 1·093 but is expressed with about ten times as great accuracy as the latter. For purposes of using it in a calculation along with 1·093 it should be rounded off to 0·962. According to the rules given in the chapter on small magnitudes $1·093 \times 0·9624$ would equal $1·0 + 0·093 - (1 - 0·9624) = 1 + 0·093 - 0·0376 = 1·0554$. Explain why this result is unjustifiable, and show the fallacy in the process of obtaining it.

139. Possible Error after a Calculation.

The possible error of the **sum** *of two or more measurements is equal to the sum of their individual possible errors.*

A table is found by measurement to be 44·3 cm. higher than a bench which has a height of 42·5 cm. from the floor. If the third significant figure of each measurement may have a possible error of half a unit what limits can be assigned for the height of the table from the floor. Ans.: between 86·7 cm. and 86·9 cm.

The possible error of the **difference** *between two measurements is equal to the sum (not the difference) of their possible errors.*

A table is 51·1 cm. lower than a shelf which is between 137·9 cm. and 140·1 cm. above the floor. How high must the table be? Ans.: from 88·75 cm. to 89·05 cm.

Would it be correct to answer the last example, "from 88·8 cm to 89·0 cm." ?

The possible error of the **product** *of the two measurements* $m_1 \pm e_1$ *and* $m_2 \pm e_2$ *will be found by ordinary multiplication to be* $m_1e_2 + m_2e_1$ under ordinary circumstances in which the error is small (§ 62) compared with the measurements themselves, so that e_1e_2 is negligible (§ 64). It will be instructive, however, to re-write the measurement and its possible error, $m \pm e$, in the form $m(1 \pm e/m)$ before performing the multiplication. The fraction e/m will now represent the relative error (§ 49). The product will be $m_1(1 \pm e_1/m_1) \times m_2(1 \pm e_2/m_2) = m_1m_2(1 \pm e_1/m_1 \pm e_2/m_2)$; *i.e. the possible percentage error of a* **product** *is equal to the sum of the possible percentage errors of its factors.*

A rectangular surface measures 50 cm. × 20 cm., each measurement being correct to two significant figures. Calculate $(50 \pm 0·5) \times (20 \pm 0·5)$, and show that the possible error of the area is 35 cm.^{2}; then find the relative value of 0·5 to 50 and that of 0·5 to 20, and see whether their sum is the same as the ratio of 35 cm.^{2} to 50 cm. × 20 cm.

Similarly, *the possible percentage error of a* **quotient** *is equal to the sum of the possible percentage errors of the divisor and dividend;* for $m_1(1 \pm e_1/m_1) \div m_2(1 \pm e_2/m_2)$ is equal to $(1 \pm e_1/m_1 \pm e_2/m_2)m_1/m_2$.

These rules are equally applicable in cases where one of the two quantities employed has no error whatever. If a 50-foot tape measure has a possible error of a tenth of a foot (0·2 per cent.) then there may be an error of three-tenths of a foot in using it to measure a distance of 150 feet. If the measured diameter of a circle is uncertain by 3 per cent. then the calculated circumference will be liable to an error of 3 per cent.

Example.—The sides of a rectangle were measured with a millimetre scale. The dimensions are length 5·26 cms., width 3·18 cms. Find the area.

Since area = length × breadth, the elementary student is apt to say

$$\text{Area} = 5{\cdot}26 \times 3{\cdot}18 = 16{\cdot}7268 \text{ sq. cms.}$$

This result is not correct. The sides have not been measured sufficiently accurately to warrant a six-figure accuracy. The reading 5·26 is an approximation and anything between 5·255 and 5·265 would be called 5·26 by an elementary student. Similarly, the limits of 3·18 are 3·175 and 3·185. Taking the lower limits the area might be

$$5{\cdot}255 \times 3{\cdot}175 = 16{\cdot}684625 \text{ sq. cms.}$$

and taking the upper limits

$$5{\cdot}265 \times 3{\cdot}185 = 16{\cdot}769025 \text{ sq. cms.}$$

The difference between these is 0·084400.

This being so it is absurd to give the result obtained by multiplying 5·26 × 3·18 to a greater accuracy than 16·73 ± 0·04 sq. cms.

Errors in one measurement may counterbalance errors in the other, but this cannot be depended on. Also there are the inaccuracies in the ruler and the trueness of the rectangle to be considered.

140. "Probable" Error.

The magnitude of a "possible error" is not actually as definite as it is usually assumed to be. There may be a high degree of probability that a measurement has no error greater than half a unit in the last decimal place that is written, but this can never be so high as to amount to an absolute certainty.

The most that can be said is that a certain large range for the error of a measurement is very improbable, narrower limits of error are less improbable, that an error is within a still smaller range may be probable rather than improbable, that the error has at least a certain minute magnitude may be so highly probable as to be almost a certainty, etc. Without taking up a quantitative discussion of the relative likelihood that the magnitude of a particular error will fall within given limits we need notice here only that the most convenient degree of probability that we can specify is that of a halfway point between probability and improbability. *For a certain limited class of error* it is possible to state that a particular error is "*just as likely as not*" to be within certain numerical boundaries, *i.e.* that the error's chance of being smaller than the stated limit is just equal to its chance of being larger than the limiting value. Boundaries of this sort are sometimes called *probable errors*, and will be considered in a later chapter (see Ch. XV. *et seq.*). At present all that needs to be noticed in regard to them is that they are not "probable" in the sense of being highly probable values nor are they even "errors" except in a restricted sense of the word, so that the accepted name is apt to prove a misleading one.

141. Accuracy Required in Special Cases.

When only a certain degree of accuracy is needed there is nothing to be gained by attempting a greater degree, and there is usually a considerable loss of time and labour if calculations are carried out to more significant figures than can be utilized. For a great many scientific, engineering, and commercial calculations three-figure accuracy is sufficient, and in such cases the slide rule can be used more readily than a table of logarithms. Four-figure accuracy is attained by the use of four-place logarithm tables, and is sufficient for most of the calculations of physics and chemistry, although five-figure accuracy is needed for certain measurements, *e.g.* determinations of mass and density, the acceleration of gravity by Kater's pendulum, barometric pressures, etc. Six-figure accuracy is sufficient for all surveying except the most precise geodetic measurements. Seven-figure accuracy is also needed for certain astronomical calculations.

142. Further Illustrative Examples.

A consideration of the formula will sometimes help a student to determine the degree of accuracy with which he must work.

(1) Thus in obtaining the circumference (c) of a circular rod from a measure of the diameter (d), since

$$c=\pi d,$$

a 1 per cent. error in d will produce a 1 per cent. error in c.

(2) Again in determining the value of an area (A) of a circle from a measure of the diameter (d), since

$$A=\frac{\pi d^2}{4}$$

a 1 per cent. error in d will produce a 2 per cent. error in A. To get the area to the same accuracy as the circumference above it would be necessary to make the diameter measurement twice as accurate. (By "twice as accurate" we mean "half the percentage error.")

(3) In determining the rigidity of a wire (n) from a torsion experiment

$$n=2G\frac{l}{\theta}\cdot\frac{1}{\pi a^4}$$

where G is the moment of the couple which applied to the end of a wire of length l and radius a produces at that end an angle of twist θ. Here it is obvious that if we consider the effects of G, l, θ, a, separately, a 1 per cent. +ve error in G gives a 1 per cent. + error in n, similarly for l; a 1 per cent. +ve error in θ gives a −ve 1 per cent. in n, and a +ve 1 per cent. error in a gives a −ve 4 per cent. error in n.* Hence, if it is desired that any one measurement should not produce more inaccuracy than either of the others, it

* $(1+\frac{1}{100})^4=1\frac{4}{100}$ nearly.

would be necessary to measure a to one quarter the percentage error of the others.

(4) In the case of a simple pendulum of length l, the period t is given by

$$t=2\pi\sqrt{\frac{l}{g}} \quad \text{whence} \quad g=4\pi^2\frac{l}{t^2}$$

hence, if the error carried into the result by an error in t should not exceed the error produced by an error in l, it follows that t should be measured to one-half the percentage error with which l is measured.

(5) The moment of inertia of a cylinder of mass M, length l, and radius r, is given by

$$M\left(\frac{l^2}{12}+\frac{r^2}{4}\right)$$

If $l=6r$ the first term is twelve times the second term, hence it is not important to measure r as precisely as l. Each term contributes equally to the total error when the percentage error in r is 12 times the percentage error in l.

143. Detailed Consideration of a Common Experiment.

In experiments where the final quantity is got by straight multiplication and division it is fairly easy to get the final percentage accuracy. Where, however, additions and subtractions have to be performed the working is more complicated.

As an example we shall consider in detail an experiment with which most elementary students are familiar, viz. the *Determination of the Specific Heat of a Metal.*

To find the specific heat of a metal, the best way is to raise the lump of metal to a fairly high temperature (uniform throughout its mass) and then to immerse it in cold water contained in a thin copper vessel called the calorimeter. The metal, water, and calorimeter take up a final constant temperature, and by equating the heat given out by the hot metal to the heat absorbed by the water and the calorimeter the specific heat of the metal may be found. For example if

M = the mass of the lump of metal in gms.
S = the specific heat of the metal in calories per gm. per degree.
W = the mass of water in gms.
w = the mass of the calorimeter in gms.
s = the specific heat of metal of which the calorimeter is made. (In the case of copper this is 0·094 calorie per gram per degree.)
$T°$ C. = initial (high) temperature of metal.
$t°$ C. = initial (low) temperature of water and calorimeter.
$\theta°$ C. = final temperature of metal, water, and calorimeter; then

$$\text{heat given out by metal} = \text{water equivalent of the lump of metal} \times (T-\theta)$$
$$= MS(T-\theta) \text{ calories}$$

$$\text{heat absorbed by water} = W(\theta-t) \text{ calories}$$

$$\text{heat absorbed by calorimeter} = \text{water equivalent of calorimeter} \times (\theta-t)$$
$$= ws(\theta-t) \text{ calories}$$

$$\text{whence } MS(T-\theta) = (W+ws)(\theta-t)$$

from which S may be found.

Experiment.—In the experiment the metal is weighed and is heated to $T°$ C. by being placed in a pot of boiling water and kept there some time, say 10 minutes. (If your thermometer reads as far up as 100° C. take the temperature of the water. If not assume it=100° C.)

Meanwhile the calorimeter is weighed and *just sufficient* water to cover the metal is poured in and this is then weighed. A sensitive thermometer is placed in the water which is stirred and after 5 minutes or so its temperature is read. Make an estimate of the temperature to one-tenth of a degree. The calorimeter should not be placed near the flame and boiler, or its temperature will gradually rise. When ready for transferring the metal bring the calorimeter near the pot and quickly remove the metal from the boiling water to the calorimeter, at the same time trying, by a judicious shake, to remove the excess of hot water clinging to it. Remove the calorimeter from the neighbourhood of the boiling water, stir the water by moving about the metal, watch the thermometer carefully, and read the highest temperature recorded. Again estimate to one-tenth of a degree. Work out the results as illustrated above. *Do not use a formula but work on first principles.*

Hints for Calculation.—In working out the result do not overcharge your arithmetic with unnecessary figures. This is a waste of time. You must pay great attention to the relative accuracies of the data you have obtained. It is of no use to take temperatures approximately and then expect to get a result correct to three significant figures. First look at your weighings. You have made them to the nearest gram. Then work out the value of the water equivalent of your calorimeter to the nearest gram. If $w=132$ gm. and $s=0{\cdot}094$ calorie per gm. per degree, then the water equivalent $=ws=132\times0{\cdot}094=12$ gm. and not $12{\cdot}408$. In obtaining this product you should use contracted methods of multiplication or a slide rule. Again, if your block of metal weighs 1,238 grams you will find it quite near enough for the purposes of this experiment to take it as 1,240 grams.

Your water temperatures have been read to one-tenth of a degree. They may be one-tenth of a degree out. Note what percentage difference is produced in the rise of temperature of the water if the original temperature were read 0·1° C. low and the final temperature 0·1° C. high or vice versa. This temperature difference is usually the factor that is in greatest error and of course conveys its error into your result.

Again, if T is taken as 100° C. without worrying about its exact value, *i.e.* whether it is 99·9° or 100·1° C., should the fall of temperature of the metal be expressed to 0·1° C. or to the nearest degree? Try if half a degree error in $(T-\theta)$ causes as much error as one-tenth of a degree in $(\theta-t)$.*

A typical set of figures is here given: $M=1{,}238$ gm., $w=132$ gm., $m=259$ gm., $T=100°$ C., $t=13·9°$ C., $\theta=28·6°$ C.

Writing out the equation

heat given out = heat absorbed,

we get

$$1238\times S\times(100-28·6)=(259+132\times0·094)\ (28·6-13·9)$$

This should be rewritten as

$$1240\times S\times(100-29)=(259+12)\ (14·7)$$

$$\text{whence } S=\frac{271\times14·7}{1240\times71}$$

First get an approximate value of S. Thus

$$S=\frac{280\times15}{1200\times70}=\frac{4}{80}=0·05$$

Now work out the answer by approximate methods without worrying about the position of the decimal point, and we get

```
271          71        88)398(45
147         124           352
---         ---           ---
271          71            46
108          14
 19           3
---         ---
398          88
```

The result is therefore 0·045 calorie per gm. per degree. A slide rule or logarithms may be used to save time.

Work out in this way your value of the specific heat of the metal obtained from the first experiment. Repeat the experiment with a different quantity of water. If the two values agree fairly well calculate their average and ***make an estimate of the percentage accuracy*** of your result.

* If the 28·6° is 0·1° high and the 13·9° is 0·1° low, the true difference is 14·5°, which introduces an error of about 1½ per cent.

Corrections for Heat Exchange.—The loss of heat from the hot metal as it is being transferred to the calorimeter is unavoidable. It is partly counterbalanced by the amount of hot water carried over by the hot metal. Indicate what effect these errors are likely to have on your value of the specific heat.

The calorimeter may gain heat from the room or lose heat to the room. The heat exchange may be diminished by starting at a temperature below that of the room and finishing up at a temperature an equal amount above that of the room. There are better methods of correcting for this exchange, but the procedure would take up too much time at this stage.

These two errors may cause as great or a greater error than those produced by accidental errors in reading the weights and the temperatures.

Exercises XII

1. An expression like "four-and-a-half-figure accuracy" is sometimes seen. Write down your opinion of what this phrase means, expressed as a percentage; and state your reasons for choosing this number.

2. Explain why the possible error of a difference is equal to the sum (instead of difference) of the errors of its components.

3. If a wooden block measures 250 cm.^{3}, with an uncertainty of 5 cm.^{3}, and weighs $2{\cdot}00 \times 10^2$ gm., how many gm./cm.^{3} does the uncertainty of its density amount to? What can you say of the effect of the uncertainty of the weight?

4. Each volume of a ten-volume encyclopedia has 2·0 inches thickness of leaves between two covers that are each 0·2 inch thick. If the number of pages per volume is liable to vary by 2 per cent. and the thickness of the cover by 1 per cent., what is the maximum length of shelf-space that may be needed for the set?

5. If a calculated measurement M is equal to $10m_1+20m_2$ what will its possible error E amount to in terms of the possible errors e_1 and e_2 of m_1 and m_2 respectively?

6. The quotient in § 139 should, by algebra, be $(1 \pm e_1/m_1 \mp e_2/m_2)m_1/m_2$. Explain why the inversion of the double sign is unnecessary in this connection.

7. The sides of a rectangle are measured. The readings to the nearest millimetre are 17·4 cms. and 9·7 cms. Find (*a*) the area, (*b*) the greatest possible area, (*c*) the least possible area. (*d*) Give finally the area with a statement of the probable deviation or error.

8. The sides of a rectangle are measured to the nearest hundredth of an inch, and given as 17·43 inches and 9·68 inches. Find (*a*), (*b*), (*c*), (*d*) as in Question 7.

9. A rectangle is measured. Length 24·7 cms., possible error 0·05 cm. Breadth 17·4 cms., possible error 0·05 cm. Find the area and its possible error.

10. The dimensions of a rectangular block are found to be $5{\cdot}26 \times 3{\cdot}18 \times 4{\cdot}97$ cms. Find its volume. Give some indications of its accuracy.

11. A bench is 124·6 cms. from the floor. Possible error 0·1 cm. A block resting on the table bears a mark 37·3 cms. above the table. Possible error 0·05 cm. Find the height of the mark above the floor.

12. A body weighs in air $5{\cdot}585 \pm 0{\cdot}005$ gm., and in water $4{\cdot}27 \pm 0{\cdot}01$ gm. Find the greatest and least possible values for its specific gravity. State its approximate value with an indication of its likely accuracy.

CHAPTER XIII

THE PRINCIPLE OF COINCIDENCE AND THE AVERAGING OF RESULTS

144. Effect of Magnitude upon Accuracy.

With a scale of millimetres measure a length of 1 inch as accurately as possible, writing down the number of whole centimetres, additional whole millimetres, and estimated additional tenths of a millimetre which it contains. Measure a length of 2 inches in the same way and divide the result by two. Calculate one-tenth of the carefully measured length of an interval of 10 inches and compare with the two other results. Notice that a large quantity can usually be measured more accurately than a small one. If the method of measurement gives the same absolute accuracy in two different cases (for example, no error greater than 0·005 cm.) the ratio of error to measurement will naturally be a smaller relative error just in proportion as the measurement itself is greater. It is for this reason that the metre was chosen for a standard of length (§ 16) instead of the centimetre, and the kilogram for a standard of mass instead of the unit, a gram.

145. Measurement by Estimation.

In order to make a more accurate determination of the length of an inch lay one metre stick on the table with the metric graduations upward and place the other beside it with the graduation in inches and tenths of an inch upward. See that the scales are in close contact and note down the number of centimetres and hundredths that happens to be opposite the mark 1 inch on the other scale. Note also the indication of the mark 31 inches. Move one scale along the other at random and repeat the observations, unless otherwise directed, until ten sets have been taken. The distance measured should always be the same, 30 inches, but it should be measured at other places, such as the mark 4 and the mark 34. *Never use the mark* 0 *at the very end of the stick, as it is often inaccurate on account of wear.*

Tabulate the results in a *form* like the following:

MEASUREMENT OF AN INCH BY THE METHOD OF ESTIMATION.
A length of 10 inches measured each time.

—	1st line.	2nd line.	Interval.	
			cms.	ins.
Cm. scale . .	20·05	45·50	25·45	
Inch scale . .	1·00	11·00		10·00
Cm.	50·00	75·47	25·47	
Inch	2·70	12·70		10·00
Cm.	10·06	35·46	25·40	
Inch	1·50	11·50		10·00
. . . . . .	. . . . . .	. . . . . .	. . . . . .	. . . . . .
Sum of 10 *such readings*			254·02	100·00
Average			25·402	10·000

Length of an inch: 2·5402 cms.

146. Measurement by Coincidence.

In the method of measurement *by estimation* one or more equal intervals are measured in the customary manner, tenths of the smallest scale-divisions being determined by a mental estimate. For the method of measurement *by coincidence* a series of equal quantities is necessary. If these can be directly compared with a series of equal scale-units it will usually be possible to observe that

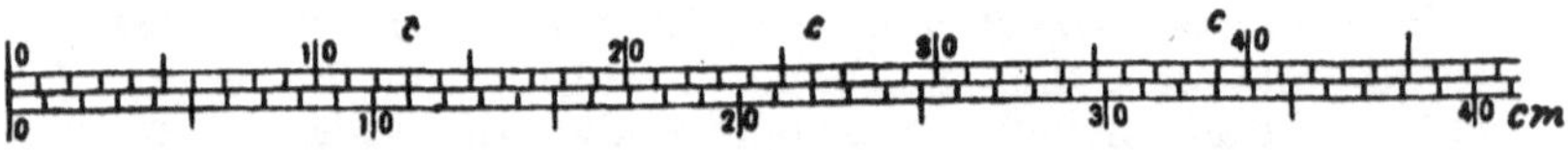

FIG. 54.—THE METHOD OF COINCIDENCES.

Each of the unknown intervals is seen to be a trifle less than 1 cm., but the fact that there is no distinguishable difference between 13 intervals and 11 cm. shows that the length of the unknown interval must be 11/13 of a centimetre.

some whole number of the equal quantities is of precisely the same magnitude as some other whole number of scale-divisions. For example, the diagram (Fig. 54) shows in three different locations (*c*) that 13 of the unknown intervals in the upper line are equal to 11 of the centimetre units along the lower scale. Accordingly, the unknown interval must be 11/13 cm.

Set the two scales so that any whole number of inches near one end of one metre stick is exactly opposite some whole number of centimetres on the other. Hold the two sticks firmly together and look again to see that the coincidence is exact. Do not be satisfied unless you are unable to say whether the upper mark is a little toward the right or toward the left of the one below it. Without allowing the two scales to slip find another place, at least 30 cm. distant from this point and preferably further, where there is another exact coincidence, this time between any centimetre or millimetre graduation and any graduation of inches or tenths. Try to decide whether the coincidence is exact or whether the imaginary central axis of one line lies a little beyond the other, and if the coincidence is faulty choose a better one elsewhere.

Record the results as in the specimen table, re-set the two scales, and repeat until ten determinations have been made. Shift the position of the first coincidence, as was done in the method of estimation; also, take care that the interval between coincidences is not of the same length each time.

MEASUREMENT OF AN INCH BY THE METHOD OF COINCIDENCES.

—	1st line.	2nd line.	Interval.	cm. in 1 in.
Cm.	90·000	35·400	54·600	
In.	4·000	25·500	21·500	2·5395
Cm.	15·000	71·400	56·400	
In.	6·000	28·200	22·200	2·5405
.	.	.	.	.
.	.	.	.	
.	.	.	.	.
Average				2·5401

It is not absolutely accurate to assume that the coincidences are always exact to a hundredth of the smallest graduation of the scale, but the lack of perfect coincidence is usually perceptible even if the discrepancy is much less than a tenth of a scale-division. It will very rarely exceed a hundredth of a smallest subdivision if the work is carefully done.

147. The Vernier.

Rule a straight line along a strip of cardboard 5 or 10 cm. wide and about 20 cm. long (Fig. 55). As shown in the diagram, lay off a scale of *centimetres* along the lower side of the line; along the upper side lay off a series of ten intervals, making each one 0·9 cm. long, so that their combined length will be just 9 cm. Rule a perpendicular line, *AB*, a little to the left of the common zero point, and mark the words "scale" and "vernier" as indicated. See that your work corresponds in every respect to Fig. 55, except that the letters *A*, *B*, and *C* are not marked, and the scale of centimetres may be longer or shorter than shown here. Observe that the mark 1 on the vernier is one-tenth of a centimetre to the left of the mark 1 on the centimetre scale, 2 on the vernier is two-tenths to the left of 2 on the scale, etc. Put the card on a larger piece of cardboard, to avoid marring the table, and cut from *C* to *B*, but no further, also from *A* to *B*. The result will be a cardboard model of a ***vernier caliper***. Notice that the zero of the vernier acts as a pointer that indicates 0 cm. when the jaws (*AB*) of the caliper are no distance apart; consequently it must give the correct reading on the centimetre scale when the jaws are slid apart to any given distance.

Holding the body of the vernier caliper stationary on the table move the vernier slightly to the right so that its line number 1 coincides with the 1-centimetre mark; then move it further until 2 and 2 coincide; then 3 and 3, etc. When the tenth division of the vernier has been brought into coincidence with a line on the

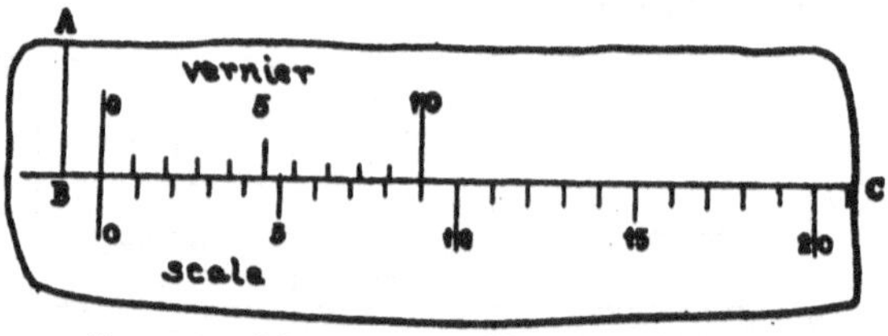

FIG. 55.—MODEL OF A VERNIER CALIPER.
The diagram can be drawn on cardboard, cut out, and used for fairly accurate measurements.

centimetre scale it will be found that the zero of the vernier stands at the mark 1 cm. and the jaws of the caliper are separated by a distance of a single centimetre. From the construction of the apparatus it will be clear that when any division of the vernier (say number 3) coincides with any division of the lower scale the jaws of the caliper must be separated by a corresponding number (3) of tenths of a centimetre more than some whole number of centimetres.

Separate the jaws of this model vernier caliper by a distance of 8½ cms., as nearly as you can by estimation; then look at the vernier and notice that while its zero is beyond the mark 8 of the centimetre scale the division 5 of the vernier coincides with some division (no matter which one) of the main scale. The reading is accordingly 8·5 cm. In the same way read the indicated length when the caliper is set at random, and repeat the process until you are perfectly familiar with it.

148. Use of the Vernier Caliper.

Hold the two parts of the caliper in alignment by pinching the line *BC* between the forefinger and thumb of the left hand, holding the lower part tightly but allowing the upper part to slide; at the same time hold the same line between the forefinger and thumb of the right hand, allowing them to slide on the lower part while grasping the upper part firmly. This will allow the model to be accurately applied to any object that is to be measured. Use the apparatus to make twelve measurements of the *thickness* of a wooden block at different positions around its edge. Write down each measurement as soon as it is obtained, and do not be disconcerted by the fact that many of them may be identical. Find the average thickness of the block, carrying the result out to one more decimal place than the individual measurements, and keep it for later reference.

If a vernier caliper is arranged to give tenths of a scale-division it is possible to make a reading of half a tenth, or a twentieth, in case two adjacent lines on the vernier show equal and opposite deviations from coincidence with two lines of the scale. In this way an ordinary barometer scale of twentieths of an inch, provided with a 25-division vernier that gives 500ths, can be read accurately to single thousandths of an inch.

149. Slide-Rule Ratios.

It has already been seen that setting a slide rule for the ratio π and picking out a point of exact coincidence will usually give a numerical value for π that has more than three-figure accuracy, *i.e.* greater accuracy than can be obtained by the ordinary process of "estimation" with the slide rule. Numbers like the 26 and 66 on the back of the slide rule for the ratio of inches and centimetres are so chosen as to give the correct value to at least as many significant figures as can be read with the apparatus used. Some slide rules give the same equivalent as 50 in.=127 cm. This has the disadvantage of not being quite as easy to set on the *A* and *B* scales as on *C* and *D*, but for use with a very finely graduated instrument or with one on which the scales are 20 inches long instead of 10 it has the advantage of greater accuracy, for $66{\cdot}0000 \div 26{\cdot}0000 = 25{\cdot}3846$ while $127{\cdot}0000 \div 50{\cdot}00000 = 25{\cdot}40000$. The former ratio is correct to three significant figures only, while the latter is correct to five. In many cases it is, however, just as easy to set on the ratio 1 inch=2·54 cm.

Check the following approximate ratios on the slide rule:

$$\sqrt{3}=\frac{5{\cdot}2}{3},\quad \sqrt{5}=\frac{9}{4},\quad \sqrt{6}=\frac{4{\cdot}9}{2},\quad \sqrt{7}=\frac{8}{3}$$

150. The Problem of Averaging in Experimental Work.

In many experiments in physical science where one quantity is expected to increase regularly with another it is possible to obtain a good average value of the rate of increase as follows :—

Let one quantity x increase by regular amounts, reading the value of y for each value of x and proceed until an even number of readings have been taken. Then divide your readings into two sets *A* and *B*, and subtract the first x of set *A* from the first x of set *B*, and the first y of set A from the first y of set B, and so on. Three examples will make the method plain.

(1) *Determination of the Elongation of a Spiral Spring.*—

(1) Load, in addition to that of pan.	(2) Reading of pointer on mm. scale. Load increasing.	(3) Load decreasing.	(4) Mean of cols. (2) and (3).	(5) Extension produced by an addition of 500 gms.
gms.	cms.	cms.	cms.	cms.
0	37·85	37·85	37·85	
100	41·82	41·85	41·84	
200	45·86	45·89	45·88	
300	49·87	49·90	49·89	
400	53·86	53·91	53·88	
500	57·85	57·91	57·88	20·03
600	61·85	61·91	61·88	20·04
700	65·81	65·90	65·86	19·98
800	69·80	69·90	69·85	19·96
900	73·80	73·80	73·80	19·92
		Average		19·99

Working in this way all the readings are taken equally into account. You may check this result by plotting a graph between extension and load. The graph will show whether the extension is proportional to the increase of the load, and its slope will give you the value of the ratio.

(2) *Determination of the period of vibration of a pendulum.*— Set the pendulum swinging. Stand a yard or so off and note by a watch, or start a stop-watch at, the moment when the bob makes a *transit*, that is, passes in front of the mark.

To do this it is best to take the time at every fifth vibration from the noughth to the fifty-fifth vibration, arranging your observations as shown in the table, which is a useful scheme for any experiment on timing. By subtracting the time of the noughth vibration from the time of the thirtieth we get time for thirty swings. By taking away the fifth from the thirty-fifth we get the time for another set of thirty, and so on. In the final column we thus get several times occupied by different sets of thirty swings.

Number of swing.	Time by watch	Number of swing.	Time by watch.	Time for 30 swings.
	h. m. s.		h. m. s.	m. s.
0	3 5 46	30	3 6 59	1 13
5	59	35	7 12	13
10	3 6 12	40	24	12
15	23	45	37	14
20	36	50	49	13
25	48	55	8 2	14
		Average		1 13·2

By taking the average and dividing this by thirty we get the period. In the case given the period

$$=\frac{73\cdot2}{30}=2\cdot44 \text{ seconds}$$

(3) *Determination of the Acceleration of the Trolley in a Fletcher's Trolley Experiment.*—In this experiment a trolley runs down an inclined plane with uniform acceleration and a pen, vibrating at right angles to the motion of the trolley, writes a fine wavy trace on the upper surface of the trolley (Fig. 56). The central line is then

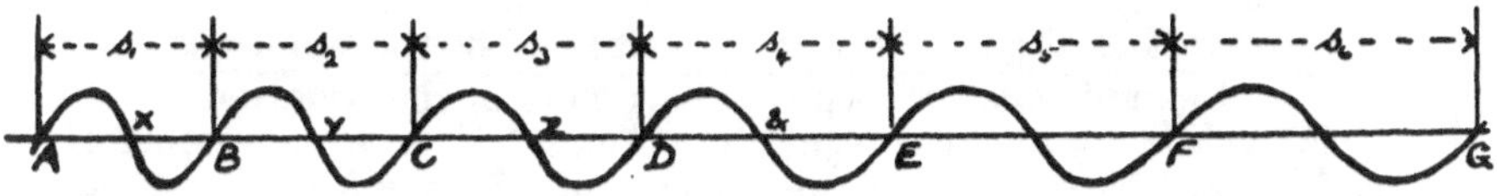

FIG. 56.

drawn and the distances *AB*, *BC*, *CD*, etc., measured. Denote these by s_1, s_2, s_3 . . . It can be shown that the acceleration of the trolley $=n^2\times$ average increase in the wave-length of the trace from one wave to the next, where n = frequency of vibration of the pen.

Now if we work out s_2-s_1, s_3-s_2, . . . and add up these five differences and take the average, we are in effect throwing away the readings s_2, s_3, s_4, s_5, and just taking one-fifth of the difference between s_1 and s_6, for $\frac{1}{5}\{(s_2-s_1)+(s_3-s_2)+\ \ldots\ (s_6-s_5)\}=\frac{1}{5}(s_6-s_1)$. Therefore to use all the readings and to get the best result, divide the table in two, subtract s_1 from s_4, s_2 from s_5, s_3 from s_6 and average these, and finally divide by 3.

Thus suppose

$$\begin{aligned} s_1&=1\cdot8 \text{ cms.}\\ s_2&=2\cdot5 \quad ,,\\ s_3&=3\cdot1 \quad ,,\\ s_4&=3\cdot8 \quad ,,\\ s_5&=4\cdot4 \quad ,,\\ s_6&=5\cdot1 \quad ,, \end{aligned}$$

then

$$\begin{aligned} s_4-s_1&=2\cdot0 \text{ cms.}\\ s_5-s_2&=1\cdot9 \quad ,,\\ s_6-s_3&=2\cdot0 \quad ,,\\ \hline \text{Average}&=1\cdot97 \quad ,, \end{aligned}$$

Therefore, increase in wave-length $=0\cdot66$ cm. and the acceleration of the trolley

$$=n^2\times0\cdot66\text{—cm. per sec. per sec.}$$

If $n=5$, this becomes $16\cdot4$ cms. per sec. per sec.

151. Generalisations.

We might put the method of arriving at the results more generally as follows.

(A) y expected to be proportional to x: say $y=bx$.

(1) x increases regularly. Take an even number of readings (x_1, y_1), (x_2, y_2) . . . (x_n, y_n). Here the x's and y's may be measured from any arbitrary zeros.

Then by the half-table subtraction

$$y_{\frac{n}{2}+1} - y_1 = b\left(x_{\frac{n}{2}+1} - x_1\right)$$
$$y_{\frac{n}{2}+2} - y_2 = b\left(x_{\frac{n}{2}+2} - x_2\right)$$

and so on.

The multipliers of b on the right hand are equal, therefore best value of

$$b = \frac{\text{average of quantities on the left hand of the equations}}{\text{the multiplier of } b \text{ on the right hand of the equations}}$$

(2) x does not increase regularly. Again take an even number of readings and subtract as before. The multipliers of b are not equal this time, and best value of

$$b = \frac{\text{sum of quantities on left hand of the equations}}{\text{sum of multipliers of } b \text{ on the right hand of the equations}}$$

(3) Since $y = bx$ we might divide each y by each x and find the average of all the quotients. This method gives equal weights to all the quotients, and those obtained from the small values of x and y are in practice less accurate than those from the larger values. Hence this method is not recommended.

(4) If x and y are measured from their true zeros:

$$y_1 = bx_1$$
$$y_2 = bx_2$$
$$y_3 = bx_3$$

Therefore we may add up the left-hand sides of the equations to form a new left hand and the right-hand sides to form a new right hand, thus getting

$y_1 + y_2 + y_3 + \ldots = b(x_1 + x_2 + x_3 + \ldots)$, or as is usually written $\Sigma y = b\Sigma x$,

whence a good value of b may be obtained by division.

(B) y expected to be related to x by the equation $y = a + bx$. Required to find a and b. The observational equations are:

$$y_1 = a + bx_1$$
$$y_2 = a + bx_2$$

and so on.

By using either method A(1) or A(2) the a's cancel out and we get b as before. Then adding up the left-hand and right-hand sides of the equations, we get

$$\Sigma y = na + b\Sigma x$$

where n is the number of observational equations.

Then $$a = \frac{1}{n}(\Sigma y - b\Sigma x)$$

Methods A(1) and A(2) are known as the *half-table methods*. They with Method A(4) and B, are also known as the *Method of Zero Sum*. This name has been given by N. R. Campbell,* who remarks that if we add or subtract a group of measured quantities the errors of measurement will in the long run cancel out.

Example 1.—Related quantities x and y have been observed. Their values are set out in the table. A knowledge of the problem tells us that they should be strictly proportional to each other (*e.g.* y might be the masses and x the corresponding volumes of lumps of the same material). Simple values of x and y have been taken to illustrate the method of extracting the value of b in the equation $y=bx$.

	x	y	Differences.		y/x
	2	1			0·500
	5	3			0·600
	7	4			0·571
	11	6	9	5	0·556
	13	7	8	4	0·538
	15	8	8	4	0·533
Sums .	53	29	25	13	

The numbers in the columns headed differences belong to method A(2). From the sums at the bottom we see $b=\frac{13}{25}=0{\cdot}52$.

Using method A(4) we get the numbers at the bottom of the first and second columns, whence $b=\frac{29}{53}=0{\cdot}548$.

Using method A(3) we get the quotients in the last column. Their average is $b=0{\cdot}549$.

Another value of b could be obtained from a graph, using a black thread and drawing the straight line through the origin. This gives $b=0{\cdot}54$. The line of best agreement with the points has $b=0{\cdot}52$, but it does not go through the origin.

Example 2.—Related quantities x and y have been observed. Their values are set out in the table and the relation between them is known to be of the form $y=a+bx$.

	x	y	Differences of x	Differences of y
	1	3		
	3	4		
	8	6		
	10	7		
	13	8	12	5
	15	9	12	5
	17	10	9	4
	20	11	10	4
Sums .	87	58	43	18

* N. R. Campbell, *Philosophical Magazine*, vol. 39, 1920, p. 177; and vol. 47, 1924, p. 816.

Using method B

$$b=\tfrac{18}{43}=0{\cdot}42$$

and

$$a=\tfrac{1}{8}(58-\tfrac{18}{43}\times 87)$$
$$=2{\cdot}7$$

$$\therefore\ y=2{\cdot}7+0{\cdot}42x.$$

The values of a and b could also be obtained by a graph and the black thread. Do this and so confirm the values already obtained.

Exercises XIII

1. Consider carefully what is meant by the expression "the principle of coincidence," and then write a definition of it in your own words. See that it is a definition of a *principle*, not of a process or of a condition.

2. What is the significance of the fact that the second point of "coincidence" in Fig. 54 is not as precise as the first, nor the third as good as the second?

3. Suppose the time of vibration to be just 7/8 as long for one pendulum as for another. How could you verify the fact by the method of coincidences?

4. If you take just seventeen steps while a man who is walking beside you takes just eighteen, how long are his steps in terms of yours, which are considered of unit length? How does the frequency of his steps compare with that of yours?

5. Explain why the method of coincidence is more exact than the method of estimation.

6. If the numbers in the column headed "*interval*," § 146, are trustworthy to five significant figures, how can you account for the fact that even the fourth significant figure fluctuates in column of centimetres per inch?

7. If an ordinary vernier can be read to half-tenths why can it not be used for other fractions of a tenth?

8. Could the upper scale in Fig. 54 be used as a vernier for the centimetre scale? If so, would it indicate elevenths of a centimetre, or thirteenths, or neither? How would its numbers be arranged?

9. Make a cardboard model of an angular scale and vernier to read to tenths of a degree. A radius of about 6 inches should be taken.

10. Study the verniers on barometers (1/10 mm. and 1/500 in.), on goniometers and spectrometers ($\frac{1}{10}^\circ$, 1 minute, 20° seconds).

11. Use the vernier calipers you have made (or a steel pair) to measure up bodies of different geometrical shapes (blocks, cylinders, spheres). Find their volumes. State their percentage accuracies. Weigh them, deduce their densities and state their percentage accuracies.

12. In determining the time of vibration of a Borda's pendulum by the coincidence method, it was observed that the clock pendulum (which had a two-second swing) made 172 (complete) vibrations to the Borda's 171. Find the time of swing of the Borda pendulum and the error made by miscalculating the coincidence by one vibration.

13. Describe the best way of taking an average in an experiment (1) on the stretching of a spring or wire, or the bending of a beam, when a number of equal weights are available; (2) on the period of vibration of a pendulum or other vibrator.

14. A falling tuning fork writes its trace on a whitewashed glass plate. The frequency of the fork is 148 per second. Successive lengths of 10 waves were found to be 4·10, 8·65, 13·05, 17·70 cms. Show how to find the best value of the average increase of wavelength per vibration and deduce the acceleration of gravity.

15. I have a pair of calipers graduated in inches and twenty-fifths of an inch. I wish to attach to it a vernier enabling me to read to thousandths of an inch. Show how this may be done, giving careful diagrams.

The scale and vernier should be numbered so as to make the operation of reading as easy as possible. Give the position of the vernier when the instrument is adjusted to measure the length of a rod 3·472 inches long.

16. Give instances of the best way of carrying out observations so that the results obtained may be accurately averaged. Take, for example, the determination of the force required to stretch a spiral spring unit length and the determination of a period of vibration. What is the use of graphs ?

17. A moving platform passes underneath a pen vibrating at right angles to the line of travel. On measurement it is found that there is a uniform increase in the wavelength of the curve traced by the pen. Show that the motion of the platform is one of uniform acceleration. If s is the average increase in wavelength from one wave to the next and n the frequency of vibration of the pen show that the acceleration is n^2s.

How would you carry out your wavelength measurements and calculations in order to get the best value of s ? How would you determine n ?

18. Try the methods of § 151 on some of the exercises in Exercise IX.

CHAPTER XIV

MEASUREMENTS AND ERRORS—STATISTICAL METHODS

152. Direct and Indirect Measurements.

An *indirect measurement* is one that is obtained from another measurement by means of a calculation. The ordinary processes of measuring length or weight are called *direct*, because the unknown length is placed beside a standard series of multiples and fractions of the unit of length and is directly compared with it, and an unknown weight is directly balanced against a series of known weights until its exact equivalent is determined. An example of an indirect measurement is the usual method of determining density. The volume of an object and its mass are determined directly, or the volume is calculated from other measurements, and its density is then obtained by making a calculation of the ratio of mass to volume.

Are measurements of area usually direct or indirect? Can they be made in the other way?

How could the density of a liquid be determined by a direct measurement?

How could the density of a solid be determined by a direct measurement?

153. Independent, Dependent, and Conditioned Measurements.

Measurements are also classified as *independent*, *dependent*, and *conditioned*, and may belong to any one of these classes whether they are direct or indirect.

Two or more measurements are said to be *conditioned* if there is a theoretical relationship between them, which must always hold good. The three angles of a plane triangle furnish an illustration of this class. Their sum must always be equal to π, or $180°$.

Measurements are said to be *dependent* if one of them is allowed to influence or bias the observer when making a later measurement of the same quantity. With accurate determinations this effect is so hard to avoid, even if the observer has the best of intentions, that it is always advisable to guard against it by some such device as hiding the scale of an apparatus from view until after the indicating mark has been set in the position that has to be read. Measurements are also dependent if any essential step in making them is not repeated in successive determinations but is assumed to have its effect remain unchanged during the series. Thus, in making independent measurements by the method of coincidences (§ 146) the scales of length were not held in one position while several coincidences were found, but were reset after each determination.*

It is customary to use the term *independent* only for measurements that are at the same time neither dependent nor conditioned.

If the density of an object should be determined by a direct measurement would its mass, density, and volume be independent, dependent, or conditioned?

154. Harmony and Disagreement of Repeated Measurements.

If the same object is measured several times in succession the measurements will in general differ from one another, but the differences will tend to be small under either of two different circumstances: (*a*) if the quantity is of a sharply defined character, or (*b*) if the method of measurement is coarse or crude. Thus, if a length of woollen cloth were compared with an accurate millimetre scale it would probably be found difficult to measure it the same twice in succession; but if the same length of steel rail were clamped on rigid supports and measured at a constant temperature it might be hard to obtain two measurements that would differ. The length of one object would be a poorly defined quantity; that of the other, a sharply defined quantity.

Although sharpness of definition of the quantity to be investigated means that repeated measurements will tend to harmonize closely, it should be carefully noted that accuracy of the methods or means of measurement has just the opposite result; it is the rough methods of measurement that make repeated determinations identical with one another, and the refined methods that show discrepancies. Two similar 1-lb. weights may appear to have precisely the same mass on a rough balance and yet differ when weighed on a more carefully constructed one. If the heavier weight should then be filed down just enough to make it equal to the lighter one a test with a delicate chemical balance might show not only that the two were still unequal but even that one of them alone would not weigh the same amount twice in succession.

The general statement can be made, then, that an accurately defined quantity or a coarse method of measurement will result in a series of determinations being harmonious or identical, while a poorly defined quantity or an accurate method of measurement will cause the results to disagree or diverge.

* In a paper on the surface tension of water by the method of measuring the number of ripples in a chosen distance Lord Rayleigh stated: "It was found advisable to work with an unknown distance, otherwise the recollection of previous results interfered with the independence of the estimates."

Which measurements do you think would be most apt to show variations among themselves, the measurements of a wooden block made with the cardboard caliper of § 147, or measurements of the same block made with a steel vernier caliper?

It is a general truth that, no matter how sharply defined a quantity may be, the use of the most precise methods will result in successive equally careful measurements of it differing perceptibly from one another, although the differences may be very small. Thus, the most accurate determinations of the length of a national prototype metre would be apt to differ by several tenths of a micron.

If equally careful measurements of the same quantity fail to agree and there is no reason for preferring one rather than another it necessarily follows that *the true magnitude of a measured quantity is always unknown.* Accordingly the various approximate measures are summarized, for actual use, by stating their *average* (the sum of n numbers, divided by n) or some other representative value, instead of by choosing one of them at random.

155. Errors of Measurement.

The *error* of a measurement is the amount by which it differs from the true value of the quantity which is measured. If the true value is always unknown the error must likewise be unknown. Such errors, however, can be discussed theoretically, and in this way much can be learned about the best manner of dealing with them.

Errors are usually classified as *constant* and *accidental*, but what are known as *mistakes* really belong in a separate class by themselves. *Constant errors* affect all the measurements of a series in the same manner or in the same direction. *Accidental errors* are *small* errors that make one measurement a trifle too large and another too small, but do not tend to bias the average result. *Mistakes* are occasional errors that are due to a lack of mental alertness on the part of the observer.

CLASSIFICATION OF ERRORS.

CONSTANT:

- THEORETICAL: usually calculable, such as the faulty length of a linear scale, due to its expansion from heat.
- INSTRUMENTAL: due to faulty graduation or adjustment of an instrument.
- PERSONAL: some persons have a constant tendency to estimate the instant of an occurrence a little too early; others a little too late.

ACCIDENTAL:

- INSTRUMENTAL: due to varying external influences, "play" of moving parts, inconstant sensitiveness, etc.
- PHYSIOLOGICAL: the senses of sight, touch, etc., have a limit of sensitiveness, and this limit is not always the same.
- PSYCHOLOGICAL: very likely the deductions about the outside world that result from the effect of sense-impressions on the mind do not correspond to the latter so closely as to be absolutely free from irregular variations when dealing with minute quantities.

MISTAKES:
- MANIPULATIVE: doing the wrong thing.
- OBSERVATIONAL: observing the wrong thing.
- NUMERICAL: recording the wrong numbers. (It is especially important to guard against focussing the attention so closely on the minute part of a measurement, *e.g.* the estimation of tenths of the smallest scale-division, that a mistake is made in recording the figures that express the larger part of it.)

156. Accidental and Constant Errors.

The difference between constant errors and accidental errors can be easily understood from the analogy of shots fired at a target. In Fig. 57 the average position or centre of clustering has a *constant* tendency downward and to the right of the bull's-eye, while the individual shots have *accidental* tendencies which carry them to one side of this average position as much as to the other. Any single shot can be considered to have a total error which is the "vector sum" or geometrical combination of its own accidental error plus

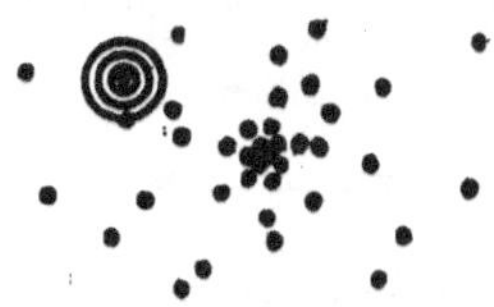

FIG. 57.—ILLUSTRATION OF ERRORS OF MEASUREMENT.

The shots (measurements) aimed at the bull's-eye (true value) show a general drift (constant error) to the right and downward, and individual deviations (accidental errors) that tend to extend equally in all directions from the centre about which they cluster.

the constant error of the whole group. Physical measurements are target shots in which the centre of clustering can be found but in which the position of the bull's-eye is unknown.

Constant errors can be avoided or reduced to a minimum by the help of theoretical knowledge (effect of gravity in deflecting the shots downward), and by changing observers, methods, and conditions (repeating the shots when the wind blows to the left instead of to the right), and above all by judgment, experience, care, and alert hunting for all possible sources of error.

Accidental errors are much more easily investigated, the chief problems from the standpoint of physical science being the determination of the centre of the cluster and the measurement of the degree of scattering or dispersion which takes place around it.

GENERAL SOURCES OF ERROR

Linear scales not parallel; lack of verticality; etc.
Faulty standardization of standards of comparison.
Uneven or irregular subdivision of standards.
Inaccurate "coincidence" or "bisection."
Faulty estimation of tenths or other subdivisions.

Parallax in reading scales or the position of a pointer.
Assumed zero of a scale.
Friction (as in a balance) and "play" or "back-lash."

State one *source of error* that was present when you performed each of the experiments mentioned in the following list. For example, in (*e*), one source of error would be the fact that the periphery of the disc is not an exact circle. This would give rise to a constant error in the determination of π if the diameter should be measured each time along one marked radius, but would cause an accidental error if many different diameters should be measured. Mark them in such a way as to show which ones gave rise to constant errors, and which ones caused accidental errors. Were there any mistakes ?

(*a*) The measurement of your span.
(*b*) The measurement of an irregular area, first method.
(*c*) The measurement of the density of an irregular solid.
(*d*) The measurement of the sine of an angle that has a given tangent.
(*e*) The experimental determination of π.
(*f*) The calculation of e^{-x^2} for different values of x.
(*g*) The use of the formula $1/(1+\delta)=1-\delta$.
(*h*) The "black-thread" determination.

157. Errors and Variations.

The theory of *accidental errors* runs closely parallel with the theory of variation in natural objects, so that a statistical investigation of the properties of several objects of the same kind, or *variates* as they are technically called, will be found to illustrate many of the facts of accidental variation which could otherwise be observed only by the more tedious study of the deviations of repeated measurements of the same object.

158. Measurement of Variates.

Examine a vernier caliper, and if it has an extra scale that reads backwards, or one scale for internal measurements and another for external ones, decide which scale reads the internal distance between the jaws of the caliper, and which point marks its zero when the jaws are closed. See that you understand the vernier and have no trouble in reading it at any setting.

Measure the length of 100 seeds or other variates of the same kind.* Make a table of preliminary measurements of ten variates in the order in which they were measured to find their general range, and then make a table of lengths *arranged in numerical order* like the above, recording each different length and the number of times that it occurs. Do not merely write the length when one

* Convenient variates for class use are beans of the kind known as kidney bean, scarlet runner bean, or French bean (botanical term *phaseolus*).

seed is measured and add a pen-stroke when another of the same length is found, but see that *each* variate is represented by a mark in the second column of your table. If an unexpectedly large or small value is found after the first column has been written down it may be noted anywhere at the beginning or end of the table, as shown here.

cm.	Frequency.
1·56	/
1·58	/
1·59	
1·60	/
1·61	//
1·62	/////
1·63	///// ///// /
1·64	///// ///// ////
1·65	///// //
1·66	///
1·67	///// /
1·68	//

EXAMPLE OF A FREQUENCY DISTRIBUTION.

Notice that the expected lengths are tabulated in numerical order before the actual measurements are made.

Plot a graphic diagram (Fig. 58 and Fig. 59) in which the variates have their measurements in centimetres laid off along the x-axis and the frequency of each length is represented by the height of a corresponding ordinate. Do not begin the scale of x-values at zero, but allow one square for each hundredth of a centimetre. The y-scale in any statistical diagram should measure one square for each unit of frequency, since there are rarely enough measurements to require a condensed scale and nothing would be gained by an expanded scale where there are no fractional values to be plotted. Connect the points of the diagram by a broken line and notice that the " curve " is highest in the middle and slopes downward, more or less uniformly, towards the ends. A graph of this sort is called a *frequency polygon.*

Make another graphic diagram from the same table, but instead of representing successive ordinates by the successive vertical ruled

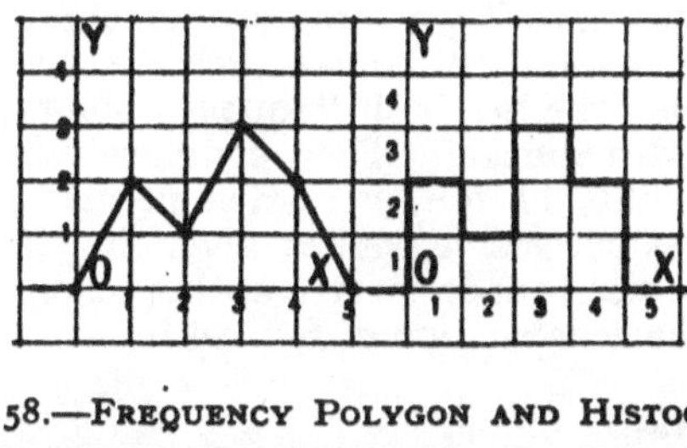

FIG. 58.—FREQUENCY POLYGON AND HISTOGRAM.

The ordinary graphic diagram made with broken lines is called a *frequency polygon* if the ordinates represent the frequency of occurrence of the abscissæ. When a *length* along the x-axis is used instead of a *point*, and a vertical strip instead of a line, the diagram is called a *histogram*. The diagrams show two slightly different ways of representing precisely the same facts, namely, $\begin{cases} x = 0, 1, 2, 3, 4, 5. \\ y = 0, 2, 1, 3, 2, 0. \end{cases}$ Notice particularly the difference in the placing of the scale numbers.

lines of the cross-section paper let them be represented by the successive white strips that lie between the vertical lines (Fig. 58), and use a vertical scale of one square for each measurement as before. A graph constructed in this way is called a *histogram.** Notice that *the area enclosed is numerically equal to the entire number of measurements or other statistics that are represented.* Theoretically it is better to work with a histogram than a frequency polygon, for the latter supposes all the variates in any one class to be concentrated at the centre of that class.

Make a histogram from your table of the measurements of 100 variates (Fig. 59).

159. Frequency Distributions—Recapitulation.

A tabulation of a set of measurements that shows how many times each observed value occurs is said to give the *frequency distribution* of the measurements, and a graphic diagram in which the ordinates give the frequencies of the measurements that are represented by the abscissæ is called a *frequency polygon* if drawn

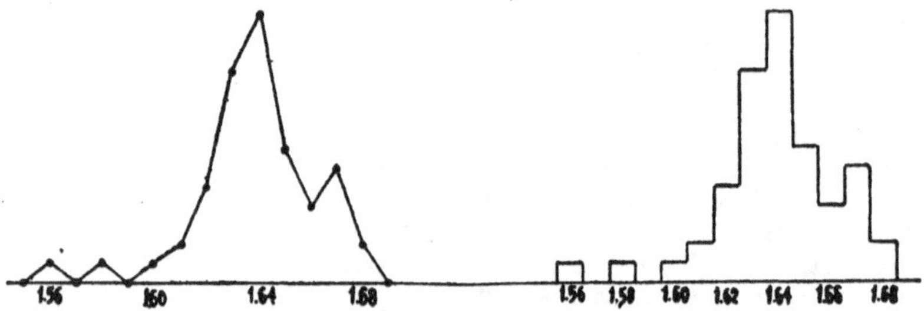

FIG. 59.—EXAMPLE OF FREQUENCY DIAGRAMS.
These correspond to the table in § 158.

with the usual broken line or a *histogram* if constructed by building up rectangles on successive segments of the base line.

Examine Fig. 59 and test it with a ruler held horizontally in order to determine whether both frequency diagrams represent the same frequency distribution. Then compare, without measuring, the apparent heights of the successive points on the frequency polygon with the successive values given in the table in § 158. Has any mistake been made in attempting to represent that table graphically by the construction of Fig. 59 ?

160. Class Interval.

When a set of variates was measured according to § 158 the determinations were rounded off to the nearest hundredth of a centimetre by the inability of the apparatus to measure them more accurately. If the five seeds, for example, which measure 1·62 cm. in the printed table could have been measured with a much higher

* This term was invented by Professor K. Pearson of University College, London.

degree of accuracy there can be no question that each of them would have a different length from any of the others.

Under such circumstances the ordinary type of frequency diagram could never be obtained, for the frequencies would invariably be something like the series

0, 1, 0, 0, 1, 0, 1, 0, 1, 1, 1, 0, 1, 0, 1, 1, 0, 1, 1, 0, 1, 0, 1, 0,

and the diagram would no longer have its characteristic " mound-like " shape with the ordinates high in the middle of the graph and dwindling away toward each end.

The shape of a frequency diagram, accordingly, may depend upon the numerical distance between the successive abscissæ that are used, the least difference between recorded measurements. In the table of § 158 this amounts to one-hundredth of a centimetre, but if those lengths had been rounded off to the nearest half millimetre the frequency distribution would have become

1·55	1·60	1·65	1·70 *centimetres,*
1	9	41	2 *cases,*

in which little is left of the original shape of the graph except a suggestion of the general " mound-like " character.

The least difference in successive recorded measurements is 0·05 cm. in the last illustration, 0·01 cm. in the original table and Fig. 59, and is called the *class interval.*

If your frequency polygon tends to the nondescript type seen when the class interval is too small the measurements of the hundred variates should be regrouped. Make a new *table* in which the values from 1·55 to 1·65 are all called 1·6, those from 1·65 to 1·75 are considered as 1·7, etc. If several measurements have been recorded as 1·65 put about half of them in the class 1·6 and the other half in the class 1·7. From this table plot separately both the frequency-polygon and the histogram that illustrate it graphically.* Do not use an expanded scale in any statistical graph either for frequency or for class interval.

161. Types of Frequency Distribution.

If the class interval is made very small and the measurements are very numerous, both forms of diagram can be considered as losing their abrupt changes until they merge into two identical curved lines, the notches that were originally present having become indefinitely small. Frequency distributions are not invariably " mound-shaped," but may be classified, according to the general shape of the graphic diagram, as (*a*) *symmetrical* or cocked-hat-shaped, (*b*) moderately *asymmetrical*, (*c*) very asym-

* If preferred, a class interval of 0·05 cm. may be used instead. This will avoid splitting any original class into two.

metrical or *J-shaped*, (*d*) bilocular or *U-shaped*, (*e*) *rectangular* (Fig. 60).

The symmetrical type (*a*) is seen in physical measurements. Notice that in it (1) the average (*i.e.* middle) *x*-value is the most frequent; (2) measurements that are a given amount above or below the average are less frequent than it but of equal frequency with each other; and (3) extremely divergent values do not occur. The asymmetrical type (*b*) is similar except that the most frequent value does not lie in the middle of the distribution and on one side of it the frequency falls away more rapidly than on the other. Moderate asymmetry is common in all statistical data, *e.g.* in the age distribution of deaths by measles. An example of the J-shaped

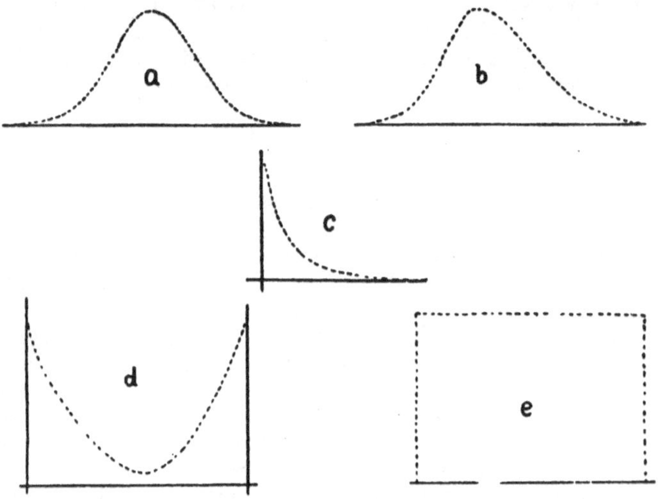

FIG. 60.—TYPES OF FREQUENCY DISTRIBUTION.

(*a*) Symmetrical. (*b*) Positively asymmetrical (skewed toward the right) (*c*) Extreme asymmetrical, or J-shaped. (*d*) Bilocular, or U-shaped. (*e*) Rectangular.

type (*c*) may be seen in the distribution of wealth among any population; the frequency of individuals with little wealth is very great, and with increasing wealth the number of cases falls off until it reaches a vanishing point. It is also seen in the distribution of petals among certain kinds of flowers, and when deaths of infants are plotted against their ages (in months). The curious U-shaped type (*d*) is seen where there are tendencies toward both extremes, or the centre is a position of unstable equilibrium, *e.g.* in specific death rates at different ages; and the rectangular type (*e*) occurs in purely mathematical cases, such as the actual error of a tabulated logarithm, which is never more than $\pm 0{\cdot}5$ of the unit in the last decimal place and has all intermediate values with equal frequency.*

* Other examples of these frequency distributions will be found in Ch. XXI.

162. The Probability Curve.

The measurement of 100 seeds will probably show a moderate degree of asymmetry, since such objects do not grow beyond a definite size but do fall short of it in many cases. This is called *negative asymmetry*, and the curve is said to be skewed toward the left. *Physical measurements, however, tend to be above the average just as often as they are below it*, and so give the symmetrical form of frequency distribution. As the measurements are made more and more numerous the frequency polygon approaches more and more closely to the form of the so-called *probability curve*, $y=e^{-x^2}$, which was calculated in the lesson on logarithms and plotted in the lesson on graphic representation. In some cases it will appear drawn out relatively flat and in others will be very high and narrow (Fig. 35), but the curve is the same in all cases, except that the scales of x-values and y-values are condensed or spread out to different degrees.

Carry out the binomial expansion that is given below at least as far as $n=15$, noticing that each two successive terms have to be added in order to obtain the term below and between them. Lay off a series of equal intervals on the x-axis. On these erect a series of ordinates proportional to the successive terms of the last polynomial, *in order*. A smooth curve through the tops of the ordinates will give a very good approximation to the probability curve.*

$$
\begin{aligned}
(1+1)^0 &= 1\\
(1+1)^1 &= 1+1\\
(1+1)^2 &= 1+2+1\\
(1+1)^3 &= 1+3+3+1\\
(1+1)^4 &= 1+4+6+4+1\\
(1+1)^5 &= 1+5+10+10+5+1\\
(1+1)^n &=
\end{aligned}
$$

The reason for the agreement of the polynomial coefficient curve with the probability curve is as follows: If a single coin is tossed the chance that it turns up "heads" is $\frac{1}{2}$, and the chance for "tails" is $\frac{1}{2}$. If two coins are tossed the possible results are:

h h
h t
t h
t t

i.e. the chance of two heads is 1 in 4 or $\frac{1}{4}$, the chance of a head and a tail together is 2 in 4 or $\frac{1}{2}$, and the chance of two tails is $\frac{1}{4}$. The chances are proportional to 1 : 2 : 1.

* If we plot curves for each of these expansions we shall see that the curves for high values of the powers have their peaks relatively more developed than those for low values of the powers. This illustrates the fact that with large number of variates the closeness of fit (or agreement) to the mean (or median, see § 165) is better than with few variates.

If three coins are tossed the possible results are:

h	*h*	*h*
h	*h*	*t*
h	*t*	*h*
t	*h*	*h*
h	*t*	*t*
t	*h*	*t*
t	*t*	*h*
t	*t*	*t*

or the chance of three heads turning up together is $\frac{1}{8}$, of two heads and a tail $\frac{3}{8}$, or one head and two tails $\frac{3}{8}$, and of three tails $\frac{1}{8}$. The chances are proportional to 1 : 3 : 3 : 1.

In the case of four coins (write the possible configurations down for yourself) the chances of 4 heads together, 3 heads and 1 tail, 2 heads and 2 tails, 1 head and 3 tails, and 4 tails are proportional to 1, 4, 6, 4, 1, respectively.

These numbers are the coefficients in the expansion of $(1+1)^4$, similarly for the numbers given in the earlier cases.

We see now that if n coins are tossed the chance of m heads turning up (and of course $n-m$ tails) is

$$(\tfrac{1}{2})^n \times \text{ the } (n-m+1) \text{ term in the expansion of } (1+1)^n$$

If, therefore, we plot a curve with the numbers in the successive terms of $(1+1)^n$ as ordinates, on an equally-spaced scale of abscissæ the probability of the occurrence of any given configuration of heads and tails in the tossing is proportional to the height of the ordinate drawn at the right place, *e.g.* the chance of m heads and $(n-m)$ tails is the height of the $(n-m+1)^{th}$ ordinate from the start. Of course, there is an equal chance for m tails and $(n-m)$ heads, hence the curve is symmetrical around a central hump.

Heads.	Tails.	Relative probability.
439	560	0·0006
449	550	0·006
459	540	0·037
469	530	0·155
479	520	0·432
489	510	0·803
494	505	0·942
495	504	0·961
496	503	0·976
497	502	0·988
498	501	0·996
499	500	1·000
500	499	1·000
501	498	0·996
502	497	0·988
503	496	0·976
504	495	0·961
505	494	0·942
510	489	0·803
520	479	0·432
530	469	0·155
540	459	0·037
550	449	0·006
560	439	0·0006

When n is very large the ordinates are relatively close together and their summits lie on a smooth curve, this curve being almost indistinguishable from the probability curve $y=be^{-a^2x^2}$ b and a being suitably selected.

Quetelet worked out the expansion of $(1+1)^{999}$. Its coefficients give us the probability of the relative numbers of heads and tails in the tossing of 999 coins.

A part of his table is reproduced here.

If we plot the last column as ordinates against either of the first or second columns as abscissa we get a curve which is a very close approximation to the normal curve. (Try this.)

The table does not go beyond 560/439 because the chance of getting that arrangement is only 6 in 10,000 tosses, and of course the chance of getting a greater disparity than 560/439 is smaller still.

As an example of the agreement of the mathematical curve drawn from theory to an experimental frequency curve we quote Bessel's result on 470 observations made by Bradley on a certain stellar measurement.

Magnitude of error in parts of a second of arc.	No. of errors of each magnitude.	
	Observed.	Given by theory.
Between		
0 and 0·1	94	95
0·1 ,, 0·2	88	89
0·2 ,, 0·3	78	78
0·3 ,, 0·4	58	64
0·4 ,, 0·5	51	50
0·5 ,, 0·6	36	36
0·6 ,, 0·7	26	24
0·7 ,, 0·8	14	15
0·8 ,, 0·9	10	9
0·9 ,, 1·0	7	5
and above 1·0	8	5

It is just a little curious that in many cases theory foretells less large errors than are found in practice.

163. Representative Magnitudes.

For most scientific work, the statement of a whole frequency distribution or of each one of a long series of measurements would be a cumbrous process and reading such statements would be a tedious task. Accordingly it is customary to summarize such a set of values by stating some representative value, such as the average. In special cases such representative magnitudes as the *geometrical mean*, $\sqrt[n]{(a_1a_2 \ldots a_n)}$, or the *harmonic mean*, $n/(1/a_1+1/a_2+ \ldots +1/a_n)$, or the *quadratic mean*, $\sqrt{\{(a_1{}^2+a_2{}^2+ \ldots +a_n{}^2)/n\}}$, have been employed, but the most usual one is the *arithmetical mean* or *average*, $(a_1+a_2+ \ldots +a_n)/n$. The *median*, which is the middle term if $a_1, a_2 \ldots a_n$ are arranged in order of size, is frequently useful as a representative value, as is also the *mode* or *modal value*, which is simply the value that occurs with the greatest frequency. The word *mean* is sometimes used in a general sense for *any representative value*, and sometimes is restricted to the same significance as the word *average*.

164. The Average.

The average is obtained by adding together a set of values and dividing the sum by the number of values. Thus the average of the five values 3, 3, 4, 5, 10, is one-fifth of their sum, or 5. If certain values occur repeatedly the average is obtained by dividing the sum by the number of values, not by the number of different values. Thus, the average of the measurements given in the table of § 158 is $(2\times1{\cdot}68+6\times1{\cdot}67+3\times1{\cdot}66\ldots)/(2+6+3\ldots)$, or 8686/53 or 1·639, or in general by the formula

$$\text{Average} = \frac{f_1x_1+f_2x_2+\ldots}{f_1+f_2+\ldots}$$

where f_1, f_2 . . . are the frequencies with which the values x_1, x_2, . . . respectively occur.

165. The Median.*

The median is obtained by choosing such a value that half of the other values exceed it and half are below it. *If the numbers are arranged in numerical order* in a column the number that is halfway down the column is the median. Otherwise it may be found by crossing off the largest number and the smallest, and repeating the process until only one is left. If two numbers are left, as will be the case if there are an even number of measurements, the number halfway between them can be taken as the median, but in physical or statistical data it usually happens that the two remaining numbers are the same.

Find the median of 3, 3, 4, 5, 10. What is the median of 7·4, 6·8, 7·3, 7·3, 7·2, 7·1, 7·2? Ans.: 7·2. Find the median of the measurements that are tabulated in § 158.

Turn to the histogram in Fig. 58, and cross off one of the small squares between the curve and the base line at the extreme left and then one at the extreme right; repeat the process until only one (or two) squares are left, and show that the median is 3. Imagine that Fig. 59 is similarly divided up into small squares, each square representing one measurement, and satisfy yourself of the truth of the fact that *the ordinate that is drawn upward from the median x-value must bisect the area of the frequency curve.*

166. The Mode.

The mode is the most frequent value. Thus in the set of values 3, 3, 4, 5, 10 the mode is the number that occurs twice, namely 3. In the illustration of the measurement of variates (§ 158) the mode is 1·64, the length that was found most often.

167. Choice of Means.

If a frequency distribution is of the *symmetrical* type the average, mode, and median will all be the same.

Where there is a *moderate degree of asymmetry* the median will

* This term was invented by Francis Galton in 1875.

come nearer to the mode than the average does, as is shown in the following diagram (Fig. 61). The mode is easily seen to be the most probable value. If a seed is taken at random from the set whose measurements were given in the table it is more likely to measure 1·64 cm. than any other amount. The mode has one decided disadvantage, however, in that it cannot always be chosen from a given frequency distribution. For example, the mode cannot be obtained from the series 3, 4, 4, 5, 6, 6, 7, 8, except by assuming that the values follow the probability law, and fitting a probability curve to them as accurately as possible by a "black thread" method; the position of the top of this theoretical curve can then be determined. Where there is a long series of measurements the average is sometimes

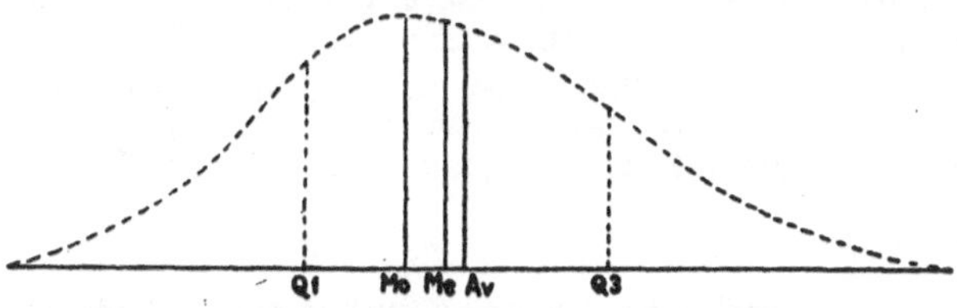

FIG. 61.—TYPICAL FREQUENCY DIAGRAM.

The mode has the highest ordinate. The ordinate of the median bisects the area under the curve. The ordinate of the average passes through the centre of gravity of the area. The median lies between the mode and the average, and is twice as far from the former as from the latter. The difference (distance) between the quartiles Q_1 and Q_3 (see § 172) is called the *interquartile range.*

calculated from the formula, indicated on the above diagram, that $\overline{me} - \overline{mo} = 2\ (\overline{av} - \overline{me})$, or that *the median lies ⅓ of the way from the average to the mode.*

The median, like the mode, is easy to determine. It can be used in two classes of cases where the average cannot be determined. One is in case measurements have been tabulated with indefinite terminal classes, *e.g.* "less than 10 mm., 7 cases; 10 to 12 mm., 4 cases; 12 to 14, 5; 14 to 16, 3; above 16, 2 cases." Here the median is the class "between 10 and 12," say 11 mm., and the mode is probably the class "12 to 14," say 13 mm., but the average cannot be found on account of nine of the numerical values not being stated; the best that could be done would be to guess that the average was a little less than the median. The other case is where quantities can be arranged in numerical order but are difficult to measure individually. Thus it may be difficult or impossible to gauge the scholarship of a student in accurate numerical terms, but if a group of students can be arranged in order of scholarship the median can be determined without difficulty. Furthermore, the median has a unique advantage over most other representative magnitudes in that it is not affected by inverting the unit of measurement. The median of a number of prices will be the same whether they are given as shillings per dozen or as dozens per shilling. Of a group of different velocities the same one will be picked out by choosing the median whether the numerical values are expressed in miles per hour or in minutes for a one-mile run. It is easy to see that this will not be the case if the average is used. The median, however, is not quite as good a representative value as the average, in case they are different, for a series of measurements that have all been made with equal care. Curiously enough, however, the more extensive a series of measurements the more likely it is to show that the individual measurements are not equally trustworthy but may be grouped in different classes according to their relative scattering. It is in such cases that the median is a much better representative figure than the average, for the average is influenced by an unduly large or small measurement just in proportion to the aberration of the latter, while as long as a measurement is above the median it makes no difference how far above it may be, its effect is no greater than that of any other single value. (Compare the average and the median of 3, 3, 4, 5, 10, with those of 3, 3, 4, 5, 30. The very erratic values ought to have the least influence instead of the most.)

168. Deviations.

Of almost as much importance as finding the best representative value for a series of measurements is the determination of how closely they cluster around it or how widely they scatter from it. This will help to furnish information in regard to the accuracy of the measurements, and hence also in regard to the accuracy of the instruments and methods employed in making them; it will also be useful in comparing and combining determinations made at different times or by different observers. The difference, $m-a$, between any single measurement and the average (or other representative magnitude) is known as the *deviation* or *variation* of that measurement and is denoted by the letter v. It is not the same as the error of the measurement, for the error may have a constant component that affects all of the measurements equally; but it may be considered as the accidental error or accidental component of the total error. Deviations give no positive indication of any constant errors that may be present.

m	v	m	v
27·35		27·36	
27·34		27·38	
27·34		27·35	
27·33		27·37	
27·34		27·32	
27·34		27·30	
27·33		27·31	
27·34		27·30	
27·35		27·35	
27·34		27·36	

If each one of a series of independent measurements is as trustworthy as any other it can be shown mathematically that their best representative value is their arithmetical mean, or average; and accordingly the average is the figure that is almost invariably used.

Copy the two adjoining columns of figures, find the average of each set,* and write after each measurement (m)

* Be on the look-out for short cuts in finding averages (see §§ 169, 170).

its deviation (v) from the average of the figures in the same column, marking it with a minus sign if the value is less than the average, and with a plus sign if in excess of the average.

Although both averages are the same it is evident that the first set of measurements must have been made by a more trustworthy instrument, observer, or method than the second. If the averages had not been of the same value the first one would undoubtedly have been entitled to more confidence, other things being equal, than the second.

Add each column of deviations and verify the fact that *the algebraical sum of the deviations from the average is always zero.* If their sum is zero their average will also be zero, so that when an "average deviation" is recorded the term means not the average of the algebraical deviations, but the average of the values that they would have if the negative signs were omitted.

169. Average by Symmetry.

The foregoing property suggests an easy method of finding the average in simple cases: If a number can be so chosen that the individual measurements are symmetrically grouped around it the sum of the positive deviations will equal the sum of the negative deviations and the number will be the required average.

What is the average of 115 and 119? Ans.: 117, because it makes the sum of the deviations (+2 and −2) equal to zero.

Find the average of 3, 6, 9, 12, 15 by the method of symmetry.

What is the average of 16, 18, 20, 22? Of 14 and 17? Of 12, 14, 17, 19? Of 126·8 and 127·4? Of 121 and 141? Of 161 and 191? Of 198, 199, 203? Of 8, 10½, 11½? What is the value, to the nearest whole number, of the average of 117, 116, 117? (Suggestion: Is the average above or below 116·5?) What is the exact average of 17, 18, 18, 19, 19, 20?

170. Average by Partition.

Before leaving the subject of the average it should be noted that there is no need of adding the entire numerical values if they contain a part in common. Thus in either of the columns of figures in § 168 the average is obtained by writing the first three figures, *which are common to all values,* and annexing the average of the last figure. This method effects a great saving of time.

171. Average from the Diagram.

The formula for finding the average (§ 164) is practically the same as the formula for finding the position of the centre of gravity of a number of particles obtained by the principle of moments (see formulæ in § 258). Hence it follows that the ordinate erected at the value of the average passes through the centre of gravity of the frequency diagram or histogram. If the frequency diagram be cut from the paper and hung from a pin we can find by means of a little plummet the position of the centre of gravity and hence the average. If the paper is stiff, we may balance the area on a knife-edge placed at right angles to the base line and the intersection of the knife-edge and the base line gives the arithmetical mean at once.

172. Quartiles.*

Just as the middle number of a series arranged in ascending or descending order of magnitude is called the median, or halfway value, so the middle one of the numbers that lie below the median is called the *lower quartile*, or quarter-way value; and the median of the numbers that lie above the median is known as the *upper quartile*, or three-quarter-way value. The quartile abscissæ are laid off on the base line of Fig. 61, at Q_1 and Q_3. It should be carefully noticed that their ordinates, together with the median ordinate, divide the whole area of the frequency diagram into four equal parts. If there is any difficulty in understanding this, remember that the area of a histogram indicates the total number of measurements (§ 158). If half of them lie below the median, by definition, and half above there can be no trouble in realizing that the median is the abscissa whose ordinate bisects the area of the figure. Of course the same thing is not true of the average, except when average and median happen to have the same value.

What are the quartiles of the series 3, 4, 4, 5, 6, 8, 11?

Find the median and quartile values of the set of numbers 28, 29, 31, 36, 31, 30, 35, 33, 32, 36, 29, 28, 32. Notice that the numbers "below" the median need not all be smaller numbers; the median is 31 and one of the lower numbers is also 31. Notice that the lower quartile is the median of the six numbers *below* the median, not of the *seven* numbers that include the median itself. Similarly, the upper quartile is not 33, but 34.

173. Median and Quartiles from the Frequency Diagram or Histogram.

In practice if the frequency diagram is drawn on squared paper, the squares included in the figure are counted (making allowance for fractions of squares) and a vertical line is drawn to bisect the area. This is done first by a trial vertical, and the squares counted on the two sides of this ordinate. If one side has a larger area than the other the ordinate is shifted towards the smaller area. The amount of shift can usually be quickly calculated. In the same way the positions of the quartiles are found.

174. Semi-Interquartile Range.

The numerical distance between the lower quartile and the upper quartile is called the *interquartile range*. In the series 7, 8, 10, 13, 19, 22, 27, the quartiles are 8, and 22, and their difference, 14, is

* Term invented by Sir Francis Galton

the range between quartiles, or inter-quartile range. Half of this, or 7, is called the *semi-interquartile range;* notice that it is the average of the two distances from median to quartiles. It is sometimes used as a measure of the scattering or clustering of a set of observations, and its theoretical importance will be seen later.

Find the semi-interquartile range of each of the two columns of figures in § 168 after re-arranging the data in numerical order. For the first column: if the five smaller values are considered as lying below the median and five above, that is, if the median is considered to occupy no numbers in the middle of the column the lower quartile will be 27·34; but if only four values are considered as lying below the median, and four above, that is, if the median is considered to include two numbers in the middle of the column the lower quartile will be 27·335. As the median is actually considered to be neither two numbers nor no numbers, but is one single number, it will be satisfactory to say that the lower quartile is neither 27·340 nor 27·335, but is halfway between these two values, say 27·338. Similarly, show that the upper quartile is 27·342. The second column should be treated in the same manner. What do you notice about the relative size of the two semi-interquartile ranges?

EXERCISES XIV

1. In addition to the sources of error that you have stated for the experiments listed at the end of § 156, write down at least ten more sources of error that were present when you performed those experiments.

2. Which form of frequency diagram do you consider to be the more "graphic," the histogram or the frequency polygon?

3. Different samples of beans were measured. The result are as follows:—

Sample 1.

Size.	Number.
cms.	
0·60—0·64*	0
0·65—0·69	1
0·70—0·74	2
0·75—0·79	6
0·80—0·84	10
0·85—0·89	14
0·90—0·94	17
0·95—0·99	22
1·00—1·04	17
1·05—1·09	7
1·10—1·14	3
1·15—1·19	1
1·25—1·24	0

Sample 2.

Size.	Number.
cms.	
1·5—1·6†	2
1·6—1·7	4
1·7—1·8	6
1·8—1·9	8
1·9—2·0	12
2·0—2·1	20
2·1—2·2	11
2·2—2·3	9
2·3—2·4	2
2·4—2·5	1

Sample 3.

Size.	Number.
cms.	
0·75+‡	0
0·80+	4
0·85+	14
0·90+	28
0·95+	32
1·00+	14
1·05+	14
1·10+	3
1·15+	5
1·20+	4
1·25+	6
1·30+	4
1·35+	1
1·40+	1
1·45+	0

* Really means 0·595 to 0·645. † Really means 1·45+ to 1·55—.
‡ Really means 0·745 to 0·795.

Plot their frequency diagrams and histograms.

Find the arithmetical averages, the median, and the modes for the three samples, also the average deviations. In each case draw an ordinate to bisect the diagrams and then two ordinates bisecting the two halves. Find the values of the medians, the quartiles and the semi-inter-quartile ranges.

Keep these curves for future use.

4. The following measurements were made of the weight of a particular crop of variates:

Weight	16	17	18	19	20	21	22	23	24	25	26	27	28	29	30	31	32	33
Frequency	0	7	18	32	47	56	57	52	46	39	32	25	18	12	7	4	2	0

Plot the frequency curve and histogram.

Find the average, median, mode, and average deviation.

5. Complete the following rule for finding the upper quartile of 12 measurements, and submit it to your instructor for inspection: "Arrange the measurements in order of magnitude and find the median by taking the average of the sixth and seventh. Then . . ."

6. When the tabulated values of § 158 are re-grouped in larger class intervals (§ 160) there are seven measurements of 1·65 cm., which theoretically should be separated into two equal groups for inclusion with the classes 1·60 cm. and 1·70 cm. Would it be objectionable to add exactly $3\frac{1}{2}$ squares to each of the strips of the histogram, or would it be better to add 3 to one and 4 to the other? Explain why.

7. Why does it generally happen that the median of a set of measurements fails to be exactly twice as far from the mode as from the average?

8. Draw roughly a curve like Fig. 61, but having negative instead of positive asymmetry. Place the mode and the median where you think they should go, and then locate the average.

9. Make a tabular statement of (*a*) the advantages of the median over the average, and (*b*) the advantages of the average over the median, as a representative value.

10. *Prove* that for any three numbers (say, *a*, *b*, *c*) the algebraical sum of the deviations from the average must be equal to zero.

11. If a train travels the first, third, fifth, seventh, . . . miles at a rate of 20 miles per hour, and the second, fourth, sixth, eighth, . . . miles at 30 miles per hour, show that its average speed is not the average of the two rates, 25 miles per hour, but is their harmonic mean (§ 163) instead.

12. What would you suggest for a *numerical measure of asymmetry* in a diagram like Fig. 61?

13. Toss a coin 100 times (and more if you have time) and count the number of times it comes head. Repeat with 2, 4, etc., coins and draw the frequency curves.

14. Show that if 1024 throws are made with 10 coins the following results should occur

No. of heads turning up	0	1	2	3	4	5	6	7	8	9	10
Frequency	1	10	45	120	210	252	210	120	45	10	1

15. Throw dice, select cards, etc., and verify the laws of chance, *e.g.* the chance of throwing any number with one die is 1 in 6; the chance of selecting a spade card from a well-shuffled pack is 1 in 4.

16. In an experiment on the grinding and sifting of sand, Martin obtained

the following figures for the numbers of particles of different sizes (30 divs. =0·001″). Plot the frequency curve. Comment.

Mean size measured in microscope divisions.	Number of particles measured.
1	65
2	110
3	94
4	63
5	48
6	36
7	21
8	21
9	14
10	11
11	0
12	7
13	0
14	3
15	3
16	0
17	2
	498

17. The degree of cloudiness of the sky at a certain European town was estimated over a space of ten years. Classifying into 10 degrees ranging from a zero of " no cloud at all " to a maximum of 10 meaning " clouded all day," the following results were obtained. Plot the frequency polygon and comment on its shape.

Degree	0	1	2	3	4	5	6	7	8	9	10
Frequency	751	179	107	69	46	9	21	71	194	117	2089

18. For the measurements in the right-hand half of the table in § 168 find the mean and the average of the deviations. Express this average as a percentage of the mean.

19. The fruit of the Great Rag Weed (*Ambrosia trifida*) of Western Canada was examined by a class of 83 students. The number of spines around the central spine were counted (central spine not included in count). Results:—

No. of spines	0	1	2	3	4	5	6	7	8	9	10	11	12	13
No. of fruits	72	38	101	339	1986	4902	2714	598	219	105	48	20	4	2

Plot the frequency curve, and find the Arithmetic Mean, Median, Mode, the Quartiles, and the Semi-inter-quartile Range.

CHAPTER XV

DEVIATION AND DISPERSION

175. Characteristic Deviations.

Just as the use of an average or other representative magnitude obviates the necessity of stating the separate measurements from which it is derived, so a statement of all the deviations from the average can be replaced by a single *characteristic deviation*. A set of measurements can then be summarized by two numbers, and it is customary to write such a result in the form $a \pm d$,* where the first number gives the average value of the quantity measured and the second expresses the limiting distances above and below the average which mark off in some way an amount of deviation that is characteristic for the set of measurements.

176. Total Range.

The simplest way of summarizing the deviations of a series of measurements is by stating their *total range*, or the algebraical difference between the smallest one and the largest one. Obviously the extreme range of the measurements themselves will also give the same result. Thus, in the two columns of measurements in § 168, show that the total range is 0·02 for the first and 0·08 for the second. Unfortunately, this is not only the easiest but also the worst method of obtaining a characteristic deviation, for the numerical value given by the total range depends upon only two of the measurements of the series; and those two are the least satisfactory ones, because a repetition of the series of measurements would be likely to show a considerable fluctuation in the largest and smallest values, while the most closely clustered values would hardly be changed at all.

What is the numerical value of the total range of the data in the table of § 158? In what kind of units should it be expressed?

177. Average Deviation (a.d.).

A better index of the amount of scattering is given by the *average deviation*. This is the average of the positive arithmetical

* The letter d will be used in this book to denote a particular kind of *characteristic deviation*. A single deviation of any individual measurement will be indicated by the letter v (*variation*).

values of all of the deviations. The true algebraical average of the deviations cannot be used if they have been calculated from the average measurement, because, as has been shown (§ 168), its value is always zero.

The average of the positive values of the deviations is always smallest when the deviations have been calculated from the median measurement, and it is in connection with the median that the average deviation is generally used. For a symmetrical distribution it is of course the same whether calculated from mean or median.

Find the average and the median of the numbers 3, 3, 4, 5, 10. Then find the deviation of each of these numbers *from the median*, being careful not to omit one of the deviations in case it happens to be zero. What is the value of the average deviation when calculated from the median? Find also the five deviations *from the average*, and determine the average of their (positive) arithmetical values. Is the average deviation also smaller when measured from the median than when measured from the mode?

178. Standard Deviation (σ).

The typical deviation that is most frequently used for statistical purposes is the *standard deviation*. This is the square root of the quotient of the sum of the squares of the deviations from the average divided by one less than the number of statistical values.* If there are n quantities whose deviations from their average are respectively $v_1, v_2, v_3, \ldots v_n$, the standard deviation, σ, of those quantities is given by

$$\sigma = \pm\sqrt{\frac{v_1^2+v_2^2+v_3^2+\ \ldots\ +v_n^2}{n-1}}$$

The standard deviation is a special case of the *mean-square deviation* (or *root-mean-square deviation*), being in fact the mean-square deviation from the average, or approximately the quadratic mean of the deviations from the average. It is also sometimes called the *mean deviation* and the *mean-square* deviation, but as the term *mean deviation* is also used for what we have defined as the *average deviation* it is much better to avoid the use of the word *mean* (§ 163) altogether in connection with a deviation.

179. Dispersion (p or pe).

The measure of deviation which is used for physical measurements in almost all cases is that particular form of characteristic

* To be strictly accurate, the standard deviation of the statistician is $\sqrt{\Sigma v^2/n}$. The reasons for using $n-1$ will be found in the larger treatises here it need only be noticed that the denominator is diminished because the numerator is smaller than the sum of the squares of the true errors see Ch. XIX). Of course, if n is fairly large, the distinction is unnecessary.

deviation which is called the *dispersion*. It is approximately two-thirds of the standard deviation, or, more exactly,

$$p = \pm 0{\cdot}6745\sqrt{\frac{v_1^2 + v_2^2 + \ . \ . \ . \ + v_n^2}{n-1}}$$

180. Significance of the Dispersion.

An important characteristic of this typical value is that for a series of measurements which is extensive enough to follow the law of the probability curve *the dispersion can be shown mathematically to express exactly the same limits above and below the average as are given with respect to the median by the semi-interquartile range.** If measurements follow the law of the probability curve the median and the average will coincide; and the upper and lower quartiles will include within the space between them just half of the total number of measurements, as was seen in connection with Fig. 61. In other words, the right-hand half of the frequency diagram will be bisected by the ordinate corresponding to the positive value of the dispersion, and the corresponding abscissa will be the median value of all the positive deviations. Since the left-hand half of the curve is likewise bisected by the negative value of the dispersion, and the graphic diagram is symmetrical with respect to the y-axis, it follows that the dispersion is the median of the absolute arithmetical values of all the deviations, or that *any single deviation is as likely to be less than the dispersion as it is to be greater.* The dispersion, with its plus-and-minus sign marks off a small distance above and below the average, and it is just as likely as not that any single measurement chosen at random will be within these limits.

181. Advantage of the Dispersion.

Where the number of measurements in a series is relatively small, say not greater than 10, the dispersion obtained by calculation and the semi-interquartile range obtained by picking out the quartile values will not usually have exactly the same value, on account of the measurements not being sufficiently numerous to follow the laws of probability very closely. Even in this case, however, the *value given by the formula* is preferable to half the difference between the quartiles, because the former is obtained from all the measurements of the series and the latter from only two. Where there are as many as ten measurements of a physical magnitude it is usually found that the two values will not differ by more than 15 or 20 per cent., and for rough preliminary measurements the semi-interquartile range is almost as satisfactory as the dispersion and can be obtained much more readily.

* Show that the median and semi-interquartile range of the 53 measurements given in the table of § 158 are 1·64 cm. and 0·01 cm. Could these two numbers be used as a rough approximation for *average* and *dispersion*?

182. Calculation of the Dispersion.

m	v	v^2
cm.		
1·82	0·033	0·001089
1·85	0·063	0·003969
1·85	0·063	0·003969
1·78	0·007	0·000049
1·82	0·033	0·001089
1·84	0·053	0·002809
1·78	0·007	0·000049
1·62	0·167	0·027889
1·81	0·023	0·000529
1·70	0·087	0·007569
10)17·87	Σv=0·536	Σv^2=0·049010
AM=1·787	ad=0·054	

$$\log 0{\cdot}049010 = \bar{2}{\cdot}6903$$
$$\log 9 = 0{\cdot}9542$$
$$2)\bar{3}{\cdot}7361$$
$$\bar{2}{\cdot}8680$$
$$\log 0{\cdot}6745 = \bar{1}{\cdot}8290$$
$$\bar{2}{\cdot}6970 = \log p$$
$$\therefore\ p = 0{\cdot}04978$$
$$= 0{\cdot}050 \text{ nearly.}$$

The table shows in the first column the measurements (m) of ten variates taken at random from the 100 that were measured before, also their average (AM), and in the second column their individual deviations (v) and the average deviation *a.d.* The third column shows the squares of the numbers in the second column, and their sum. According to the formula in § 179 the sum of these squares is to be divided by 9, the square root of this quotient is then to be found and multiplied by about $\frac{2}{3}$ in order to obtain the dispersion of this set of measurements.*

The value of d is 0·04978 cm., or 0·050 nearly. This means that the limits 1·737 cm. and 1·837 cm. are the boundaries between which about half of the measurements should lie (see § 180). Verify this from the tabular values of m.

183. Rule for Accuracy of the Average.

In the above example the squares of v have been taken out in some cases to five significant figures. In determining how many figures of the average are to be retained as significant it is best to follow the rule that *at least half of the deviations should have in their*

* After plenty of practice the deviations may be written down without their decimal points, *i.e.* as whole numbers. Similarly for the dispersions. A great saving of time is thus effected. The squaring of the v's may also be hastened by use of a table of squares (see the Appendix).

significant places a number greater than 3. In the illustrative example notice that keeping three decimal places in the average has made more than half of the deviations greater than 0·030. In such a case the average could obviously be rounded off to *two* decimal places, 1·79, and at least half the deviations would still be greater and 0·03. This will be found to simplify the calculation and the result will not be essentially different.

Find the dispersion of the same ten variates by writing a column of deviations from the value 1·79, and calculating the value of 0·6745 times the square root of one-ninth of the sum of their squares. Use logarithms but *do not* refer to the illustrative example in § 182 for each step of the process ; refer to the formula (§ 179) instead.*

Notice that your final result agrees with the one previously obtained as far as three significant figures. This is ample accuracy, since two significant figures are all that is usually wanted for the value of a dispersion.

184. Simplification of the Work and Use of the Table of Dispersions.

The squaring, root extraction and long multiplication and division should, of course, never be done by the tedious arithmetical process. Even the logarithmic process used in the illustration can be avoided.

The squaring of the deviations should be done mentally or with the slide rule and, after dividing by $(n-1)$, the square root also should be extracted with the rule. In taking the square root, pay attention to the position of the decimal point, for any arrangement of significant figures has two square roots (see § 80). Thus we could write the final steps of the example above as :

$$\Sigma(v^2)=0·04901$$

$$\frac{\Sigma(v^2)}{9}=0·00545$$

$$\sqrt{\frac{\Sigma(v^2)}{9}}=0·074$$

$$0·675\sqrt{\frac{\Sigma(v^2)}{9}}=0·050$$

the last two operations being done on the rule (also the second if it is not easy to divide mentally by $(n-1)$).

Some slide rules carry a mark at 0·6745 on the C scale. This makes the working still easier.

Once you get used to calculating dispersions you may be able to dispense with the tedious work of working out the decimal point and the ciphers in the columns headed v and v^2 and retain only the

* Referring to the steps of a typical example is almost always the easiest way of *obtaining a required result* ; it is a very poor way of *learning a method of procedure.*

significant figures. If this is done in the example above we get $\Sigma(v^2)=49010$. Dividing by 9 gives 5446; the square root of this is nearly 70 and $\frac{2}{3}$ of 70 is 48.

We may now use the table of squares in the Appendix. In this table if N is a number in the third column corresponding to a number n in the first column

$$N=\frac{(n\pm\frac{1}{2})^2}{0{\cdot}6745}$$

or

$$(n\pm\tfrac{1}{2})=0{\cdot}6745\sqrt{N}$$

or any value of N between the values listed in the third column would correspond to the nearest value of n in the first column. Taking the number 5446 from the example above, the corresponding value of n is seen to be 50, which agrees with the previously obtained 49·78 as far as the two significant figures, which is all that is required. A place for 5446 would also be found opposite 16 in the table, but the rough preliminary check-calculation showed the answer should be about 48, so that there is no doubt that the required answer is 50 rather than 16.

Practise using this table when doing the examples at the end of the chapter.

185. Sigma Notation.

The capital letter sigma (Σ) of the Greek alphabet is often used by mathematicians, prefixed to an algebraical term, to denote *the sum of all such terms*; thus the expression for the average, namely, $(a_1+a_2+a_3\ \ldots\ +a_n)/n$ is abbreviated to the equivalent form $(\Sigma a)/n$. In the same way the formula

$$0{\cdot}6745\sqrt{\frac{v_1{}^2+v_2{}^2+\ \ldots\ +v_n{}^2}{n-1}}$$

is more easily and compactly written in the form

$$0{\cdot}6745\sqrt{\Sigma(v^2)/(n-1)},$$

and the use of $\Sigma(v^2)$ will have been noticed previously in the example of the calculation of a dispersion.

Rewrite the formulæ for the harmonic mean and the quadratic mean (§ 163), using the sigma notation. Write the formula for $(x_1+x_2)^2$, using the same notation as far as possible.

If five measurements of the quantity x give 3, 3, 4, 5, 10, what is the value of Σx? If the average is 5 what is the value of Σv? Of $\Sigma(v^2)$? Of $\Sigma(x^2)$?

186. Dispersion of an Average (P. or P.E.).

The average length of 10 variates has already been calculated. If 10 more were measured their average would be somewhere nearly the same as the first average. If a considerable number of such averages had been determined it would be a simple matter to determine *their* dispersion, and the result would naturally be a smaller number than the dispersion of any of the sets of individual measurements, since an average is a more trustworthy figure than a single determination. In fact it can be shown mathematically (*vide infra*) that if ten equally good measurements are averaged the single measurements will show a variation which is greater than the variation of such averages in the proportion of $\sqrt{10}$ to 1, and similarly that *the dispersion of averages will be* $1/\sqrt{n}$ *as great as the dispersion of individual measurements if the latter are averaged in groups of* n. This means that it is not necessary to calculate several averages in order to find their dispersion, for it can be determined from a knowledge of the number of measurements that go to make up a single average and from the dispersion of these individual measurements. Thus the *dispersion of the average* (P), as it is called, of nine measurements is at once seen to be one-third as large as the dispersion for single measurements (p), since the square root of 9 is 3.

If a set of 16 measurements are free from constant errors how much more accurate is the average than one of the individual measurements?

The formula for the dispersion of an average is easily written, for

if $$p=0{\cdot}6745\sqrt{\frac{\Sigma(v^2)}{n-1}}$$

then $$P=0{\cdot}6745\sqrt{\frac{\Sigma(v^2)}{n(n-1)}}=\frac{p}{\sqrt{n}}$$

the second formula being $1/\sqrt{n}$ as large as the first.

In the same way the average deviation of the arithmetical mean may be written

$$AD=\frac{ad}{\sqrt{n}}$$

Peters' Approximate Formulæ.—Instead of going to the trouble of squaring the deviations and finding the dispersions in the usual way, it is sometimes convenient to use the following expressions which are approximately true:

dispersion of an individual
$=0{\cdot}845\times$average deviation of an individual

dispersion of the mean$=0{\cdot}845\times$average deviation of the mean.

187. The Statement of a Measurement.

It is customary to write the result of an accurate physical measurement in the form of two numbers separated by a plus-or-minus sign. The first number is the *average;* the second one is the *dispersion of the average*, not the dispersion for single measurements.

Tabulate the first eleven of the twelve measurements of the wooden block made with the cardboard model of a vernier caliper (§ 147). Pick out their quartiles and find the semi-interquartile range, to be used as a rough value of the dispersion of the individual measurements, and divide it by the square root of *n*. Write the thickness of the block in the form $AM \pm P$. Divide *p* for your measurement of the 10 seeds of § 182 by $\sqrt{10}$ in order to obtain *P* for them, and state their measurement in the same form.

Model cal.	Steel cal.
3·7	3·75
3·7	3·74
3·7	3·75
3·8	3·76
3·7	3·76
3·7	3·75
3·7	3·76
3·7	3·75
3·7	3·75
3·7	3·74
3·7	3·74

COMPARISON OF ROUGH MEASUREMENTS AND PRECISE MEASUREMENTS.

Make eleven more measurements of the wooden block, this time with the vernier caliper that gives tenths of a millimetre and treat them in the same way as the previous eleven. A typical set of results is shown in the margin. Two important facts are illustrated by this table: (1) It is the accurate method that shows discrepancies between repeated measurements, and the rough method that shows more uniform agreement. (2) There will be nothing fallacious about the statement that a dispersion "is zero" if care is taken to ***point off that zero***. The semi-interquartile range of the first column is 0·0 while that of the second is 0·01 cm., but 0·0 cm. is the only correct way of rounding off to one decimal place the number which is expressed in the second decimal place by the figures 0·01 cm.

Characteristic Values.—The characteristic values of a set of measurements are the Arithmetical Mean (*AM*), the Mode (*Mo*) the Median (*Med*), the average deviation of an individual reading or variate (*a.d.*), the average deviation of the mean (*A.D*), the probable error of an individual (*p.e*), the probable error of the mean (*P.E*), and sometimes the lower and upper quartiles (Q_1 and Q_3) and the semi-interquartile range (*S.I.Q.R*).

188. Relative Dispersion.

It has already been shown that in order to see how serious an error really is, it should be referred to or divided by the true magnitude of the quantity measured. In the same way, instead of using the actual value of the dispersion, it is often found more useful to obtain the ***proportional dispersion***, or ***relative dispersion***, or ***fractional dispersion***, as it is also called; the ratio of dispersion to representative magnitude.

In the two measurements just stated, the length of a seed and the thickness of the wooden block, divide the dispersion of the average by the average itself in order to obtain the ***relative dispersion of the average***. Express this either as a decimal or as a percentage.

Notice that it is smaller for the thickness of the block, which is fairly uniform, than for the length of a seed, which varies considerably. The absolute dispersion of the average in centimetres, however, may be larger for the block than for the seeds if a large dimension of the block is measured while the seeds that are used are small.

189. Comparison of Characteristic Deviations.

Turn back to your notes on the use of logarithms and find the graphic diagram of $y=e^{-x^2}$ (Fig. 34). Mark off the following values on the base line, $p=0{\cdot}4769$, $a=0{\cdot}5642$ and $\sigma=0{\cdot}7071$. These represent respectively the dispersion, the average deviation, and the mean square error or standard deviation, and are roughly proportional to 10 : 12 : 15 ; a better approximation to their ratios than is given by 10 : 12 : 15 may be found with the aid of the slide rule. Draw the corresponding ordinates, and notice that the last one meets the curve at the *point of inflection*, that is, at the point where it is momentarily straight as it changes from convex upward to convex downward.

Although the biologists often prefer to use the "standard deviation" or "mean square error" to the "probable error," the only real meaning given to the "mean square error" is the geometrical fact quoted above. It may, of course, be looked upon as an arbitrary measure of the scattering of the variates about the mean, but it lacks the 50–50 meaning which is the strong point in the expression "dispersion."

190. Chance of a Variate lying between or within a Stated Distance of the Mean or between Stated Limits.

Take the symmetrical form of frequency curve and regard it simply as the limiting shape of the normal histogram when the number of variates is large.

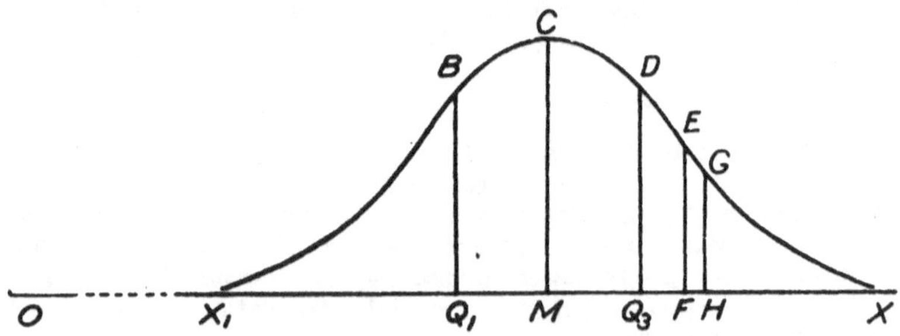

FIG. 62.

In Fig. 62 the origin is away to the left of the frequency curve, OM = the arithmetical mean, OQ_1 = the lower quartile, OQ_3 = the upper quartile, the distance $MQ_1 = MQ_3 =$ the semi-interquartile range = the dispersion.

The area under the curve X_1CX is numerically equal to the total number of variates. Half of the area between X_1CX and the base line lies between the curve, the ordinates BQ_1, DQ_3, and the base line, *i.e.* half of the variates lie within the range " $A.M. \pm$ dispersion." Hence, if we regard the curve as practically coming down to the base line at X and X_1, we may say

$$\frac{\text{Number of variates having deviations less than the "dispersion"}}{\text{Total number of variates}} = \frac{\text{area } BCDQ_3Q_1}{\text{area } X_1CX}$$

Since the curve is symmetrical this ratio is also equal to

$$\frac{\text{area } CDQ_3M}{\text{area } CXM}$$

In the same way if EF is another ordinate, we may say

$$\frac{\text{Number of variates having deviation less than } MF}{\text{Total number of variates}} = \frac{\text{area } CEFM}{\text{area } CXM}$$

and if GH is another ordinate

$$\frac{\text{Number of variates having deviations between } MF \text{ and } MG}{\text{Total number of variates}} = \frac{\text{area } EGHF}{\text{area } CXM}$$

or the

$$\text{Chance that a variate has a deviation between } MF \text{ and } MH = \frac{\text{area } EGHF}{\text{area } CXM}$$

also the

$$\frac{\text{Chance that a variate has a deviation less than } MF}{\text{Chance that a variate has a deviation greater than } MF} = \frac{\text{area } CEFM}{\text{area } EXF}$$

The dispersion, average deviation, and standard deviation are approximately in the ratio of 10 to 12 to 15, so that the range—" Mean $\pm$ average deviation " will include a larger percentage of the variates than the range " Mean $\pm$ dispersion." The range " Mean $\pm$ standard deviation " contains a still larger percentage.

The table shows the grouping of the variates for a large number of variates which agree with the normal symmetrical curve :

Range.	Percentage included.	Range.	Percentage included.	Range.	Percentage included.
$AM \pm p$	50·0	$AM \pm ad$	57·3	$AM \pm \sigma$	68·3
$AM \pm 2p$	82·3	$AM \pm 2ad$	89·0	$AM \pm 2\sigma$	95·5
$AM \pm 3p$	95·7	$AM \pm 3ad$	98·4	$AM \pm 3\sigma$	99·7
$AM \pm 4p$	99·3	$AM \pm 4ad$	99·9	$AM \pm 4\sigma$	99·99
$AM \pm 5p$	99·9				
$AM \pm \frac{1}{2}p$	26·4	$AM \pm \frac{1}{2}ad$	31·0	$AM \pm \frac{1}{2}\sigma$	38·3

and from this table it can be shown that the chance of lying within the range $AM \pm 2p$ to the chance of lying outside the range is 82·3 to 17·7, *i.e.* 4·6 to 1 and so on as in the next table.

Range.	Chance of lying within to lying without.
$AM \pm p$	1 : 1
$AM \pm 2p$	46 : 1
$AM \pm 3p$	21 : 1
$AM \pm 4p$	142 : 1
$AM \pm 5p$	1310 : 1

Similar tables may be drawn up for the average deviation and standard deviation.

191. The Probable Error.

By this time the student ought to be thoroughly aware of the fact that the dispersion is not properly an error, but a deviation. If he also realizes that deviations within its limits are no more probable than improbable there can be no objection to his using the term *probable error* that is always employed by physicists in speaking of this characteristic deviation. In this book the term *dispersion* has been used in order to avoid repeatedly informing the student that it is an error and repeatedly suggesting that there is something very probable about it. The "most probable" error is of course zero. The "dispersion" will hereafter be often spoken of as the probable error, and of course it will be understood that it is used in two forms, the probable error of the individual measurements, denoted by p or $p.e.$ and the probable error of the average, denoted by P or $P.E.$

The value of the dispersion or probable error is also likely to be in error. It can be shown by mathematics that the probable limits of the probable error are

$$P.E.\left(1 \pm \frac{0·4769}{\sqrt{n}}\right)$$

where n is the number of variates.

For example, if $n=100$ it is an even chance that the probable error lies within 4·8 per cent. of its calculated value.

If the approximate formula

$$P.E.=0{\cdot}8453\frac{\Sigma v}{n\sqrt{(n-1)}} \text{ is used the limits are}$$

$$P.E.\left(1 \pm \frac{0{\cdot}5096}{\sqrt{n}}\right)$$

which is slightly greater than the other value quoted above.

EXERCISES XV

1. If each deviation, v, is a number of thousandths explain why the mathematical operations of finding the dispersion cause p also to be given in thousandths of a unit.

2. If the results of a series of measurements are found to be 8·16 cm.$\pm$0·033 cm. would it be advisable either (*a*) to write 8·160$\pm$0·033, or (*b*) to write 8·16$\pm$0·03, instead of using the first form? Explain why.

3. Re-arrange the following table so that each colloquial statement is associated with its proper numerical equivalent:

"a few hundred"	20$\pm$5
"nearly a gross"	250$\pm$50
"a dozen or so"	70$\pm$10
"upward of three score"	130$\pm$10

4. Express each of the following in the form of a representative magnitude and a measure of scattering:

"over a dozen"	"six or eight"
"nearly a hundred"	"a few"
"about a thousand"	"several"
"in the neighbourhood of 15 or 20"	"some."

5. If "average deviation" means the average of the (positive) arithmetical values of the deviations, what would probably be meant by the expression "median deviation"? Has any characteristic deviation already been named which has practically the same significance?

6. Arrange the magnitudes 28, 29, 31, 36, 31, 30, 35, 33, 32, 36, 29, 28, 32, in order to find the median. Write down under them the deviations from the median. Count a quarter of the way in from each end to find the deviations of the quartiles and find the semi-inter-quartile range. Divide this by the square root of the total number to get the probable error of the median.

7. The following measurements were made of the lengths of a particular crop of variates.

Length	16	17	18	19	20	21	22	23	24	25	26	27	28	29	30	31	31	33
Frequency	0	7	8	32	47	56	57	52	46	39	32	25	18	12	7	4	2	0

Plot the frequency polygon on square paper and find graphically the Mode, Median, and Arithmetical Mean.

(To find the latter cut out the polygon, hang it loosely from a pin and using a plumb-line find the position of the centre of gravity.)

Check by the ordinary arithmetical method of finding the Mean.

Find also the average deviation of an individual reading.

8. (*a*) What are the principles that should guide you in making the calculations of the percentage error of a numerical result obtained in an experiment?

(*b*) The following ten readings were obtained for the focal length of a convex lens. They are equally reliable and free from constant error (Houston, "Light"):

f, cms. 19·91, 20·07, 20·11, 20·14, 20·16, 20·27, 20·21, 20·14, 20·22, 20·31

What are meant by (1) The arithmetical mean? (2) The average deviation of an individual reading? (3) The average deviation of the mean? (4) The probable error or dispersion of an individual reading? (5) The probable error or dispersion of the mean? Illustrate the usefulness of these terms.

(*c*) Find the values of these expressions in the case of the sample of readings quoted and state the final result of the focal length of the lens.

9. Ten equally reliable measurements of the length of a line give the following results (in inches):

11·62	11·70	11·78	11·78	11·81
11·82	11·82	11·84	11·85	11·85

Find the characteristic values.

10. The following measurements were made of the length of a line:

Length, cms.	Number of times occurring.	Length, cms.	Number of times occurring.
2·71	1	2·76	5
2·72	1	2·77	4
2·73	2	2·78	2
2·74	4	2·79	1
2·75	6	2·80	1

Find the arithmetical mean, the median, and the mode. State what you think is the value of the length with an indication of the limits between which it may lie.

11. A number of equally reliable readings are made on some physical quantity. Describe the method by which the best characteristic value is obtained and how the dispersion of the readings from this quantity is indicated. Illustrate your answer by using the following readings obtained as values of the density of a certain gas:

37·05, 36·83, 36·76, 36·94, 37·12.

12. Measurements made by the Cambridge Anthropometric Association on the heights of 4,415 undergraduates of the University of Cambridge of British extraction yielded the following results:

Height.	Number.	Height.	Number.
ins.		ins.	
61	0	70	675
62	20	71	510
63	50	72	310
64	90	73	160
65	180	74	100
66	340	75	50
67	530	76	10
68	640	77	0
69	750		

Note.—Any height between 62½ and 63½ inches is called 63 inches.

Plot the frequency curve. Find the Arithmetical Mean, the Median, the Mode, the Quartiles, and the Semi-inter-quartile range.

13. Re-cast the data of the last question in this form.

Number below $61\frac{1}{2}$ inches = 0
" " $62\frac{1}{2}$ " = 20
" " $63\frac{1}{2}$ " = 70
" " $64\frac{1}{2}$ " = 160
" " $65\frac{1}{2}$ " = 340, etc.

Plot a curve between inches as ordinates and the numbers in the second column as abscissæ. The curve obtained was called by Galton an ogive. It would be obtained if we arranged all the men in order of height along a straight line and drew a smooth curve along the tops of their heads.

It is steep at its ends, showing great differences in heights from man to man at the two extremes and fairly flat in the centre showing that here is a large number having little variation from the mean.

14. The tables contain experimentally determined values of the lengths of a wire *A*. Find the characteristic values: 7·82, 7·85, 7·85, 7·78, 7·82, 7·84, 7·78, 7·62, 7·81, 7·70.

15. Ten measurements of the specific gravity of a substance made with equal precision gave the following results:

1·162	1·164	1·167	1·180	1·154
1·173	1·159	1·162	1·163	1·145

Find the characteristic values.

16. In a certain first year university examination of a class formed of two classes taken together the following results were obtained:

Between	Class 1.	Class 2.	Combined class.
10 and 19 %	2	0	2
20 " 29 %	3	5	8
30 " 39 %	5	11	16
40 " 49 %	4	10	14
50 " 59 %	8	10	18
60 " 69 %	4	7	11
70 " 79 %	4	1	5
80 " 89 %	1	2	3

Plot the frequency curves. Discuss the nature of the third curve.

17. In a certain first-year examination the students obtained marks as follows:

4 between 20 and 29
11 between 30 and 39

and 20, 16, 8, 7, 5, 2 in the successive ranges of 10 marks. Plot the frequency curve and deduce the characteristic values.*

18. Test the accuracy of the statement

dispersion of an individual = 0·845 × average deviation of an individual

with the problems given above.

---

* An interesting paper on the subject, "The Education of Examiners," appeared in *Nature*, vol. lxx, 1904. The curves of good examiners approximate to the curve of errors, resembling the curve obtained by a bad shot firing at a target. Causes which may interfere with the formation of a good curve are: (1) a small number of candidates, (2) a preliminary selection.

19. Trouton (*Proc.* Roy. Soc., 1912) measured the quantity of ether taken up by 100 gm. of water at atmospheric pressure with the following results:

1·060, 1·040, 1·048, 1·032, 1·052, 1·063, 1·054, 1·087, 1·064, 1·045

Find the characteristic values.

20. Ferguson (*Phil. Mag.*, 1923) in measuring the surface tension of benzene used a formula $T=\frac{gr}{2}H$. His values of H at 20·2° C. are unity +0·142, 0·163, 0·171, 0·174, 0·178, 0·181, 0·189, 0·189, 0·191, 0·192, 0·192, 0·193, 0·193, 0·202, 0·203, 0·203, 0·208, 0·209, 0·209, 0·211, 0·214, 0·219, 0·221, 0·223, 0·233, 0·234, 0·245, 0·255. Find the mean and its probable error. Group the readings in classes 1·12–1·14, 1·14–1·16, etc., and plot the frequency curve. If $g=981·2$ cm. sec.$^{-2}$ and $r=0·0480$ cm., find the value of T.

21. The density of ordinary zinc at 16·3° C. as determined by Egerton and Lee, in 1923, is:

7·1397	7·1401
7·1398	7·1408
7·1400	7·1404
7·1398	7·1394
7·1398	7·1404
7·1401	7·1399
7·1404	7·1394

Find the mean and its probable error.

22. Morse found the molecular weight of cane sugar from osmotic pressure experiments. Values: 327·9, 336·8, 344·9, 341·1, 343·5, 340·4, 347·7, 344·8, 338·8, 340·1, 339·8. Find the mean value and its probable error? Use Peters' rule.

23. Honigschmid (see *Jour.* Amer. Chem. Soc., 1921) gets the following values for the atomic weight of bismuth (Cl=35·457): 209·020, 0·024, 0·022, 0·010, 0·040, 0·022, 0·015, 0·019, 0·031, 0·017, 0·013, 0·012, 0·019, 0·013. Find the mean and its probable error. Use Peters' rule.

24. In an interference experiment designed to test the effect of the earth's rotation on the velocity of light, Michelson and Gale (*Nature*, April, 1925) measured a "fringe-difference" between rays which had traversed a certain optical path in different directions. 269 observations were made, averaged usually in groups of 20 in the order taken. Thirteen such means are given in the table:

	Displacement in fringes.	Number of observations.
1	0·252	20
2	0·255	20
3	0·193	20
4	0·246	20
5	0·235	20
6	0·207	26
7	0·232	20
8	0·230	20
9	0·217	20
10	0·198	20
11	0·252	20
12	0·237	20
13	0·230	23

Find the A.M., ad., A.D., P.E. Comparing the result with a calculated result of 0·236±0·002, what inference should be drawn?

CHAPTER XVI

THE WEIGHTING OF OBSERVATIONS

192. Necessity of Weights for Observations.

A representative value is often wanted for measurements which are not all equally trustworthy. The accepted values for such constants as the maximum density of water, the mechanical equivalent of heat, the length of the true ohm of mercury, the velocity of light *in vacuo*, have all been derived from measurements by different observers at various times, and in general by different apparatus and methods. Any of these varying factors will produce varying results, and one determination can sometimes be accepted with more confidence than another, and so will be entitled to greater " weight " when it is necessary to decide upon a representative value.

193. Density by Different Methods.

An example of the effect of different methods on the determination of a physical magnitude may be given by the measurement of the density of a metal block. If the mass is known this can be accomplished either by mensuration, or by measuring the displacement of water, or by a measurement of buoyant force in water. According to the Principle of Archimedes the apparent loss of weight of a body immersed in a fluid is the same as the weight of an equal volume of the fluid. If the volume of a metal block is v, its weight w, and its apparent weight in water w', the density can be found (1) as the ratio of the weight, w, to the loss of weight, $w-w'$, supposing that the density of water is unity; or it can be determined as (2) the ratio of the weight, w, to the measured volume of water that is actually displaced on immersion, say v'; or (3) the block can be measured with a caliper and the density calculated as m/v.

The experiments may be performed several times with fairly simple apparatus by the different members of a class and their results reported to the instructor for classification and comparison. The practical details will be found in books on practical physics.

194. Weights for Repeated Values.

The simplest case of weighting different observations is when separate numerical values have each been obtained a definite number of times. Suppose, for example, that the density of a block of aluminium has been determined both as 2·6 and as 2·7, in the total of five measurements, the smaller value having been found on four occasions while the larger value was obtained only once. The best representative figure from these data certainly would not be the number 2·65, halfway between 2·6 and 2·7, but ought to be a number situated four times as far from the least frequent measurement, 2·7, as from the most frequent one, 2·6; in other words, it should be the number 2·62. Moreover, this is easily seen to be the same result as would be obtained by taking the average of the five individual measurements. (Try it.) The rule in such a case is obviously to *give each numerical value a weight proportional to the number of times of its occurrence.*

Find the *weighted average* of the values of a measured length if it was found to be 2·345 cm. in each of six trials, 2·350 cm. in twelve trials, and 2·355 in nine trials.

195. The Weighted Average.

The weighted average is found in any case by considering that certain values have been obtained more frequently than others. In the case just discussed this was a fact, in other cases it is only a supposition made to fit the known or estimated intrinsic value of the observations.

If a difficult measurement had been made by an experienced student and found to be 0·35, while the same experiment gave the value 0·41 when performed by a beginner, it might be decided somewhat arbitrarily to give the first number twice the weight of the second. The process of finding the weighted average, $(2\times0{\cdot}35+1\times0{\cdot}41)/3$, would then be equivalent to supposing that the better measurement had been obtained on two occasions but the poorer one only once. If a measurement of some quantity had been found to be 1·36 when made under unfavourable circumstances, and 1·41 when made under circumstances that were more favourable to experimentation it might be considered best to assign the respective weights of 1 and 1·5 to the two values. The weighted average would then be $(2\times1{\cdot}36+3\times1{\cdot}41)\div5$, or 1·39, a figure which will be seen to be nearer to the better value than to the poorer one in exactly the ratio of 1 to 1·5.

196. Arbitrarily Assigned Weights.

The objectionable feature of such an arbitrary assignment of weights is very obvious. The relative weights depend too much upon the judgment of the individual computer; furthermore, it is often difficult to avoid being influenced by the fact that certain determinations vary more or less widely from the expected value, instead of keeping one's judgment focussed on the quality of the experimental work.

Which do you consider the better method of determining density, by buoyancy, or by displacement? Choose what you consider the best ratio for their relative accuracies and find the corresponding weighted average, but be careful not to give extra weight to either measurement on account of its coming close to the third determination made by calculating the volume obtained by mensuration (see § 153).

If the average of six students working with method (1) is 2·74, of another six working with method (2) is 2·70, and of another six working with method (3) is 2·76, and if the assigned weights are method (1) —3, method (2) —1, method 3 —2. Find the weighted average.

197. Weight and Dispersion.

Determinations of any carefully measured magnitude are usually stated in the form of an *average* and *its dispersion*, $A \pm P$. Subject to the condition that the influence of constant errors can be neglected, it can be shown mathematically that the best value for a measurement is obtained by *weighting each determination of an average in inverse proportion to the square of its dispersion*. Thus, if one determination has a dispersion of 0·0040 and another has a dispersion of ·012 the former should be given nine times as much weight as the latter. This can be expressed in a general formula, if P is used to denote the dispersion of an average, by saying that the *weighted average* of

$$A_1 \pm P_1,\ A_2 \pm P_2,\ A_3 \pm P_3,\ \ldots$$

is

$$\frac{A_1/P_1^2 + A_2/P_2^2 + A_3/P_3^2 + \ldots}{1/P_1^2 + 1/P_2^2 + 1/P_3^2 + \ldots}$$

or

$$W.A. = \Sigma(A/P^2)/\Sigma(1/P^2) = \Sigma(wA)/\Sigma(w)$$

but it is much better to learn the principle involved than to memorize the formula.

Example.—There are two means worked out for a particular quantity, (1) 518·23 with a $P.E. = 0·67$, and (2) 518·31 with a $P.E.$ 0·22. Obtain the final mean

$$\frac{P.E._1}{P.E._2} = \frac{67}{32} = \frac{3}{1}$$

therefore give No. 2 nine times the weight of No. 1. This gives a final average of 518·30.

If there are n direct observations having the weights $w_1, w_2, w_3 \ldots w_n$, and we write $w_1 + w_2 + \ldots + w_n = \Sigma w$, the probable error of the weighted average is given by

$$0·6745\sqrt{\frac{\Sigma(wv^2)}{(n-1)\Sigma w}}$$

where $v_1, v_2, \ldots v_n$ are the deviations from the weighted average.

The probable error of the weighted average in the problem just stated

$$0{\cdot}6745\sqrt{\frac{1\times0{\cdot}0049+9\times0{\cdot}0001}{10}}=0{\cdot}6745\sqrt{0{\cdot}00058}=0{\cdot}016$$

198. Limitations of the Above Formula.

Attention should again be directed to the fact that weighting according to dispersions takes no account of the fact that constant errors may be present in the given data. The dispersion summarizes only the accidental errors, and if the constant errors are greater than these the weighted average is no better than the simple arithmetical average (Fig. 63).

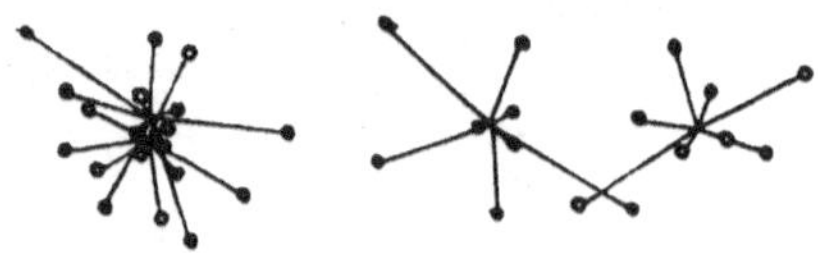

FIG. 63.—TARGET DIAGRAMS ILLUSTRATING ERRORS.

In the left-hand figure the two groups have nearly the same centre, that with the smaller dispersion naturally being the more trustworthy. In the right-hand figure at least one of the groups shows such a large constant error that the relative difference in the two sets of accidental errors is unimportant.

Tabulate the determinations, made by the various members of the class, of the density of aluminium as found by the effect of buoyancy. Calculate the typical value in the form $A \pm P$.

Find in the same way the average and dispersion of the density as determined by displacement. Use the slide rule, not the tables of logarithms.

Calculate the weighted average of these two data.

There are several sources of constant error in each of the two above methods of determining density. State at least four that are common to both methods, and at least one that influences one form of experiment but not the other. For example, if bubbles cling to the block when it is immersed (it is usually difficult to avoid them altogether) the apparent weight will be lessened, always causing the calculated density to be too low; they will also cause too much water to overflow, again making the calculated density lower than it should be. Consider also the effects of such things as inequality of the beam-arms of the balance, capillary attraction where the thread cuts the surface of the water, etc.

199. Exception to the Rule.

The method of weighting observations in inverse proportion to their dispersions is used for separate and independent data whose relative accuracy is assumed to be shown by their dispersions. Where two or more series of observations, however, are known to have been made with equally trustworthy apparatus, methods, and observers they should be weighted merely according to the number of measurements which each comprises, notwithstanding that their dispersions might indicate a very different result.* To do otherwise would be to repudiate the principle of the average, which depends upon the fact that all observations are supposed to be equally trustworthy. On the other hand, when different observations are known to be unequally trustworthy, even if they occur in the same series, weight may be given to the fact that some are closely clustered about an apparent central position while others diverge erratically. A great advantage of the *median*, as a representative magnitude, is that it is not unduly influenced by a very large or a very small measurement and hence it automatically gives less weight to the more aberrant measurements of a given group.

Which is to be preferred, the average or the median, for a determination, like the one just made, of density by buoyancy? Why?

Exercises XVI

1. In determining the specific gravity of an aluminium block why is it hung with a single thread instead of a string? Would there be any advantage in using wire instead? If wire should be used what kind of wire would be best, and what size?

2. Is it necessary to calculate dispersions in order to weight averages in accordance with § 197. What simpler calculation can be used to give exactly the same result?

3. Read § 180 again, and write in your note-book a statement of the most important fact that it contains. Do not follow the form of any of the printed statements, but try to write out the fact from an essentially different point of view.

4. The volume of a flask was measured by three students. A obtained 1020 c.c., B got 1030 c.c., C got 1010 c.c. Giving A's result a weight 5, B's a weight 3, B's a weight 2, find the average.

5. A gets for a measurement $1{\cdot}21 \pm 0{\cdot}11$. B gets $1{\cdot}31 \pm 0{\cdot}22$. Find the best average and its probable error.

6. Three observers measured the frequency of a tuning-fork by a stroboscope method. Their results were $58{\cdot}53 \pm 0{\cdot}07$, $58{\cdot}47 \pm 0{\cdot}15$, $58{\cdot}40 \pm 0{\cdot}06$. Find the most probable value of the frequency and its probable error.

7. The density of lead at 20° determined by Egerton is $11{\cdot}3437 \pm 0{\cdot}0001$ by Richards $11{\cdot}3370 \pm 0{\cdot}0009$. Find the weighted mean. Would you judge that a constant error was present in either or both results?

8. Morley determined the density of oxygen in three sets of experiments. His results are: $1{\cdot}42879 \pm 0{\cdot}000034$, $1{\cdot}42887 \pm 0{\cdot}000048$, $1{\cdot}42917 \pm 0{\cdot}000048$. Find the weighted mean according to § 197.
Morley himself weighted the last result twice as much as either the first or the second. Find his mean.

---

* Similarly, a single set of measurements known to be uniformly and equally good must be simply *averaged*; to give more weight to the individual measurements that have the smaller deviations would be a procedure akin to substituting the median for the average (but see also the end of § 167).

9. H. A. Rowland (*Proc.* Amer. Acad. 1879) believed that Joule's determinations of the mechanical equivalent of heat should have the weights indicated in parentheses after the several results. Find the best value of J. from Joule's work.

442·8 (0) 427·5 (2) 426·8 (10) 428·7 (2)
429·1 (1) 428·0 (1) 425·8 (2) 428·0 (3)
427·1 (3) 426·0 (5) 422·7 (1) 426·3 (1)

10. The following values for the frequency of a tuning-fork were obtained by a stroboscope on two successive days by the same student. The experiment was new to him on the first day.

1st day	258·6	257·3	255·4	256·6	256·7	256·7	256·5	256·3	255·1
2nd day	256·0	256·8	256·3	256·5	256·0	256·3	256·9	256·4	256·7

Discuss these results. Did the observer improve in accuracy?
Find the weighted mean

11. Michelson (*Nature*, Dec. 1924) quotes the following values for the velocity of light (in kilometres per second):—

Investigator.	Method.	Distance used, km.	Weight.	Velocity.
Cornu . . .	Toothed wheel	23	1	299,950
Perrotin . . .	,, ,,	12	1	299,900
Michelson . .	Revolving mirror	0·6	2	299,895
Newcomb . .	,, ,,	0·5	3	299,860
Michelson . .	,, ,,	35·4	3	299,820

Find the weighted mean and its probable error. Michelson gives his second result as 299,820 ± 30. How does that agree with the weighted mean?

12. Do you think that the first statement of § 187 is too strong? Would not the statement A.M. ± *p.e.* be more correct than A.M. ± P.E., and not so likely to lead a reader to overestimate the worth of the result?

CHAPTER XVII

THE REJECTION OF DOUBTFUL OBSERVATIONS, CRITERIA OF REJECTION

200. Observational Integrity.

When successive re-determinations of a quantity have been made in the course of an experimental investigation it is to be supposed that they have all been made with an equal degree of care.

It is important to remember that an observation should never be rejected simply because it is not in satisfactory agreement with the other determinations of the series. If the experimenter realizes that one of his measurements was made under some kind of a handicap or under such conditions that a faulty result would be likely it is permissible to cross out the corresponding value in his notes and to omit it in the final consideration of the data, but there must be some definite and satisfactory reason for discarding it other than the fact of its divergence from the expected value. The temptation, often felt by the beginner, to omit or "re-determine" * a discordant result may be very perceptible, but absolute freedom from prejudice (see *dependent measurements*, § 153; also § 42) should be cultivated to such a point that the experimenter is habitually able to feel a certain disinterestedness in the outcome of a measurement after he has first taken pains to ensure its being as trustworthy as possible. His attitude should be, "I have done all that can be done in the way of preparations for making this measurement accurate and independent; now let the result turn out as it will."

201. Importance of Criteria.

Even with all care to make successive measurements equally accurate it often happens that one or more of them show unduly large deviations from the average. In order to prevent these values from having an abnormal influence on the representative value certain rules have been formulated for determining whether they shall be retained or discarded, for if an observer merely used his own judgment in deciding the question the result would depend too much upon his own individuality and temperament, and different

* A re-determination is not intrinsically objectionable, but it should be made *in addition* to the other determination, not *in place* of it.

observers would obtain different results from data identically the same, just as in the case of the arbitrary assignment of weights (§ 196). In fact, the rejection of a measurement is nothing more nor less than giving it a weight equal to zero.

202. Chauvenet's Criterion.

One of the easiest to understand of the various devices for testing doubtful observations is known as *Chauvenet's criterion of rejection,* according to which rejectability is determined as a function of deviation, dispersion, and number of measurements. An unduly large deviation is an argument for rejection, especially if the dispersion is relatively small; furthermore, a deviation so large

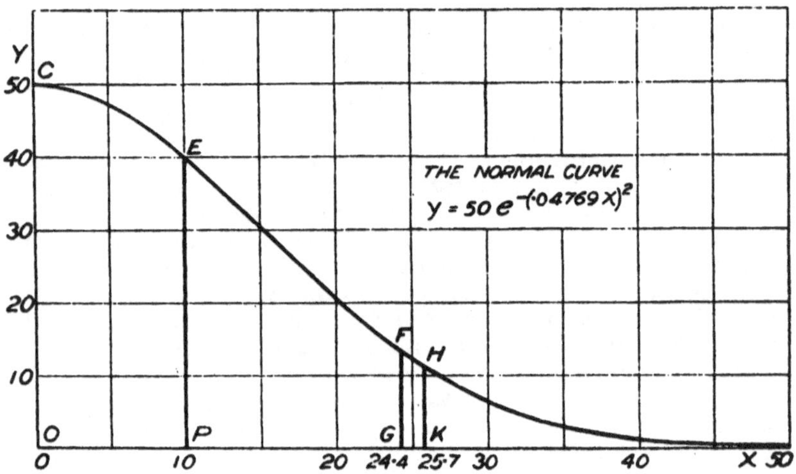

FIG. 64.

that it would not be expected to occur more than once out of a hundred cases might not seriously affect the average of a hundred values, but if it happened to occur in a set of only ten measurements it would probably exercise an altogether disproportionate effect upon their average. The object of a criterion of rejection is not to indicate that a certain measurement is wrong, but merely to point out that it is liable to be misleading if it occurs among a small group.*

The following exercise is for the purpose of showing the theory of the effect that the combined influences of deviation, dispersion, and number of measurements exert upon the determination of

* Remarkably wide deviations may be expected to occur once in a while if the number of measurements is extremely large; notice that there is *some* space between the curve $y=e^{-x^2}$ and the x-axis even at a great distance from the y-axis (§§ 61, 99).

the advisability of keeping or rejecting any single measurement of a group.

Draw a graphic diagram (Fig. 64) * from the accompanying table. Then draw the ordinates $x=10$ and $x=50$ from the base line up to the curve. The result will be the right-hand half of a normal frequency diagram, the x-values corresponding to deviations and the y-values to the frequency of their occurrence. Remember that the total area of such a curve corresponds to the number of deviations (§ 158); in the same way the area between the curve and base line and bounded on the left and right by any two ordinates represents the number of observations whose numerical deviations lie between those two limits (§ 190). As the dispersion is the same as the median deviation (§ 180) it is evident that the ordinate which bisects the area (this is the ordinate $x=10$, for the scales used in *this* diagram) must have its abscissa numerically equal to the dispersion.†

x	y
0	50·0
1	49·9
2	49·6
3	49·0
4	48·2
5	47·2
6	46·0
7	44·6
8	43·3
9	41·6
10	39·9
12	36·0
14	32·1
16	28·0
18	24·0
20	20·2
22	16·6
24	13·8
26	10·8
28	8·4
30	6·5
32	4·8
34	3·6
36	2·6
38	1·8
40	1·2
42	0·9
44	0·7
46	0·5
48	0·3
50	0·2
52	0·1
54	0·1
56	0·0
58	0·0
60	0·0

VALUES OF $y=50e^{-(0{\cdot}04769x)^2}$

Suppose another ordinate FG is drawn at a position so far to the right that it includes between the y-axis and itself nine-tenths of the total area under the curve and leaves only one-tenth of the area beyond it to the right, then the corresponding abscissa would similarly have a value that would be exceeded by only one-tenth of the total number of deviations, and if any one deviation were chosen at random there would be only one chance in ten that it would be larger than the corresponding x-value. It can be proved mathematically ‡ that in order to include nine-tenths of the area the ordinate must be drawn 2·44 times as far to the right of the y-axis as the line which bisects the area and corresponds to the dispersion.§

On your diagram draw the ordinate FG that includes nine-tenths of the area and make sure that its abscissa fulfils the condition stated above. If the total number of measurements were ten how many would most

* Fig. 64 shows only the large squares of the graph paper. The student's diagram will have the smaller squares also.

† Check this by counting the small squares in the areas EXP, CXO of Fig. 64.

‡ Certain statements are intended to be taken on faith by the student. This one is just as true as the statement that the frequencies of accidental errors follow the law $y=e^{-x^2}$; all of them can be proved.

§ Check this by counting the small squares in the areas FXG, CXO of Fig. 64.

probably be represented by the area to the right of the ordinate? How many if the number of measurements were 50? How many if the number were 4? How many if 6? The last two questions should be answered to the nearest whole number.

Since the ordinate for $x=2{\cdot}44\ p$ includes nine-tenths of the area and the limit $2{\cdot}44 \times$ *dispersion* includes nine-tenths of the deviations, it might be said theoretically and rather figuratively that if there were only five measurements in a certain series the number of measurements whose deviations were greater than this limit would most probably be just *one half*. In other words, the limit would be just on such a border line that if it were decreased we should expect it to exclude one measurement rather than no measurements, and if it were increased we should expect it to exclude no measurements rather than one measurement.

It follows, then, that *no* one of a series of five measurements theoretically ought to have a deviation of *more* than 2·44 times the dispersion. This being the case it is only natural to consider that one is justified in discarding any one of the measurements of a series of five if its deviation does exceed this *limit*. Just as the ordinate *FG*, Fig. 65, at $2{\cdot}44\ p$ excludes one-tenth of the area (*i.e.* excludes half a measurement if there are five in all), so the ordinate *HK* at $2{\cdot}57\ p$ excludes one-twelfth of the area, which would correspond to half a measurement if the total number of measurements were six. Accordingly no measurement in a set of six should theoretically have $v > 2{\cdot}57\ p$ if the set is supposed to follow the law of frequency distribution for accidental errors. *Chauvenet's criterion* is simply an extension of this delimitation to other values of n as well as 5 and 6. The column, l, of the following table shows the limiting x-value

n	l	n	l	n	l	n	l
1	1·00	11	2·97	21	3·35	32	3·59
2	1·71	12	3·02	22	3·38	34	3·62
3	2·05	13	3·07	23	3·40	36	3·65
4	2·27	14	3·11	24	3·43	38	3·68
5	2·44	15	3·15	25	3·45	40	3·70
6	2·57	16	3·19	26	3·47	49	3·81
7	2·67	17	3·22	27	3·49	64	3·95
8	2·76	18	3·26	28	3·51	81	4·06
9	2·84	19	3·29	29	3·53	100	4·16
10	2·91	20	3·32	30	3·55	671	5·00

***Chauvenet's Criterion*.—If the most divergent measurement out of a series of n determinations has a deviation more than l times as great as the dispersion of the individual measurements it should be rejected.**

for which the corresponding part of the area (see Appendix) included under the curve $y=e^{-x^2}$ is $1-1/2n$ and the excluded part is half of $1/n$, as before, this value being expressed in terms of the dispersion.

It should be carefully kept in mind when considering any criterion of rejection that we are interested in the individual measurements, and, accordingly, the dispersion to which the criterion applies is the *dispersion of the individual measurements*, not the dispersion of the average. Chauvenet's criterion, then, is the test of whether any deviation is greater than l times the dispersion of the individual values of a series of n measurements, where l corresponds to n in the way shown in the table.

203. Use of the Table of the Probability Integral.*

This table of figures has been calculated to show the chance of a reading occurring whose deviation is within some prescribed limit.

It is calculated on the basis that the dispersion (or probable error) is 0·4769, *i.e.* that it is an even chance whether any selected reading has a deviation greater or smaller than 0·4769. In terms of areas it means that the ordinate drawn at $x=0{\cdot}4769$ bisects the right-hand half of the area under the curve $y=e^{-x^2}$.

As an example of its use let us suppose that in some particular problem the dispersion (or probable error) is 3·00, and we wish to know what is the chance of occurrence of a reading whose deviation is greater or less than 5·00.

3·00 is the dispersion which corresponds to $x=0{\cdot}4769$ of the table, therefore 5·00 corresponds to

$$x=\frac{5{\cdot}00}{3{\cdot}00}\times 0{\cdot}4769=0{\cdot}7948$$

The "probability integral" for this value of $x=0{\cdot}739$. That is 73·9 per cent. of the area of the right-hand half of the curve occurs to the left of the ordinate 0·794 and 26·1 per cent. of the area is beyond this ordinate; or 73·9 per cent. of the variates have deviations which are less than 5·00 and 26·1 per cent. have deviations greater than 5·00. The chances are 73·9/26·1 in favour of a deviation $<5{\cdot}0$ and 26·1/73·9 in favour of a deviation $>5{\cdot}0$.

The table given for Chauvenet's criterion is calculated from the probability integral table.

Take a set of 10 readings. As shown in the text above, there should be no reading which has a deviation so great that its ordinate has on the outside of it only $\frac{1}{20}$ of the area under one-half of the frequency curve. That is, 19/20 or 0·95 of the area should be between this ordinate and the highest ordinate. The table tells us this value 0·95 corresponds to an x-value of 1·385, which is 1·385/0·4769, or 2·91 times the dispersion of an individual reading.

The little table on the right hand of page 325 containing the probability integral table gives this result in a slightly quicker way.

* The table is given in the Appendix.

204. Example.

In an experimental determination of the specific heat of lead shot the values in the adjoining table were obtained by a class of students. Test them by Chauvenet's criterion to determine whether any measurement falls outside of the theoretical limits, but if two or more such values are found *reject only the most divergent one*,* find a new average for those that remain, and apply the criterion to them in turn.† Repeat the process, if necessary, until no more values can be discarded, and then state the best value obtainable from the figures, with its "probable error." Try also the example in the next section.

0·022
0·0309
0·031
0·032
0·0347
0·035
0·036
0·038
0·045
0·050

SPECIFIC HEAT OF LEAD SHOT.

205. Graphic Approximation to Chauvenet's Criterion.

Where a graphic diagram is to be used for only a single series of numbers instead of for sets of values of two varying quantities it is advisable to use a horizontal scale and lay off the individual measurements as small dots or circles (Fig. 65) unless they are sufficiently numerous to allow a good histogram to be drawn.

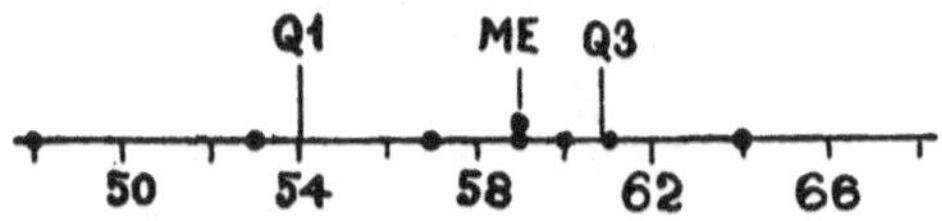

FIG. 65.—DISTRIBUTION OF A FEW MEASUREMENTS.

When the majority of frequencies are very small each measurement may be represented by a dot instead of the formal square of the histogram. The centre of gravity of such a system of equal-sized dots is more readily apparent to the eye than is the position of the line that would be needed to bisect the area of the frequency polygon or histogram.

The following figures are the experimental values of the slope of the "black-thread" diagram of § 108 as obtained by a class of students: 0·60, 0·57, 0·64, 0·53, 0·48, 0·59, 0·59, 0·61. Make a graphic diagram of these values and mark the median and quartiles. Do not measure the semi-interquartile range nor multiply it by l, but mark off its length on the edge of a strip of paper and apply this to your diagram, laying it off to the right and to the left of the median ‡ as many times as may be indicated by the criterion. In this way a rough application of the criterion can be made graphically and the long calculation can be avoided. Determine from the diagram whether

* If one large measurement and one small one are beyond the limits the rejection of the more divergent one may result in shifting the average far enough toward the other one to bring it within bounds. Where it is obvious that this cannot happen it is unnecessary to adhere to the letter of the rule.

† The greatest deviation, namely that belonging to the 0·022 reading, is not as great as 2·91 × *p.e.* Therefore there is no rejection. Final value 0·0355 *P.E.* 0·0017.

‡ It is better to use the mid-quartile point than the median in case the two are not the same.

any value should be rejected, and then verify the result by the usual form of calculation.

Use the graphic method for applying Chauvenet's criterion to the following set of barometer readings : *

BAROMETER READINGS.

29·986 inches	29·977 inches
29·982 ,,	29·984 ,,
29·990 ,,	29·982 ,,
29·984 ,,	29·986 ,,
29·984 ,,	29·988 ,,
29·980 ,,	29·984 ,,
29·986 ,,	

What advantage has the arithmetical method over the graphic method ?

Write down, in your own words, just what it is that is represented by (1) the probable error of a single measurement, $0·6745 \sqrt{\Sigma v^2/(n-1)}$, and by (2) the probable error of the average, $0·6745 \sqrt{\Sigma v^2/n(n-1)}$.

206. Irregularities of Small Groups.

The probable error, or "dispersion," cannot be considered as having much meaning in cases where the total number of measurements is less than ten, and even with ten measurements it should be treated with a certain amount of caution. A number of values less than ten will hardly ever give a histogram of their frequency-distribution which is recognizably similar to the graph of $y=e^{-x^2}$, the curve which all unbiassed measurements will be found to follow if they are sufficiently numerous.

207. Justification of the Criterion.

For the same reason, it is hardly worth while to use a criterion of rejection for less than ten measurements. The example given above with only five is intended merely for an illustration of the method of using the criterion, and the still smaller values in the table are only of theoretical importance. Chauvenet's criterion is not to be considered as showing that any one measurement is a mistake, but only as indicating that a very large deviation is such a rarity that it would have an unduly large influence upon the average if it were allowed to remain along with the other values of a very limited series of measurements (§ 202 and footnote).

208. Wright's Criterion.

Another criterion of rejection, which is sometimes employed, is that of Wright.† According to this *the arbitrary rejection of a single*

* Reject 29·975. Final result by arithmetic 29·9845 *P.E.* 0·0006.

† T. W. Wright, "A Treatise on the Adjustment of Observations" (Van Nostrand).

measurement may be considered if its deviation is more than five times the probable error.

Turn back to the graphic diagram of the table in § 202, and notice how small a part of the area of the curve lies to the right of the ordinate, $x=50$, which represents five times the probable error. Turn to the table of values for Chauvenet's criterion and note how many measurements would need to be made before "half a measurement" would be likely to diverge from the average five times as far as the probable error.

In the measurements to which you have already applied Chauvenet's criterion how many would have been rejected if Wright's criterion had been used instead? If a deviation is great enough to be practically sure of rejection by one of the two criteria will it ordinarily be rejected by the other? If a maximum deviation is small enough to avoid rejection by one criterion will it be practically certain to escape rejection by the other? Explain why What are the relative advantages of the two criteria?

Other limiting values, which give practically the same result as Wright's criterion, are *four times the average deviation,* and *three times the standard deviation.*

Stevens recommends that the doubtful reading be omitted and the mean and average deviation found. If the distance of the doubtful reading from the mean is greater than four times the average deviation it should be rejected.

The whole problem of rejection of discordant readings has been the subject of much argument. Bessel and Airy maintained that if the readings are equally reliable to begin with (and if they are not there is no justification for applying the theory of least squares) we have no right to reject any single reading. A rejection of a reading with a large variation has in some cases led to a diminution of the accuracy of the final result.

Exercises XVII

1. Write a definition of Chauvenet's criterion in your own words. Notice that the last sentence of § 202 is not a satisfactory definition.

2. Can a set of three measurements comprise such values that one of them will be rejected by Chauvenet's criterion? Give an example to illustrate your answer.

3. Can a set of two measurements comprise such values that one of them will be rejected by Chauvenet's criterion? Give an example to illustrate your answer.

4. What is meant by the term "probable error or dispersion"?
The focal length of a lens was determined by ten equally reliable experimenters. The results are as follows (all constant errors having been corrected for):

19·91 inches
20·07 ,,
20·11 ,,
20·14 ,,
20·16 ,,
20·27 ,,
20·21 ,,
20·14 ,,
20·22 ,,
20·31 ,,

Apply Chauvenet's criterion. If any reading or readings should be rejected do so and find the final characteristic values, deviations, dispersions.

5. Ten apparently equally reliable readings of the length of a rod are made with a micrometer, the last figure being an estimation. Result as follows:

1·278 inches
1·270 ,,
1·267 ,,
1·263 ,,
1·262 ,,
1·264 ,,
1·266 ,,
1·259 ,,
1·265 ,,
1·268 ,,

Find the arithmetical mean, the deviations and probable errors. Apply Chauvenet's criterion to see whether any reading should be rejected. If so reject it and find the new characteristic values. State finally how many readings come within the range given by the probable error of an individual reading and how many come outside that range.

6. A number of equally reliable students made determinations of the latent heat of steam at 100° C. Their values were 511, 527, 531, 534, 536, 547, 541, 534, 542, 551 calories per gm. Apply Chauvenet's criterion to see if any reading or readings should be rejected. State your final results.

State what "constant" errors are likely to exist in this experiment. Do you think there remains a constant error in the above results?

7. The late Lord Rayleigh obtained by the method of ripples, the following values for the surface tension of a clean water surface: 74·5, 73·7, 74·0, 73·2, 74·7, 74·7, 74·7, 74·2, 75·2, 74·3, 74·3, 74·3, 74·3, 74·1, 73·2, 73·5 dynes per cm. Find the characteristic values.

Should any of the readings be rejected?

8. Apply Chauvenet's criterion to the following data:

Length of a wire 7·82, 7·85, 7·85, 7·78, 7·82, 7·84, 7·78, 7·62, 7·81, 7·70 cms

and find the final values of the characteristic measures.

9. The Indian Geodetic Service measured the height of Mount Everest with their theodolites and base lines and obtained these six results:

28991·6 feet
29005·3 ,,
29001·8 ,,
28998·6 ,,
29026·1 ,,
28990·4 ,,

Average . . 29002·3 ,,

Should any reading(s) be rejected? If so do so, and find the final values of the characteristic measures.

10. Trouton measured the quantity of ether taken up by 100 gm. of water at atmospheric pressure. The results of 10 experiments were:

1·060	1·040	1·048	1·032	1·052
1·063	1·054	1·087	1·064	1·045

Should any reading (*s*) be rejected? If so do so, and find the final value of the characteristic measures.

11. Cavendish determined the density of the earth, making 29 experiments with the following results. Density in gms. per c.c.

Between	Number of times this value was obtained.
4·8 and 4·9	1
4·9 ,, 5·0	0
5·0 ,, 5·1	1
5·1 ,, 5·2	1
5·2 ,, 5·3	4
5·3 ,, 5·4	5
5·4 ,, 5·5	4
5·5 ,, 5·6	5
5·6 ,, 5·7	5
5·7 ,, 5·8	2
5·8 ,, 5·9	1
5·9 ,, 6·0	0

Plot the frequency curve. Comment.
Should any of the readings be rejected? Find the final mean.
Compare this result with the correct result 5·53. Comment.

CHAPTER XVIII

INDIRECT MEASUREMENTS

209. Importance of Indirect Measurements.

An indirect measurement is one that is calculated from one or more direct measurements instead of being directly observed by the experimenter.

In a certain sense almost all measurements are indirect, especially those that require the highest degree of accuracy. In making a careful determination of the weight of an object the direct determinations are of the turning points of the pointer that swings back and forth as the beam of the balance oscillates (§ 119). From these points the resting position of equilibrium is calculated, and from this position and a similar one obtained when the balance is empty a calculation is made of the discrepancy between the weight of the object and that of the standard weights that are balanced against it, using a previous determination of the shift of resting point that is caused by a very small standard weight.

The term, *indirect*, however, is commonly applied to measurements like that of the density obtained by dividing mass by volume, one or more carefully determined measurements being treated by calculation in such a way as to give a required " indirect " result.

In some cases values calculated from other data are a necessity. The density of a solid may be obtained by direct measurement if it can be tested by immersing it in fluids whose densities are known, and this method is found to be very convenient for testing small objects that do not have a very great density, such as precious stones.* For denser objects, however, some indirect method is necessary. The usual one is to calculate $d=m/v$, v also being obtained indirectly from the loss of weight on immersion in a fluid of known density.

In some cases much time and laborious calculation can be saved by calculating the average and probable error, $a \pm p$, for each quantity that is measured directly and then investigating the resultant probable error of the indirect measurement in accordance with theoretical considerations. For example, it would take much

* A solution of mercuric potassium iodide, *sp. gr.* = 3·11 or less, is used by jewellers for this purpose. Densities as high as 4·8 can be determined by using the double nitrate of silver and thallium.

longer to calculate a density from each one of a set of twenty-five determinations of mass and volume and then average the results than it would to average the masses and the volumes separately and perform a single division. When using the latter method it is possible to find what the probable error of the density will amount to by investigating the probable errors of mass and volume.

In a case like the one just mentioned it would also be possible to make twenty-five calculations of density from the twenty-five sets of measurements of mass and density and to compute the probable error of the average density. In many cases, however, such a process cannot be carried out. If there were more data at hand for the mass than for the volume the excess could not be utilized ; and, furthermore, some of the values needed in the formula may be predetermined constants, such as those mentioned in Chap. XVI (see § 192), which are given only in the form of a representative magnitude and its probable error. The question then arises as to the way in which the probable error of the indirect measurement is influenced by the size of the probable errors of the direct measurements on which it is based.

210. Probable Error of a Sum.

The simplest case is when the indirect measurement is merely the sum of the two independent direct measurements. Let the direct measurements be denoted by the small letters,

$$a_1 \pm p_1 \text{ and } a_2 \pm p_2 .$$

when stated in the form of average and probable error, and let the indirect measurement with its probable error be represented by capitals,

$$A \pm P,$$

the direct measurement being equal to the sum of the indirect measurements, so that $A = a_1 + a_2$; then it can be proved that

$$P^2 = p_1{}^2 + p_2{}^2$$

The dispersion of the sum is not as large as the sum of the two direct dispersions from which it is calculated. This is because positive and negative deviations will counterbalance each other to a certain extent. It is larger than either of the others alone, however, for the extra measurement gives an extra degree of uncertainty.

Example.—A bench has a height of $42{\cdot}50 \pm 0{\cdot}03$ cm. above the floor, and a table is $44{\cdot}35 \pm 0{\cdot}04$ cm. higher than the bench. Write the height of the table, with its probable error.

$$A = 42{\cdot}50 + 44{\cdot}35$$

$$= 86{\cdot}85 \text{ cms.}$$

$$P^2 = 0{\cdot}03^2 + 0{\cdot}04^2 = 0{\cdot}05^2$$

$$\therefore \text{Height} = 86{\cdot}85 \, P.E. \; 0{\cdot}05 \text{ cm.}$$

Similarly if
$$A=a_1+a_2+a_3+\ldots$$
$$P^2=p_1{}^2+p_2{}^2+p_3{}^2+\ldots$$

211. Probable Error of a Difference.

If the indirect measurement, A, is equal to the difference, a_1-a_2, between two direct measurements, a_1 and a_2, it is perhaps a natural inference that the square of its dispersion, P^2, should be equal to $p_1{}^2-p_2{}^2$. This is not the case, however, but

$$P^2=p_1{}^2+p_2{}^2$$

in all cases in which

$$A=a_1\pm a_2$$

Example.—The table that was measured in the last exercise is 51·10 ± 0·04 cm. lower than a shelf which is 137·95 ± 0·03 cm. above the floor. How high is the table?

Notice that the same formula applies to both of these exercises. The difference of two measurements has as large a probable error as their sum. Measuring first up and then down does not give a more precise result than measuring up and then further up.

212. Probable Error of a Multiple.

If
$$A=c_1a_1,$$

where c_1 is some constant not subject to error, then

$$P^2=c_1{}^2p_1{}^2, \quad \text{or} \quad P=c_1p_1.$$

Example.—The diameter of a circular disc is found to be 7·98 ± 0·03 cm. What is its circumference?

Notice that the *relative* dispersion of the circumference (3/800) is the same as the relative dispersion of the diameter (*vide infra*).

213. Associative Law.

In general, if

$$\mathbf{A}=\mathbf{c}_1\mathbf{a}_1\pm\mathbf{c}_2\mathbf{a}_2\pm\mathbf{c}_3\mathbf{a}_3\pm\ldots$$

then

$$\mathbf{P}^2=\mathbf{c}_1{}^2\mathbf{p}_1{}^2+\mathbf{c}_2{}^2\mathbf{p}_2{}^2+\mathbf{c}_3{}^2\mathbf{p}_3{}^2+\ldots \quad (1)$$

where p_n is the probable error of the average a_n.

This formula can be used to find the probable error of an algebraical expression when the probable error of each of its terms is known.

A wall consists of 15 courses of bricks, each of which is 56·5 ± 0·5 mm. thick, separated by 14 layers of mortar which have an average thickness of 7·5 ± 1·5 mm. Show that the probable error of the height of the wall is ± 22·3 mm., and state how much this figure would be reduced if the bricks were absolutely uniform. How much if the mortar was of uniform thickness instead?

214. Probable Error of a Product.

If two independent measurements are multiplied together the probable error of the product will follow the law expressed in the following equation:

If $\qquad A=a_1a_2 \quad . \quad . \quad . \quad . \quad . \quad . \quad . \quad . \quad . \quad . \quad . \quad . \quad . \qquad (1)$

then $\qquad P^2=p_1^2a_2^2+a_1^2p_2^2$

Likewise if

$$A=a_1a_2a_3$$

then $\qquad P^2=(p_1a_2a_3)^2+(a_1p_2a_3)^2+(a_1a_2ap_3)^2$

and similarly for any number of factors; but it is more satisfactory in practice to make use of the *relative* probable error (relative dispersion) as in the following form, which is easily deducible from the form just given.

If $\qquad A=a_1a_2a_3 \quad . \quad . \quad . \quad . \quad . \quad . \quad . \quad . \quad . \quad . \quad . \quad . \quad . \qquad (2)$

then $\qquad (P/A)^2=(p_1/a_1)^2+(p_2/a_2)^2+(p_3/a_3)^2+ \ . \ . \ .$

Prove that the last equation, as far as it goes, is equivalent to

$$P^2=(p_1a_2a_3)^2+(a_1p_2a_3)^2+(a_1a_2p_3)^2+ \ . \ . \ .$$

A rectangular block measures $20{\cdot}00\pm0{\cdot}04$ cm. in length, $10{\cdot}00\pm0{\cdot}01$ cm. in breadth, and $5{\cdot}00\pm0{\cdot}01$ cm. in thickness.* What is its volume?

215. Probable Error of a Power.

If $\qquad A=a_1^n$

then $\qquad (P/A)^2=(n^2p_1^2/a_1^2) \quad$ or $\quad P/A=np_1/a_1$

and, likewise, if

$$A=c_1a_{;}^n$$

then $\qquad P/A=np_1/a_1$

the constant not appearing in the formula if the relative probable error is used.

In the particular case for which $n=1$, notice that the *relative* probable error of c_1a_1 is the same as the relative probable error of a_1 itself (*vide supra*).

One edge of a cubical block is $10{\cdot}00\pm0{\cdot}01$ cm. What is its volume? What is the area of its total surface?

Does the rectangular block (above) appear to have been measured as accurately as the cubical block?. How do their volumes compare in respect to accuracy? What is there about the data of the rectangular block which make its volume determinable as closely

* Simple numbers are chosen in the examples in order that the arithmetic involved may not obscure the calculation of the probable error.

as that of the cubical block, although its measurements are less precise ?

Show that the combined volume of the two blocks is 2000· ± 4·2 cm.[3]

216. Distributive Law.

In general, whether the exponents are positive, negative, or fractional, if

$$\mathbf{A = c_1 a_1^m a_2^n a_3^r \ldots}$$

then

$$\mathbf{(P/A)^2 = (mp_1/a_1)^2 + (np_2/a_2)^2 + (rp_3/a_3)^2 + \ldots} \quad (2)$$

This formula can be used to find the probable error of any single term of an algebraical expression when the probable errors of its factors are known.

The formula for the volume of a cylinder is

$$v = \pi l r^2$$

If the measurements of l and r have respective probable errors of p_l and p_r find the value of P, the probable error of the calculated volume.

Test the accuracy of the statement in § 186, viz. that $P = \frac{p}{\sqrt{n}}$ by letting p denote the probable error of each of the n single measurements and finding the probable error of their average.

217. Recapitulation.

In the following formulæ the relative dispersion (P/A, or p/a) is represented by R or r

If	Then
$A = a_1 \pm a_2 \pm \ldots$	$P^2 = p_1^2 + p_2^2 + \ldots$
$A = c_1 a_1$	$P = c_1 p_1$
$A = c_1 a_1 \pm c_2 a_2 \pm \ldots$	$P^2 = (c_1 p_1)^2 + (c_2 p_2)^2 + \ldots$

PROBABLE ERRORS OF INDIRECT MEASUREMENTS.—In these cases the probable error (p) for each average (a) is most convenient for calculation. The capital letters refer to the indirect measurement.

If	Then
$A = a_1 a_2 a_3 \ldots$	$R^2 = r_1^2 + r_2^2 + r_3^2 + \ldots$
$A = a_1^n$	$R = nr$
$A = c_1 a_1^n$	$R = nr$
$A = c_1 a_1^m a_2^n a_3^s \ldots$	$R^2 = (mr_1)^2 + (nr_2)^2 + (sr_3)^2 + \ldots$

PROBABLE ERRORS OF INDIRECT MEASUREMENTS.—In these cases the use of the *relative* probable error (r) simplifies the calculation.

Notice the complete formal correspondence between (*absolute*) *probable errors* after adding, multiplying by a coefficient, and assembling terms, and *relative probable errors* after multiplying, raising to a power, and assembling factors, respectively.*

218. Graphs of Propagated Errors.

It has been seen that the probable errors of two or more direct measurements are propagated through any kind of a calculation and give the indirect measurement a probable error whose formula is of the type $\sqrt{x^2+y^2}$. Since the square root of the sum of two squares can always be represented by the hypotenuse of a right-angled triangle a graphic solution of the probable error of an indirect measurement is easily effected.

From any "origin" draw a horizontal line and a vertical line, making their lengths equal to the dispersions 3 and 4 of the measurements in § 211. Complete the right-angled triangle and note that

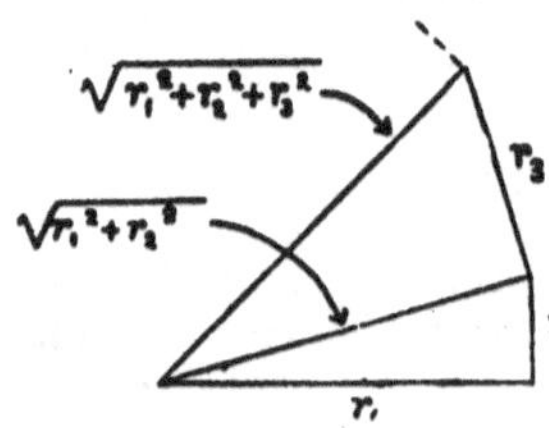

FIG. 66.—GEOMETRIC DETERMINATION OF PROPAGATED ERRORS.

the length of the hypotenuse, 5, is the "propagated" dispersion of the indirect measurement, as previously found by calculation.†

The fact that $p_1^2+p_2^2+p_3^2$ is the same thing as $(\sqrt{p_1^2+p_2^2})^2+p_3^2$, namely, the sum of *two* squares, makes this method extensible to any number of terms (Fig. 66).

Use the geometrical construction to find the square root of the sum of the squares of 4, 3, and 12. Afterward, verify your result by calculation.

219. Relative Importance of Compound Errors.

The fact that an indirect probable error which depends upon the measurement of two or more different quantities always assumes the form $\sqrt{x^2+y^2}$ means that it will be more decidedly diminished by reducing the larger of the two independent probable errors than

* The above formulæ may also be used with the average deviation or standard deviation substituted in place of p and r.

† These numbers refer, of course, to hundredths of a centimetre. It would be possible, but not advisable, to perform graphically the operation $\sqrt{0{\cdot}04^2+0{\cdot}03^2}=0{\cdot}05$ so as to obtain the answer in centimetres.

by attempting to improve the more accurate measurement. Show that $\sqrt{5^2+2^2}$ is reduced by 41 per cent. if the 5 is changed to 2·5, but only by 5 per cent. if the 2 is changed to 1.

Draw a right-angled triangle with one side several (say 8 or 10) times as long as the other. Change the long side by making it a little longer or shorter and notice that the change in length of the hypotenuse is almost in exact proportion. Change the short side and notice that the hypotenuse is hardly affected at all.

The formula for the volume of a cylinder is $v=\pi lr^2$. In determining this indirect measurement which of the two dimensions ought to be measured the more carefully? How much more carefully? Why?

If one factor in an expression can be determined much more accurately than the others it may be treated as a constant and the work thereby simplified (for example, the mass of the metal in a specific heat experiment, see § 143). The usual routine in an experiment is to so measure the factors entering into the final quantity that the amount of error produced by each in the result is the same. A little forethought given to this rule often lessens considerably the time spent on the experiment.

220. General Formula for the Calculation of the Error in an indirect Determination.

For the convenience of those who know the calculus a general formula is given here.

$$\text{Let } A=f(x, y, z, \ldots)$$

where $x, y, z \ldots$ are the quantities measured directly, and A is the deduced quantity.

Let $p_1, p_2, p_3, \ldots$ be the probable errors in $x, y, z, \ldots$ and P the probable error in M.

$$\text{Then} \quad P^2=\left(\frac{\partial f}{\partial x}\right)^2 p_1^2+\left(\frac{\partial f}{\partial y}\right)^2 p_2^2+\left(\frac{\partial f}{\partial z}\right)^2 p_3^2+\ldots$$

where $\frac{\partial f}{\partial x}, \frac{\partial f}{\partial y}, \frac{\partial f}{\partial z}$ are the coefficients of partial differentiation of $f(xyz \ldots)$ with respect to x, y, z, respectively, *i.e.* $\frac{\partial f}{\partial x}$ is obtained by differentiating $f(xyz \ldots)$ with respect to x regarding everything but x as constant.

Examples.

In § 210 where $A=a_1+a_2$, $\frac{\partial A}{\partial a_1}=1$, $\frac{\partial A}{\partial a_2}=1$.

in § 212 where $A=c_1a_1$, $\frac{\partial A}{\partial a_1}=c_1$.

In § 213 where $A=c_1a_1 \pm c_2a_2 + \ldots \; \frac{\partial A}{\partial a_1}=c_1, \frac{\partial A}{\partial a_2}=c_2, \ldots$

In § 214, Equation 1, where $A=a_1a_2, \frac{\partial A}{\partial a_1}=a_2, \frac{\partial A}{\partial a_2}=a_1.$

In § 214, Equation 2, where $A=a_1a_2a_3, \frac{\partial A}{\partial a_1}=a_2a_3, \frac{\partial A}{\partial a_2}=a_1a_3.$

In § 215 where $A=a_1^n, \frac{\partial A}{\partial a_1}=na_1^{n-1}$

$$\therefore P^2=(na_1^{n-1})^2p_1^2=(na_1^n)^2\left(\frac{p_1}{a_1}\right)^2$$

or

$$\frac{P}{A}=n\frac{p_1}{a_1}$$

In § 216 where $A=c_1a_1^m a_2^n a_3^r \ldots$

$$\frac{\partial A}{\partial a_1}=c_1ma_1^{m-1}a_2^n a_3^r \ldots$$

$$\frac{\partial A}{\partial a_2}=c_1na_1^m a_2^{n-1}a_3^r \ldots$$

$$\therefore P^2=(c_1ma_1^{m-1}a_2^n a_3^r \ldots)^2p_1^2+(c_1na_1^m a_2^{n-1}a_3^r \ldots)^2p_2^2$$

$$\therefore \frac{P^2}{A_2}=m^2\frac{p_1^2}{a_1^2}+n^2\frac{p_2^2}{a_2^2}+ \ldots$$

If you know the calculus it is much easier to apply the general formula than to remember the rules given in the earlier pages.

You will also find that with a continued product it is usually quicker to apply the fractional or relative method than the absolute method (see below).

Examples.

(1) The diameter of a cylinder is 3·16 *p.e.* 0·03 cm., and the length is 4·75 *p.e.* 0·05 cm. Find the volume and its probable error.

$$\text{Volume}=v=\frac{\pi}{4}d^2l=\frac{\pi}{4}(3{\cdot}16)^2 4{\cdot}75=37{\cdot}25 \text{ c.c.}$$

$$\frac{\partial v}{\partial d}=\frac{\pi}{2}dl=23{\cdot}5, \quad \frac{\partial v}{\partial l}=\frac{\pi}{4}d^2=7{\cdot}8$$

$$\therefore P^2=(23{\cdot}5)^2(0{\cdot}03)^2+(7{\cdot}8)^2(0{\cdot}05)^2$$

$$=(0{\cdot}705)^2+(0{\cdot}390)^2=0{\cdot}495+0{\cdot}152=0{\cdot}647$$

$$\therefore P=0{\cdot}805 \text{ c.c.}$$

The probable error due to diameter alone=0·705 c.c., due to length alone 0·390 c.c., due to the two combined 0·805 c.c.

Result: Volume=37·25, *p.e.* 0·81 c.c.

If we apply the equation giving the relative probable error and remember the last equation in the lower table in § 217, we merely write

$$\frac{P^2}{v^2}=(2r_1)^2+(r_2)^2=\left(2\times\frac{0{\cdot}03}{3{\cdot}16}\right)^2+\left(\frac{0{\cdot}05}{4{\cdot}75}\right)^2$$
$$=0{\cdot}00047$$

whence $\frac{P}{v}=0{\cdot}0217$, or probable error $=2{\cdot}17$ per cent.

Multiplying by v,

$$P=37{\cdot}25\times0{\cdot}0217=0{\cdot}805 \text{ c.c., the same as above.}$$

(2) A simple pendulum of length $l=100{\cdot}0$ cm. has a period of $t=2{\cdot}006$ seconds. The probable error in the length is $0{\cdot}05$ cm. and in the time $0{\cdot}001$ second. The acceleration of gravity is given by the formula $g=4\pi^2\frac{l}{t^2}$. Find g and its probable error.

$$g=4\pi^2\times\frac{100{\cdot}0}{(2{\cdot}006)^2}=981{\cdot}1 \text{ cm. per sec. per sec.}$$

To use the differential formula first take logarithms

$$\log g=\log 4\pi^2+\log l-2\log t$$
$$\therefore \frac{1}{g}\cdot\frac{\partial g}{\partial l}=\frac{1}{l} \quad\text{or}\quad \frac{\partial g}{\partial l}=\frac{g}{l}=9{\cdot}81$$

and

$$\frac{1}{g}\cdot\frac{\partial g}{\partial t}=-\frac{2}{t} \quad\text{or}\quad \frac{\partial g}{\partial t}=-\frac{2g}{t}=-978$$
$$\therefore P^2=(9{\cdot}81)^2(0{\cdot}05)^2+(978)^2(0{\cdot}001)^2$$
$$=(0{\cdot}49)^2+(0{\cdot}98)^2$$
$$=0{\cdot}24+0{\cdot}95=1{\cdot}19$$
$$\therefore P=1{\cdot}09=1{\cdot}1$$

or

$$g=981{\cdot}1 \; p.e. \; 1{\cdot}1 \text{ cm. per sec. per sec.}$$

By the fractional method

$$\left(\frac{P}{981}\right)^2=\left(\frac{0{\cdot}05}{100}\right)^2+\left(\frac{2\times0{\cdot}001}{2{\cdot}00}\right)^2=0{\cdot}00000125$$
$$\therefore \frac{P}{981}=0{\cdot}00112, \text{ which}=0{\cdot}112 \text{ per cent.}$$

(3) The density of a penny was obtained from its weight in vacuo and in water at 4° C.:

Weight in vacuo $=w_1=9{\cdot}456\pm0{\cdot}002$ gm.
Weight in water $=w_2=8{\cdot}401\pm0{\cdot}005$ gm.

(Note w_2 has a greater probable error than w_1, caused by surface tension around the suspending fibre.)

The density $D=\frac{w_1}{w_1-w_2}=\frac{9{\cdot}456}{1{\cdot}055}=8{\cdot}962$

$$\frac{\partial D}{\partial w_1}=-\frac{w_2}{(w_1-w_2)^2}$$

$$\frac{\partial D}{\partial w_2}=\frac{w_1}{(w_1-w_2)^2}$$

$$\therefore P^2=\frac{1}{(w_1-w_2)^4}(8{\cdot}401^2\times0{\cdot}002^2+9{\cdot}456^2\times0{\cdot}005^2)$$

$$=\frac{1}{1{\cdot}055^4}(0{\cdot}00028+0{\cdot}00224)=\frac{1}{1{\cdot}055^4}\times0{\cdot}00252$$

$$\therefore P=\frac{1}{1{\cdot}055^2}\sqrt{0{\cdot}00252}=0{\cdot}045$$

$$\therefore D=8{\cdot}962\pm0{\cdot}045 \text{ gm. per c.c.}$$

(4) A spherometer is used to measure the curvature of a spherical surface. The distance from the axes of the central leg to the feet of the outer legs$=45{\cdot}5\pm0{\cdot}2$ mm.$=a$, say. The distance the foot of the central leg is raised above the plane of the feet of the outer legs in order to touch the surface$=2{\cdot}64\pm0{\cdot}02$ mm. Find R from the formula

$$R=\frac{a^2}{2h}+\frac{h}{2}$$

also its probable error.

$$R=\frac{45{\cdot}5^2}{5{\cdot}28}+1{\cdot}32=393{\cdot}4 \text{ mm.}$$

$$\frac{\partial R}{\partial a}=\frac{a}{h}=17{\cdot}2$$

$$\frac{\partial R}{\partial h}=-\frac{a^2}{2h^2}+\frac{1}{2}=-148$$

$$\therefore P^2=17{\cdot}2^2\times0{\cdot}2^2+148^2\times0{\cdot}02^2=11{\cdot}8+8{\cdot}7=20{\cdot}5$$

$$\therefore P=4{\cdot}5$$

$$\therefore R=393{\cdot}4,\ p.e.\ 4{\cdot}5 \text{ mm.}$$

EXERCISES XVIII

1. Each volume of a ten-volume encyclopedia has 5·0 cm.±0·1 cm. thickness of leaves between its two covers. Each cover has a thickness of 0·35 cm.±0·02 cm. How much more shelf-room than 57·00 cm. is it likely to need? What is the meaning of the word " likely " in the last sentence?

2. A rectangle is 20·00±0·04 cm. long and 10·00±0·01 cm. wide. Find the area and its probable error.

3. In getting a coefficient of expansion if the percentage dispersion of the expansion, the length, and the temperature difference are 1·2, 0·1, and 1·5 respectively, find the percentage dispersion of the result.

4. Apply the rules of this chapter to prove: Probable error of the mean of n determinations of the same quantity $= \frac{1}{\sqrt{n}} \times$ probable error of an individual determination.

5. A bowl whose interior is an exact segment of a sphere is found to have a depth of $25{\cdot}00 \pm 0{\cdot}02$ cms. and a rim diameter of $50{\cdot}00 \pm 0{\cdot}30$ cms. Find its capacity from the formula

$$v = \frac{\pi h r^2}{2} + \frac{\pi h^3}{6}$$

where r is the radius of the circular rim and h is the depth of the segment. Find the probable error in the capacity.

6. Young's modulus of steel was determined by the stretching of a steel wire. Length of wire $= 585 \pm 0{\cdot}5$ cm.; radius of wire $0{\cdot}0335 \pm 0{\cdot}0005$ cm. Four experiments with loads of 16 lbs. gave extensions of 5·70, 5·50, 5·55, 5·65 mm., respectively. Assume the 16 lbs. as correct, and 1 lb. wt. $= 445{,}000$ dynes. Find Y from the formula $Y = \frac{FL}{\pi a^2 l}$ and its probable percentage error.

7. The moment of inertia of a cylinder is given by $I = M\left(\frac{l^2}{12} + \frac{r^2}{4}\right)$. $M = 1104 \pm 1$ gm., $l = 25{\cdot}57 \pm 0{\cdot}02$ cm., $r = 1{\cdot}28 \pm 0{\cdot}01$ cm. Find I and its probable error.

8. A Kater's pendulum was adjusted to have its erect and inverted periods equal. Six observers got the following readings for the length, l, and the period, t, respectively: 97·15 cms. and 1·975 sec., 97·1 and 1·979, 97·2 and 1·982, 97·2 and 1·978, 97·2 and 1·981, 97·1 and 1·976. Find the acceleration of gravity from the formula $t = 2\pi\sqrt{\frac{l}{g}}$, also its probable error.

CHAPTER XIX

LEAST SQUARES

221. The Average as a Least-Square Magnitude.

The mathematical *principle of least squares* is that when measurements are equally trustworthy their best representative value is that for which the sum of the squares of the deviations has the lowest numerical value.* It is upon this principle that the use of the average is based, for it is easy to show that the sum of the squares of the deviations of any particular set of numbers will be greater when measured from some other value than when measured from the average.

Find the average of the numbers 3, 3, 4, 5, 10 ; also their deviations from the average, and the sum of the squares of the deviations. Find the median of 3, 3, 4, 5, 10 ; and the sum of the squares of the deviations from the median. Find the sum of the squares of the deviations from the harmonic mean (call it 4·1) or from the mode, and notice that $\Sigma(v^2)$ is smaller when the deviations are measured from the average than when measured from any of the other numerical values.

222. Least Squares for Conditioned Measurements.

If we are dealing with two conditioned measurements, as in the case of the x-values and the y-values of the black-thread experiment, the principle of least squares shows that the line which expresses the *condition* or gives the law of relationship between the two variables must be so placed that the sum of the squares of the

* This principle can be proved from the "fact of experience" that deviations follow the law of the exponential equation $y=e^{-x^2}$. As an example of its application consider the marks 6, 8, and 7, on a scale of ten, which one student obtained on three examination questions, and the marks 7, 7, 7, which were obtained by another student. Find the deviations from the theoretical mark 10, and see which student has the smaller value for $\Sigma(v^2)$. It has been said that the mathematicians respect the theory of least squares because the experimental scientists have confirmed it by experience and that the latter respect it because they think the mathematicians have proved it. It is, however, certain that the arithmetical mean or average was accepted axiomatically at quite an early date as the best representative of equally trustworthy results, and this is an agreement with the principle of least squares.

distances from it to all of the experimental points shall have the smallest possible value (Fig. 67), or if little circles be described

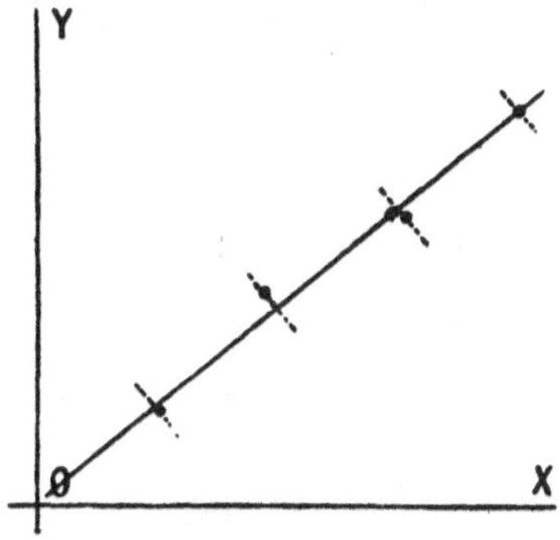

FIG. 67.—THE PRINCIPLE OF LEAST SQUARES.

The general principle is that the theoretical relationship is to be so arranged that the sum of the squares of the discrepancies of the actual measurements shall be as small as possible. If a straight line is to be used it must be so arranged that the sum of the squares of the distances to it from the various points is a minimum.

around the plotted points of size just sufficient to touch the line the latter must be drawn to make the sum of the areas of these circles a minimum.

223. The problem of obtaining the best equation between x and y or the best straight line through the plotted points when x and y are equally fallible is too complicated for the present. We leave it to the end of the chapter (§ 230).

The case taken up here is that when only the *y observations are likely to have accidental error, the x observations being strictly correct.**

If $y=a+bx$ is the best straight line through the points, we find that in general none of the pairs of values x_1y_1, x_2y_2, x_3y_3, . . . x_ny_n satisfies the equation $a+bx-y=0$, but instead of zero there will always be a small remainder (d), positive or negative; *e.g.* $a+bx_1-y_1=d_1$, $a+bx_2-y_2=d_2$, and so on.

These values of d are the *vertical* distances from the plotted points to the line.† According to the principle of least squares the sum of

$$d_1^2+d_2^2+d_3^2+ \ldots +d_n^2$$

* This happens in many physical experiments, such, for example, as when we apply definite weights to a spring or wire and measure the extension, or find what load we have to apply at some point in a machine to overcome a known load attached at some other point.

† Note that in Fig. 67 we dealt with the shortest distance from a plotted point to the line. Fig. 68 shows the vertical distances.

must be as small as possible. This means that the sum of the squares of the left-hand members of the equations, or $\Sigma(a+bx-y)^2$, must have its minimum value, and it can be shown by processes of pure mathematics, involving the calculus, that this will be the case if

$$\Sigma(a+bx-y)=0$$

and
$$\Sigma\{x(a+bx-y)\}=0$$

i.e. if
$$na+b\Sigma x-\Sigma y=0$$

and
$$a\Sigma x+b\Sigma x^2-\Sigma xy=0$$

Solving these two simultaneous equations we get

$$a=\frac{\Sigma(x)\Sigma(xy)-\Sigma(y)\Sigma(x^2)}{(\Sigma x)^2-n\Sigma(x^2)} \quad . \quad . \quad . \quad . \quad (1)$$

and
$$b=\frac{\Sigma(x)\Sigma(y)-n\Sigma(xy)}{(\Sigma x)^2-n\Sigma(x^2)} \quad . \quad . \quad . \quad . \quad . \quad . \quad (2)$$

where x and y stand for $x_1, x_2, x_3, \ldots$ and $y_1, y_2, y_3, \ldots$, the x-values and the y-values of the experimental points.

These equations enable one to determine the values of a and b, and hence to find the position (see § 108) that a straight line (black thread) must have if it is to be so located that the sum of the squares of the distances to it parallel to the y axis from all of the experimental points shall have the smallest possible value.

By similar processes a, b, and c could be found for the equation of the parabola $y=a+bx+cx^2$, or the appropriate coefficients for curves having even more complicated equations, but the processes of computation become so tedious that it is better to replace the variables, as explained in the lesson on graphic analysis, by others that will conform to the straight-line law.*

Example.—Find by least squares the equation that will fit the following experimental values of x and y (x values accurate, y in accidental error):

x	0	1	2	3	4	5	6
y	2·0	2·6	2·9	3·6	3·9	4·6	4·9

Tabulate as shown. Work neatly; take care to avoid using the wrong algebraical signs or confusing $\Sigma(x^2)$ with $(\Sigma x)^2$. Keep only three significant figures in the final result.

* The simultaneous equations for a, b, c are:

$$na+b\Sigma x+c\Sigma x^2-\Sigma y=0$$
$$a\Sigma x+b\Sigma x^2+c\Sigma x^3-\Sigma(xy)=0$$
$$a\Sigma x^2+b\Sigma x^3+c\Sigma x^4-\Sigma(x^2y)=0$$

x	y	xy	x^2
0	2·0	0	0
1·0	2·6	2·6	1·0
2·0	2·9	5·8	4·0
3·0	3·6	10·8	9·0
4·0	3·9	15·6	16·0
5·0	4·6	23·0	25·0
6·0	4·9	29·4	36·0
21·0	24·5	87·2	91·0
Σx	y	$\Sigma(xy)$	$\Sigma(x^2)$

Note $n=7$.

Then
$$a=\frac{21{\cdot}0\times87{\cdot}2-24{\cdot}5\times91{\cdot}0}{441-7\times91{\cdot}0}=2{\cdot}04$$

$$b=\frac{21{\cdot}0\times24{\cdot}5-7\times87{\cdot}2}{441-7\times91{\cdot}0}=0{\cdot}487$$

Therefore the equation is $y=2{\cdot}04+0{\cdot}487x$. Check by graphing and using the black thread. Simple numbers were chosen for the example for the sake of easy working. In practice unless the numbers were determined accurately, it would be quicker to use the graph and not bother about least squares.

224. Least Squares and Proportionality.

If two variables, x and y, are always proportional the linear equation $y=a+bx$ reduces to the form

$$y=0+bx \text{ (Fig. 68)}$$

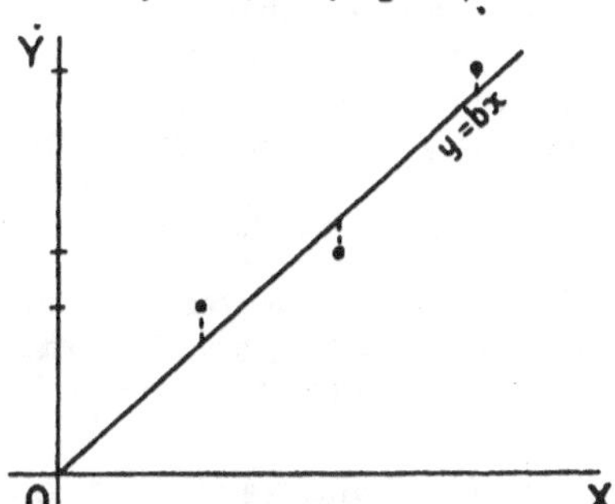

FIG. 68.

and the best value of the coefficient will be found to be

$$b=\frac{\Sigma(xy)}{\Sigma(x^2)}$$

Proof.—Put $a=0$ in the equation

$$a=\frac{\Sigma(x)\Sigma(xy)-\Sigma(y)\Sigma(x^2)}{(\Sigma x)^2-n\Sigma(x^2)}$$

we get

$$\Sigma(x)\Sigma(xy)=\Sigma(y)\Sigma(x^2) \quad . \quad . \quad . \quad . \quad (1)$$

Multiplying by $n\Sigma(x)$,

$$n(\Sigma x)^2\Sigma(xy)=n\Sigma(x)\Sigma(y)\Sigma(x^2)$$

whence $\Sigma(x)\Sigma(y) : (\Sigma x)^2 :: n\Sigma(xy) : n\Sigma(x^2)$.

But if $a : b :: c : d$, then $a-c : b-d :: c : d$; accordingly,

$$\frac{\Sigma(x)\Sigma(y)-n\Sigma(xy)}{(\Sigma x)^2-n\Sigma(x^2)}=\frac{n\Sigma(xy)}{n\Sigma(x^2)}$$

or

$$b=\Sigma(xy)/\Sigma(x^2) \quad . \quad . \quad . \quad . \quad . \quad . \quad . \quad . \quad (2)$$

After stage (1) we might have written

$$\therefore b=\frac{\Sigma(x)\Sigma(y)-n\frac{\Sigma(y)\Sigma(x^2)}{\Sigma x}}{(\Sigma x)^2-n\Sigma(x^2)}=\frac{\Sigma(y)}{\Sigma(x)} \quad . \quad . \quad . \quad . \quad (3)$$

which is simpler to use than (2).

Equation (3) may also be got straightaway from equation (1) by using (2). Of course the expressions (2) and (3) will only give the same value of b if (a) is zero. For example, if we wished to deduce the density of a material by measuring the masses and volumes of different lumps there must be a constant proportionality between the masses and the volumes, for their ratio is the density which is a constant. Therefore, if y=mass and x=volume

$$y=bx \text{ strictly}$$

With our experimental data we could use either expression (2) or (3). Remember that the formula of this section assumes the x values are absolutely correct, the y values alone being in error.

Example.—Find by least squares the equation that will fit the following experimental values of x and y (x values accurate, y in accidental error), given that there is a constant proportionality between the two quantities represented by x and y.

x	0·0	1·0	2·0	3·0	4·0	5·0	6·0	7·0	8·0	9·0	10·0
y	0·0	0·8	1·2	2·2	2·6	3·6	4·2	5·2	5·6	6·6	7·0

Tabulate as shown.

x	y	xy	x^2
0·0	0·0	0	0
1·0	0·8	0·8	1
2·0	1·2	2·4	4
3·0	2·2	6·6	9
4·0	2·6	10·4	16
5·0	3·6	18·0	25
6·0	4·2	25·2	36
7·0	5·2	36·4	49
8·0	5·6	44·8	64
9·0	6·6	59·4	81
10·0	7·0	70·0	100
55·0	39·0	274·0	385
Σx	Σy	$\Sigma(xy)$	(Σx^2)

Note $n=11$.

By Equation (2) for b we get $b=\dfrac{\Sigma(xy)}{\Sigma(x^2)}=\dfrac{274}{385}=0{\cdot}712.$

By Equation (3) for b we get $b=\dfrac{\Sigma y}{\Sigma x}=\dfrac{39{\cdot}0}{55{\cdot}0}=0{\cdot}709.$

The difference is due to the value of a being not exactly zero. (By the formula of § 223 $a=-0{\cdot}04$ and $b=0{\cdot}718$.)

225. Least Squares for a Theoretically Constant Value.

If the variation of y is supposed to be *nil, i.e.* if y is a constant, the best representative value for the fluctuating experimental determinations of it can easily be found by the method of least squares :

In the equation $y=a+bx$ if $b=0$ the equation

$$b=\frac{\Sigma(x)\Sigma(y)-n\Sigma(xy)}{(\Sigma x)^2-n\Sigma(x^2)}$$

gives $$\Sigma(x)\Sigma(y)=n\Sigma(xy)$$

Eliminating $\Sigma(xy)$ between this equation and the equation

$$a=\frac{\Sigma(x)\Sigma(xy)-\Sigma(y)\Sigma(x^2)}{(\Sigma x)^2-n\Sigma(x^2)}$$

gives $$a=\frac{(\Sigma x)^2\Sigma(y)/n-\Sigma(y)\Sigma(x^2)}{(\Sigma x)^2-n\Sigma(x^2)}$$

which is obviously the same as

$$a=\Sigma(y)/n$$

In other words, the best representative value of y is obtained by adding all n of its experimental values and dividing by n—as already stated without proof.*

226. Consecutive Equal Intervals.

If the successive unknown intervals of a scale are not perceptibly different from one another and from the intervals on an equally divided ruler, their value may be found by the method of coincidence (§ 145). If the sizes of the intervals on the scale and ruler are very different or incommensurable we may proceed as follows. Lay the ruler alongside the scale and note the reading on the ruler opposite each graduation mark on the scale, estimating to tenths if necessary. Or we may lay the scale on the bed of a travelling microscope and take the readings by the scale and vernier (or divided circle) of the microscope.

Now though the division marks on the scale are, by hypothesis, equally spaced yet we shall find in general that pairs of successive readings on the ruler (or microscope scale) do not differ by the same amount. This is due to accidental error in setting and reading.

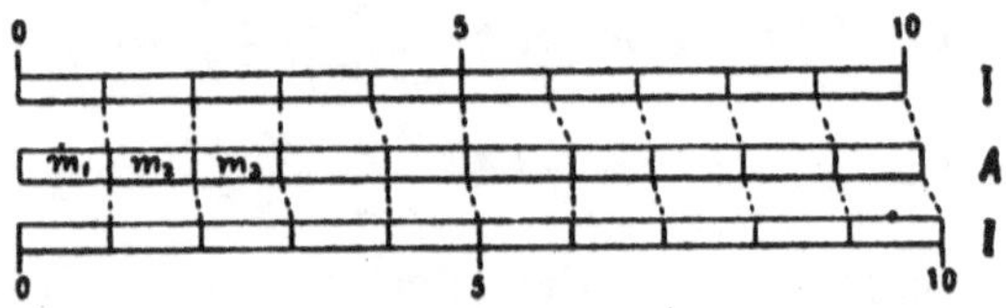

FIG 69.—BEST VALUE FOR CONSECUTIVE INTERVALS.

An imaginary uniform scale is to be so chosen that $\Sigma(v^2)$ for the irregular intervals is a minimum. A, actual scale; I, I, imaginary scales.

The problem is to find the best value of the scale spacing in terms of the graduations on the ruler.

A similar problem occurs in timing a pendulum. We take our watch and note the watch-time at the end of every fifth vibration. Pairs of successive readings will be found to differ by unequal amounts. This is due to errors of observation. What is the right value of the five-vibrations interval?

To revert to the ruler. Suppose the readings that we have taken are marked or scratched on it in some way. The principle of the method of calculation is to lay alongside our marked ruler an imaginary equally divided scale whose intervals are of such a size that the sum of the squares of the discrepancies between the graduations is a minimum. Fig. 69 illustrates this. A represents the marked ruler having (in general) unequally marked intervals of length, m_1, m_2, m_3 and I, I represent two imaginary equally-

* Remember that the principle of least squares is based on the assumption that all of the measurements under consideration are equally trustworthy. If this condition could always be fulfilled we should have little use for any representative value except the average (*cf.* §§ 167, 199).

divided scales.* The value of Σv^2 for the upper pair will be different from Σv^2 for the lower pair. We have to find that *I* scale for which Σv^2 is the least possible.

The mathematical problem is similar to that of § 223. Let us call the ruler readings $y_1, y_2, y_3, \ldots y_n$, and the corresponding readings on the imaginary scale $x_1, x_2, x_3, \ldots x_n$. Fig. 70 shows the state of affairs. We have to find the slope of the line which is

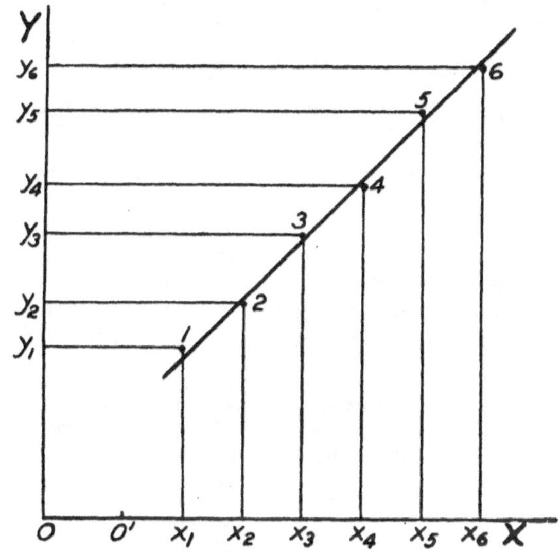

FIG. 70.

so drawn that the sum of the squares of the little vertical intercepts on the equally spaced ordinates is the least possible.

This line has for its equation $y=a+bx$, but since we wish only to find the best average value of the successive increase in y from point to point in terms of the equal increase in x from point to point we need calculate only the b. To make our numerical work simpler shift the origin of co-ordinates to a point O' such that reckoned from this origin $x_1=1, x_2=2, x_3=3, \ldots x_n=n$.

We apply the equation for b given in § 223, viz.:

$$b=\frac{\Sigma x\Sigma y-n\Sigma(xy)}{(\Sigma x)^2-n\Sigma(x^2)}$$

with the simplication that now

$$\Sigma x=1+2+3+\ldots+n=\tfrac{1}{2}n(n+1)$$
$$\Sigma x^2=1^2+2^2+3^2+\ldots+n^2=\tfrac{1}{6}n(n+1)(2n+1).\dagger$$

* The three scales may not be in agreement at the left end although the figure shows them so.

† These formulæ are proved in elementary algebra.

This gives us

$$b=\frac{\frac{1}{2}n(n+1)(y_1+y_2+\ .\ .\ .\ y_n)-n(y_1+2y_2+3y_3+\ .\ .\ .\ +ny_n)}{\frac{1}{4}n^2(n+1)^2-\frac{1}{6}n^2(n+1)(2n+1)}$$

$$=6\frac{(n+1)(y_1+y_2+\ .\ .\ .\ +y_n)-2(y_1+2y_2+\ .\ .\ .\ +ny_n)}{3n(n+1)^2-2n(n+1)(2n+1)}$$

$$=6\frac{2(y_1+2y_2+3y_3+\ .\ .\ .\ +ny_n)-(n+1)(y_1+y_2+\ .\ .\ .\ +y_n)}{n(n^2-1)}$$

Collecting the coefficients of the y's we get

$$b=6\frac{(1-n)y_1+(3-n)y_2+(5-n)y_3+\ .\ .\ .\ +(n-3)y_{n-1}+(n-1)y_n}{n(n^2-1)}$$

Finally, collecting the coefficients of $(n-1)$, $(n-3)$, etc., we get

$$b=6\frac{(n-1)(y_n-y_1)+(n-3)(y_{n-1}-y_2)+(n-5)(y_{n-2}-y_3)+\ .\ .\ .}{n(n^2-1)}$$

for the best value of the interval.

If n is odd the middle reading is neglected. It is therefore better to arrange the experiment so that an even number of readings is made.

Example.—The following observations were taken on ten points, equally spaced along a straight line. No attempt was made to get the zero of the ruler coincident with the first point.

		Differences.
y_1	120	
		160
y_2	280	
		163
y_3	443	
		164
y_4	607	
		166
y_5	773	
		165
y_6	938	
		159
y_7	1097	
		162
y_8	1259	
		166
y_9	1425	
		165
y_{10}	1590	

The best value of the interval is

$$6\left(\frac{9\times1470+7\times1145+5\times816+3\times490+1\times165}{9\times10\times11}\right)=\frac{26960}{165}=163{\cdot}39$$

The half-table method of § 151 gives 163·44. If we simply add up the successive differences and divide by 9 we get 163·33. But, of course, this comes to the same thing as subtracting y_1 from y_{10}, and dividing by 10; that is, we neglect to consider y_2 to y_9, and the time spent in getting them has been wasted.

227. Alternative Method for Consecutive Equal Intervals.*

The observed readings on the ruler are $y_1, y_2, y_3, y_4, \ldots y_n$. Let a=the true value of the first reading and l the true length of the unknown interval of the evenly divided scale. We have the following observational equations:

$$\begin{aligned} a-y_1&=0 \\ a+l-y_2&=0 \\ a+2l-y_3&=0 \\ &\ldots \\ a+(n-1)l-y_n&=0 \end{aligned}$$

The normal equations for a and l are:

$$na+\{1+2+3+\ldots(n-1)\}l-\Sigma y=0$$

$$\{1+2+3+\ldots+(n-1)a\}+\{1+2^2+3^2+\ldots(n-1)^2\}l$$
$$-\{y_2+2y_3+\ldots(n-1)y_n)\}=0$$

whence both a and l may be found.

Referring to Fig. 71, which is of the same shape as Fig. 70, we

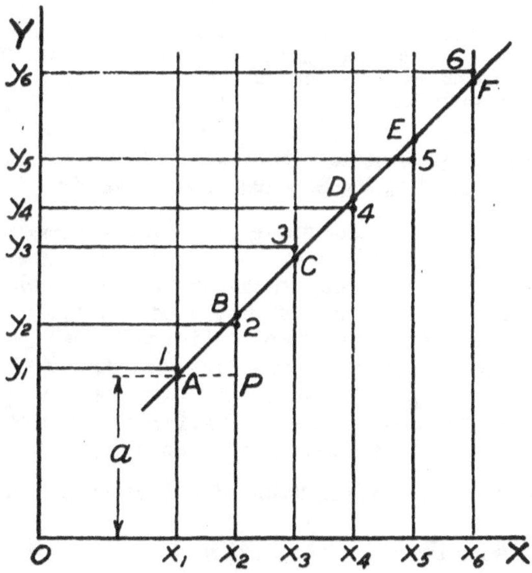

FIG. 71.

see that $x_1A=a$=the true value of the first reading, the abscissæ increase each time by l and the line $ABC \ldots F$ is inclined at 45°. Thus $a-y_1=-A1$, $a+AP-y_2=a+BP-y_2=B2$, and so on, and we adjust to make $(A_11+B2^2+C3^2+\ldots)$ a minimum.

* This section may be left until § 229 has been read.

Example.—Apply the alternative method to the data of the last section. The normal equations are:

$$10a+45l-8532=0$$
$$45a+285l-51874=0$$

whence $l=\frac{26960}{165}=163{\cdot}39$ and $a=117{\cdot}92$.

The same methods can be used for equal changes of length that are due to any uniform cause; for example, a spring elongates by uniform intervals when weights proportional to 1, 2, 3, 4, . . . are hung on it. They are also applicable to intervals of time, *e.g.* the period of a pendulum or of the indicating pointer of a galvanometer or analytical balance; and, indeed, to any variable that is proportional to another quantity which can be varied by equal intervals; thus, the π-disc (§ 44) may be rolled until any given point on its circumference has touched the flat scale at several equidistant points, or the area indicated by a planimeter may be read when the tracing point has been carried around the periphery once, twice, thrice, etc. Usually it is l alone which is required.

Exercise.—Write out the equations of the pendulum problem of example 2, § 150, and solve by the alternative method given above. Take the true time of the first transit as a, the clock readings as $y_1, y_2, y_3, \ldots$, and the true period of the pendulum as l.

228. Comparison of Results obtained by Least Squares with the Results obtained by the Methods of Average given in Chap. XIII.

The application of Least Squares calls for an expenditure of more time than the earlier methods. Unless an experiment has been carefully performed it would in general be preferable to use the simpler methods. The student should compare the methods by reworking the examples of Chap. XIII, § 150, with Least Squares, and the examples of this chapter with the earlier methods. Whether they agree or not depends in many cases on the assumptions made on the possibility of error in the values of the quantities represented by x and y. Thus, if the values of x are not in error and y should be strictly proportional to x, the value of b obtained by $b=\frac{\Sigma y}{\Sigma x}$ in § 150 should agree with the value obtained by § 224, namely $b=\frac{\Sigma(xy)}{\Sigma(x^2)}$.

229. Simultaneous Indirect Measurements.

In certain kinds of measurement several quantities (usually either two or three) have to be determined by methods which will not allow each to be measured separately but which furnish functional relations in which there is always more than one unknown involved. For example, two magnets may have the strengths of their poles determined by measuring their relative intensities as a ratio and their mutual attraction as a product. The data $x/y=a$ and $xy=b$ are then sufficient to determine both x and y. Sometimes only one unknown quantity needs to be measured, but cannot be determined except along with a different one. For example, by the use of an ammeter to measure the flow of electric current that a particular battery can drive through an unknown resistance it is easy to deduce the numerical value of the resistance, but the method must be modified before it can be used to measure the resistance of an "earthed" connection, *i.e.* of the place where the current passes from the negligibly resistant wire into the body of the non-resistant earth, for the current must return from the earth to the battery through some other resistant "ground." When there are two such points, it is possible to find the sum of their resistances but not the value of either one alone. If three separate connections to earth are available, their resistances may be called x, y, and z, and after measuring in turn the resistances $x+y$, $y+z$, and $x+z$, the value of any one of them may be found by solving the three simultaneous equations that are obtained. If there are four such points, however, more separate and independent observations are obtainable than there are unknown quantities, and the determinations will need adjustment by the method of least squares. If the resistances are called w, x, y, and z, then the following data are available :

$$\left.\begin{array}{llllll} w+x & & & =a \\ & x+y & & =b \\ & & y+z & =c \\ w & & +y & =d \\ & x & & +z=e \\ w & & & +z=f \end{array}\right\}$$

These are called *observational equations.*

In general no exact solution will be possible, and the problem is to determine such a solution that the sum of the squares of the deviations of the observed values from the *adjusted values* that are given by the assumed solution shall be as small as possible.

A graphic illustration of the effect of more equations than unknowns is given in Fig. 72, which represents a set of experimental determinations

$$\left.\begin{array}{r} 3x-y=3 \\ y-x=2 \\ 2x-y=0 \end{array}\right\}$$

By plotting these three equations on one diagram it will be seen that the three straight lines do not pass through a single common intersection, but a point (2·4, 4·9) can be found which comes fairly close to satisfying the three conditions.

The **general method** of obtaining the best representative values of the unknown quantities can be shown by the application of the principles of the differential calculus to be as follows :

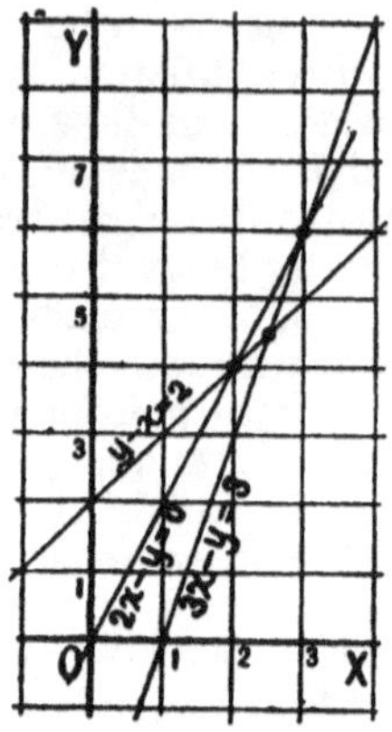

FIG. 72.—THREE OBSERVATIONAL EQUATIONS.

Since all observations contain errors the three equations are not consistent, but the best point to represent the intersection of the three lines can be found by choosing it so that the sum of the squares of the discrepancies is a minimum.

1. Multiply each observational equation by the coefficient of x in that equation, and add the results. The sum is called a *normal equation.*

The equations of Fig. 72

$$\begin{aligned} 3x-y&=3 \\ -x+y&=2 \\ 2x-y&=0 \end{aligned}$$

thus give

$$\begin{aligned} 9x-3y&=\ \ 9 \\ x-y\ &=-2 \\ 4x-2y&=\ \ 0 \\ \hline 14x-6y&=\ \ 7 \end{aligned}$$

2. Multiply each observational equation by the coefficient of y in that equation, and add the results in order to obtain a second normal equation :

$$\begin{aligned} -3x+\ y&=-3 \\ -\ x+\ y&=\ \ 2 \\ -2x+\ y&=\ \ 0 \\ \hline -6x+3y&=-1 \end{aligned}$$

3. If the equations contain third, fourth, fifth, . . . nth unknown quantities, form a normal equation for each one in the

same way, multiplying each equation by its proper coefficient of the particular unknown quantity and adding, thus obtaining n normal equations in n unknown quantities.

4. Solve the normal equations by any algebraical process. The result will be the least-square values of the unknowns.

In the example under consideration this gives

$$x = 5/2,$$
$$y = 14/3.$$

Choose any other point that you think best and show that it gives a larger value for $\Sigma(v^2)$ than this one does. Write the equations in the form

$$3x - y - 3 = v_1,$$
$$y - x - 2 = v_2,$$
$$2x - y = v_3.$$

An *approximate method* of obtaining normal equations is sometimes used in order to avoid the increased labour of applying the above processes, 1, 2, 3, 4, to observational equations that contain decimal fractions, but it is often unsatisfactory: Make the coefficient of x positive in each of the n equations by multiplying through by -1 where necessary; then add the resultant n equations to form the first normal equation. Make the coefficients of y positive in the same way, and add the n equations to obtain a second normal equation. Proceed in this way until n normal equations in n unknown quantities have been formed; then solve. It is fairly obvious that this method will be the most satisfactory in cases where all the coefficients of all the observation equations are of approximately the same size.

Apply the approximate method to the equations of Fig. 72, and note one objectionable feature which it may have.

To what extent are *graphic methods* applicable to the solution of simultaneous indirect measurements? See § 232.

230. Least Squares for Conditioned Measurements when Both x and y are likely to Error.

We now take up the problem postponed from § 223, namely, to find the best values of a and b of the equation $y = a + bx$, which agrees best with the pairs of values (x_1y_1), (x_2y_2), (x_3y_3)—n pairs in all—when both x and y are fallible. Let the x observations have a weight w compared with a unit weight for the y observations.

(1) Write the equations as $y = a + bx$. Suppose the x-values to be without error and solve for b by writing down the n observational equations, forming the two normal equations and solving as in § 229 or by, what comes to the same thing, using equation (2) of § 223. Denote this value of b by b_1.

(2) Write the equations as $x = -\frac{a}{b} + \frac{1}{b}y = c + dy$, and regarding the y values as being without error find the value of d by the same method or methods. Denote the reciprocal of d by b_2.

(3) Solve the quadratic equation

$$b^2 - \left(b_2 - \frac{w}{b_1}\right)b - w = 0$$

for b. There will be no doubt which value to take. This value is the most probable value of b.

(4) Get a by adding up the n equations, a process which yields

$$\Sigma y = na + b\Sigma x$$

and substituting for b. This value of a is the most probable.*

Example.—Take the following values of x and y and assign to the x values a weight of $\frac{1}{2}$ compared to unity for y.

x	1	3	6	8
y	2	3	4	5

To save time simple numbers have been chosen.

(1) x without error. The observation equations are:

$$\begin{aligned} 2 &= a + 1b \\ 3 &= a + 3b \\ 4 &= a + 6b \\ 5 &= a + 8b \end{aligned}$$

The normal equations are

$$\begin{aligned} 14 &= 4a + 18b \\ 75 &= 18a + 110b \end{aligned}$$

These give $b_1 = 24/58 = 0{\cdot}4138$.

(2) y without error. The observational equations now read

$$\begin{aligned} 1 &= c + 2d \\ 3 &= c + 3d \\ 6 &= c + 4d \\ 8 &= c + 5d \end{aligned}$$

The normal equations are:

$$\begin{aligned} 18 &= 4c + 14d \\ 75 &= 14c + 54d \end{aligned}$$

These give $d = \frac{24}{10}$ or $b_2 = \frac{10}{24} = 0{\cdot}4167$.

* For a fuller treatment see M. Merriman's *Method of Least Squares*, p. 127; also H. S. Uhler, *Journal of the Optical Society of America*, November, 1923, p. 1043.

(3) $$b^2-\left(\frac{10}{24}-\frac{\frac{1}{2}}{24/58}\right)b-\tfrac{1}{2}=0$$

or $$b^2+\tfrac{10}{24}b-\tfrac{1}{2}=0. \quad \therefore\ b=0{\cdot}4145.$$

(4) $$14=4a+18\times 0{\cdot}4145. \quad \therefore\ a=1{\cdot}635.$$

Therefore $$y=1{\cdot}635+0{\cdot}4145x.$$

231. Conditioned Measurements in a Triangle.

Suppose that the three angles of a plane triangle are measured independently. Their sum should be exactly 180°. In practice the measurements will add up to either a little greater or a little less than 180°, and the problem is to adjust them to get the most probable values which will add up to 180°.

Let A', B', C' be the measured values (of equal weights, suppose) and A, B, C the most probable values. We have the (approximate) observational equations:

$$\begin{aligned} A&=A' \\ B&=B' \\ A+B&=(180-C') \end{aligned}$$

Find the normal equation for A by multiplying the equations by the coefficients of A, *i.e.* by 1, 0 and 1, respectively, and adding. We get

$$2A+B=180+A'-C'$$

Find the normal equation for B. It is

$$A+2B=180+B'-C'$$

Solve these equations for A and B.

$$\begin{aligned} A&=A'+\tfrac{1}{3}\{180-(A'+B'+C')\} \\ B&=B'+\tfrac{1}{3}\{180-(A'+B'+C')\} \end{aligned}$$

and hence $$C=180-(A+B).$$

Example.—$A'=50°\ 37'$, $B'=70°\ 26'$, $C'=59°\ 3'$.

The observational equations are:

$$\begin{aligned} A&=\ 50°\ 37' \\ B&=\ 70°\ 26' \\ A+B&=120°\ 57' \end{aligned}$$

The normal equations are:

$$\begin{aligned} 2A+B&=171°\ 34' \\ A+2B&=191°\ 23' \end{aligned}$$

whence $A=50°\ 35'$, $B=70°\ 24'$, and consequently $C=59°\ 1'$.

We see that we subtract from each angle one-third of the excess of their sum over 180°.

Show that this method makes Σv^2 a minimum.

The student might say this is an obvious correction and that it is not necessary to call in the aid of Least Squares. If the measurements of the angles had not equal weight the correction would not be obvious, but Least Squares will supply additional equations for this work and the readings of the angles can be adjusted to fit the new conditions.

232. The use of a Graph for solving a Problem in which there are more Equations than Unknowns.

We have just seen one use (Fig. 72). Cases, however, frequently arise in which a graph affords an elementary student a quicker and

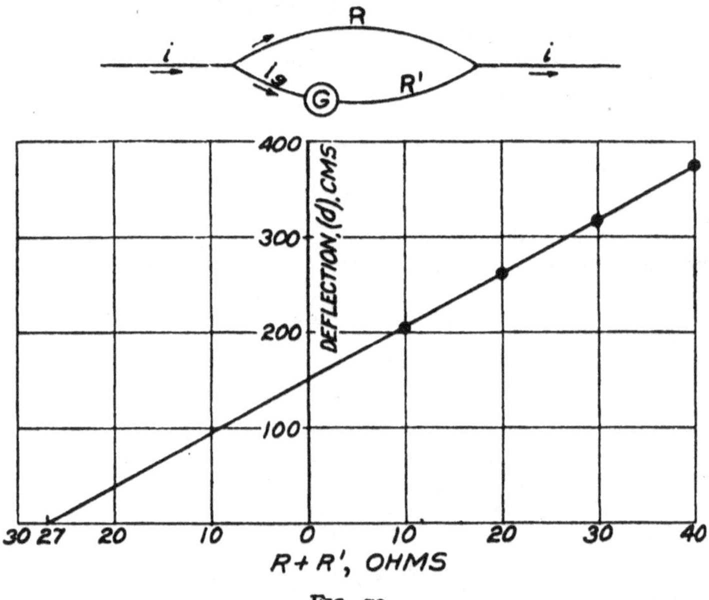

FIG. 73.

better way of getting an unknown. For example, in the following method of getting the resistance of a galvanometer a constant total current (i) was sent through a divided circuit (Fig. 73). One arm contained a constant resistance R, the other arm contained the galvanometer (resistance G), and a resistance box from which a resistance R' could be unplugged. The galvanometer current therefore $=i\dfrac{R}{R+R'+G}$.

The deflection (d) of the galvanometer was read on a scale,

$$\therefore\ d \propto \frac{R}{R+R'+G}.$$

To get G algebraically, we have the equations

$$\frac{d_1}{d_2}=\frac{R+R'_2+G}{R+R'_1+G}$$

and by combining the d's in pairs we can get G. The values obtained usually differ considerably, so that this method is poor.

If, however, we throw the equation into the form

$$R+R'+G \propto \frac{1}{d}$$

we see that plotting the values of $(R+R')$ as abscissæ against the values of $\frac{1}{d}$ as ordinates should give a straight line which will cut the axis of $(R+R')$ at a distance to the left of the origin equal to G. The advantage of the method is that the " black thread " or " least squares " will give only one value of G, and this is the best value that can be got from all the readings.

Example.—In an experiment the value of $R=10$ ohms. The values of R' and d are given in the first two columns of the table and

R'	d	$R+R'$	$1000/d$
0	48·8	10	205
10	37·8	20	265
20	31·3	30	319
30	26·7	40	375

the values of $(R+R')$ and 1,000 times the reciprocal of d (obtained with the slide rule) in the last two columns. Show that the graph gives $G=27$ ohms (see Fig. 73).*

As a test of the convenience of the graphical method try the algebraical method, taking any two readings at a time.

Exercises XIX

1. The sum of the squares of the deviations of any particular set of variates is least when the deviations are obtained from the arithmetical mean than when they are obtained from any other value of the variate (the Principle of Least Squares). Test this with the numbers 182, 183, 184, 185, 186, 187, 188.

2. Tabulate the values of x and y for the black thread experiment of § 108. Calculate the values of a and b from the formulæ and write the equation in the form $y=a+bx$, and then in the form $x/m+b/n=1$.

Compare the calculated values of the intercepts with your experimental numbers that there obtained in your work on graphic analysis (Chap. IX).

* The ordinate scale of Fig. 73 should read " Reciprocals of deflections."

3. The following results were obtained in an experiment with a machine :

E	*R*	*E*	*R*
pds.	pds.	pds.	pds.
2·5	0	13·0	60
6·5	20	15·5	80
8·5	40	18·5	100

Find by least squares, the half-table method, and by a graph the value of a and b in the equation

$$E = a + bR$$

4. In studying the density of water (§ 111) the ratio of $\sqrt{y}$ to x was found graphically to be about 0·83. Determine the accurate value of this ratio by calculating each product of the variables x and $\sqrt{y}$, summating, and dividing by the summated squares of the x's.

5. Two quantities are known to be exactly proportional to each other. Their measured values are :

x	0	1	2	3	4	5	6
y	0	1·0	1·5	2·5	3·5	4·0	5·0

Find the best value of $\frac{y}{x}$.

6. Apply least squares to get the best representative value for the subdivisions given in the difference column of the table of § 226. Also apply the half-table method.

7. Solve the observational equations

$$x + y = 10$$
$$4x - y = 19$$
$$2x + 3y = 25$$

by both methods of § 229.

8. There are three points, X, Y, Z, above a datum level. It is not possible to measure the height of Z directly, so it was measured with respect to X and Y and the following equally reliable readings of the heights were taken :

$x = 7$ feet, $y = 10$ feet, $y - x = 6$ feet, $x - z = 3$ feet, $y - z = 7$ feet.

Find the best values of the heights of the three points.

9. In the example of § 230 work out the value of a and b :
(1) If the x's are absolutely correct and the y's only are in error ;
(2) If both x and y are in error and their observations have equal weight ;
(3) If the y's are absolutely correct and the x's only are in error.

10. Given the following pairs of values of x and y, all the observations having equal weight, find the best equation connecting them.

(*a*) x	4	6	8	9
y	5	8	10	12

(*b*) x	1	3	6	7
y	2	3	5	6

Check graphically.

11. Using the values given in § 150, example I, namely:

x	2	5	7	11	13	15
y	1	3	4	6	7	8

Calculate b by least squares and compare with the previous result.

12. Using the values given in § 151, example 2, namely

x	1	3	8	10	13	15	17	20
y	3	4	6	7	8	9	10	11

Calculate a and b by least squares and compare with the values obtained in § 151.

13. Solve the following equations by the method of least squares

$$\begin{aligned} x+2\cdot65y-0\cdot33z&=6\cdot21, \\ x+2\cdot37y+0\cdot12z&=3\cdot18, \\ x+2\cdot25y-0\cdot29z&=5\cdot89, \\ x-0\cdot87y+3\cdot27z&=4\cdot28, \\ x+3\cdot38y+\quad z&=4\cdot07. \end{aligned}$$

Be careful that each product is given the right sign; check them as each equation is completed. If squared paper is not used be careful to keep the decimal points under one another. The tedium of adding positive and negative decimals can be greatly relieved by *overlining each negative digit* * and adding *en masse*.

Apply the approximate method to the last exercise and compare the ease of computation and the accuracy of solution.

14. Given n observational equations of the form shown in § 223, viz.

$$a+bx_n-y_n=0$$

and wishing to solve for a and b we form the normal equations for a and b, namely

$$\begin{aligned} na+b\Sigma x-\Sigma y&=0 \\ (\Sigma x)a+b\Sigma(x^2)-\Sigma(xy)&=0 \end{aligned}$$

respectively. Show that the values of a and b obtained by solving these simultaneous equations are those given in § 223.

15. In Searle's rotating cylinder experiment in viscosity there is an end correction k to be applied to the length l of the cylinder to fit the equation

$$y=m(l+k).$$

l and y have the following experimental values (l is set and y is measured):

l	1·30	2·35	3·40	4·30	5·30	6·40
y	837	1332	1794	2184	2655	3120

Find, by plotting y against l, the value of k and deduce the separate values of m. Compare the mean value of m with the value obtained from the slope of the graph.

16. Practise the method of least squares for consecutive equal intervals on some of the problems of Chap. XIII, for example § 150 (1).

---

* Find the value of $2\cdot4123-3\cdot6418+5\cdot7827$ column by column as follows:

$$\begin{array}{r} 2\cdot4123 \\ \bar{3}\cdot\bar{6}\bar{4}\bar{1}\bar{8} \\ 5\cdot7827 \\ \hline +4\cdot5532 \end{array}$$

17. The magnet of a Kew magnetometer was set in vibration and the time of passage through the centre was taken with a half-seconds chronometer. Results as follows :—

Magnet moving to the right.				Magnet moving to the left.			
Number of vibration.	Time of transit.	Number of vibration.	Time of transit.	Number of vibration.	Time of transit.	Number of vibration.	Time of transit.
	′ ″		′ ″		′ ″		′ ″
0	2 0·0	100	9 9·0	5	2 21·0	105	9 30·5
10	2 42·7	110	9 51·7	15	3 4·0	115	10 13·5
20	3 25·5	120	10 34·8	25	3 47·0	125	10 56·5
30	4 8·5	130	11 17·7	35	4 30·0	135	11 39·4
40	4 51·5	140	12 0·4	45	5 12·5	145	12 22·0
50	5 34·5	150	12 43·5	55	5 55·7	155	13 5·0
60	6 17·2	160	13 26·5	65	6 38·7	165	13 48·0

Find the time of vibration.

18. Humphreys and Abbot measured the velocity of the Mississippi River at different depths. Find the law connecting velocity and depth.

Feet from surface	0·0	0·1	0·2	0·3	0·4	0·5	0·6	0·7	0·8	0·9
v in feet per second	3·1950	3·2299	3·2532	3·2611	3·2616	3·2282	3·1807	3·1266	3·0594	2·9759

(Hint: The graph is parabolic. Try $v=A+Bx+Cx^2$).

19. Theory gives the equation between the maximum vapour pressure of a liquid and the temperature in the form

$$\log p=A-\frac{B}{T}-C\log T$$

where p and T are the pressure and absolute temperature, respectively, and A, B, C are constants. Using the data of Question 30, Exercises IX, find A, B, C. (Hint: Treat $\frac{1}{T}$ as an x and $\log T$ as a y.)

20. Rankine suggested the vapour pressure equation

$$\log p=A-\frac{B}{T}-\frac{C}{T^2}$$

Find A, B, C by least squares from the data of Exercises IX, Question 30.

CHAPTER XX

SYSTEMATIC AND CONSTANT ERRORS

233. Definitions.

It has been shown that errors may be either accidental or constant. There is another class of errors, often included under the term constant errors, in which the error is not actually constant, nor does it vary according to the law of probability. This is the class of *systematic errors*, or errors that undergo a more or less regular change during the course of making a set of measurements. They may be subdivided into *progressive errors*, which show a steady increase (or decrease) from one determination to the next, and *periodic errors*, which increase for a number of measurements, then decrease, and then repeat the previous cycle or period.

234. A Test for Systematic Errors.

Where systematic errors are absent a comparison of any measurement of a series with the preceding one will tend to show an increase in the numerical value about as often as a decrease ; a fact that can easily be tested by writing between each two successive values a plus sign, a minus sign, or a zero, according as the second value is respectively greater than, less than, or equal to, the first, and then comparing the number of the plus signs with that of the minus signs.

Where progressive errors have been greater than accidental errors there may be all plus signs or all minus signs as the result of applying the test. If the accidental errors are relatively large they will probably cause several of the signs to be plus and several minus, but the presence of a progressive error at the same time will cause one sign to appear more frequently than the other.

If the systematic errors are periodic there will be alternate groups of plus signs and minus signs, as is shown in the next table.

235. Example of a Systematic Error.

In an experiment in which water in a reservoir was drawn up into a tube by suction and successive readings of its height were made, values having the following decimals were obtained in order: 0·76, 0·74, 0·70, 0·62, 0·63, 0·61, 0·55, 0·56, 0·51, 0·50, 0·44, 0·44, 0·39, 0·40, 0·35, 0·35. Are the results probably affected by progressive errors, or periodic errors, or neither? Use a graphic diagram if the question is hard to answer. What effect would you expect to result from a slight leakage of air into the upper part of the tube?

236. Example of a Periodic Error.

If the pivot of the second hand of a watch is not exactly in the centre of the dial the indicated seconds will be subject to a periodic error. For example, if it is located too far to the right the hand may indicate 29 instead of 30 when it points downward, and 1 instead of 60 when it is pointing upward. That is, it will have a periodic error which will be a maximum (positive) at 60 seconds, a minimum (negative) at 30 seconds, and zero at 15 and 45 seconds (Fig. 74).

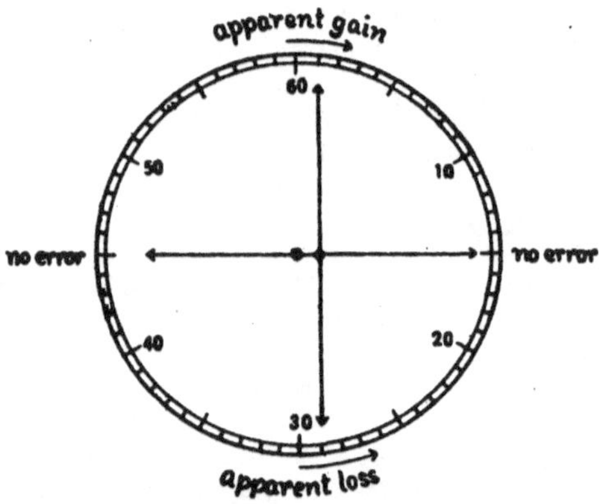

FIG. 74.—ILLUSTRATION OF A PERIODIC ERROR.

An eccentric clock-hand will appear to be ahead of time, accurate, behind, accurate, ahead, etc., in the course of its rotation. A determination of the times at which the gain and the loss are most marked will enable the direction and amount of displacement to be found.

From the illustration it is evident that the direction of displacement must be toward a point halfway between the positions at which the error is greatest and least.

Stand where you can hear the clock beat seconds and read the time indicated by your watch. Every seven seconds as indicated by the clock read the seconds and estimated tenths of a second from the watch and state the result to another student, who will take down the values in his notebook. After three or four minutes change places with him and note down the time as he reads it off. Every seven-second interval should have its time by the watch noted, for a full period of seven minutes. It is advisable to practise the procedure beforehand until you are sure that you can estimate tenths of a second with reasonable accuracy. If your estimates are predominantly 0·5 and 0·0 the results will not be satisfactory. In order to avoid losing count of the (audible) seconds while you are stating the time it may be advantageous to choose seven of your fingers and tap on the table with each in turn at one-second intervals. Concentrate your sense of sight on the watch and your sense of hearing on the beats of the clock.

See that you have the complete table of sixty values in your own notebook, and mark the observed tenths of a second with a plus sign where they increase from one observation to the next and with a minus sign where they decrease, as shown below. With most watches it will be found that the second hand is not pivoted in the exact centre of the graduated circle and the periodic error will be shown very distinctly.

hr. min. sec.		*hr. min. sec.*		*hr. min. sec.*	
4 : 37 : 65·2	+	4 : 40 : 25·8	–	4 : 42 : 45·2	–
38 : 12·3	+	32·6	–	52·0	+
19·6	O	39·3	–	59·1	+
26·6	–	46·1	–	43 : 06·3	+
33·5	–	53·0	+	13·5	+
40·3	–	60·2	+	20·6	+
47·1	O	41 : 07·4	+	27·8	–
54·1	O	14·6	+	34·6	–
39 : 01·1	+	21·7	O	41·3	–
08·2	+	28·7	–	48·1	–
15·4	+	35·6	–	55·0	+
22·6	O	42·2	–	44 : 02·3	+
29·6	–	49·1	O	09·4	+
36·4	–	56·1	+	16·6	+
43·1	–	42 : 03·2	+	23·8	–
50·0	O	10·3	+	30·7	–
57·0	+	17·5	+	37·4	–
64·2	+	24·7	–	44·1	O
40 : 11·4	+	31·6	–	51·1	O
18·6	+	38·4	–	58·1	+
				4 : 45 : 05·2	

APPARENT TIME OF SEVEN-SECOND INTERVALS.

The tabulated numbers are the times indicated by a watch at audible intervals that were known to be exactly seven seconds. The presence of a periodic error is shown by the tabular differences occurring in alternate groups of positive and negative values.

Draw a graphic diagram in which the abscissæ represent the integral part of the number of seconds in your table, and the ordinates represent the corresponding tenths of a second (Fig. 75). Draw a smooth curve to eliminate accidental errors in the determination of time.

Determine the direction and the amount of the displacement, and summarize the result by stating "the pivot of the second hand of the watch is displaced toward the figure . . . of the dial by an amount equal to the length of . . . seconds' division on the graduated circle."

Explain how the periodic error can be eliminated in case such a watch is used for determining intervals of time.

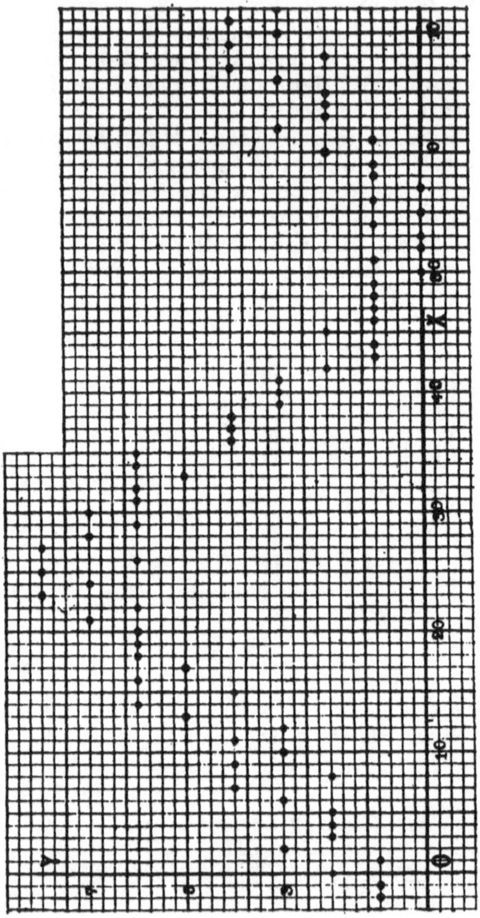

FIG. 75.—PERIODIC ERROR OF A CLOCK-HAND.

Each measurement of time that occurs in the previous table has its whole number of seconds represented by an x-value and its tenths of a second by a y-value.

237. Example of a Progressive Error.

The list of figures given in § 235 was obtained from a determination of specific gravity by Hare's method. If the lower ends of two upright tubes (Fig. 76) dip into two separate reservoirs while their upper ends are both joined to a third tube from which the air can be partially exhausted it can easily be proved that the heights to which the fluids are raised will be inversely proportional to their densities; so that if a fluid whose density is unity is raised to a height h_1, and a heavier fluid to a lesser height h_2, the density or specific gravity of the latter must be h_1/h_2. The complete list of determinations of height included readings of both columns of liquid; they were made at approximately equal intervals of time, and in the order in which they are given in the table, viz. 75·76, 73·06, 75·74, 73·04, 75·70, 73·00, 72·98, 72·95, 75·62, etc.

Pure water.	Salt solution.
75·76	
	73·06
75·74	
	73·04
75·70	
	73·00
	72·98
	·95
75·62	
	·94
·63	
·61	
	·89
	·88
·56	
·55	
	·84
	·84
·51	
·50	
	·79
	·78
·44	
·44	
	·75
	·70
·39	
·40	
	·77
	·77
	·77
·35	
·35	

EXAMPLE OF HEIGHTS IN HARE'S METHOD.

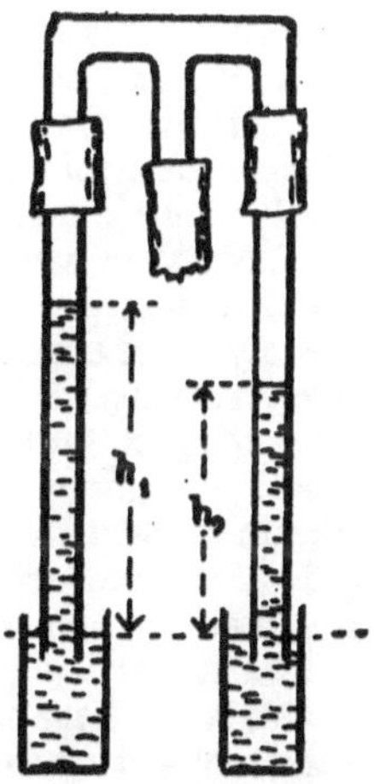

FIG. 76. — HARE'S METHOD OF BALANCING COLUMNS.

The heights of the two liquids are inversely proportional to their densities.

If the density is calculated by dividing 75·76 by 73·06 it is evident that the progressive error will make the resulting figure too large, for the height of the water had fallen somewhat below 75·76 when the reading of the salt solution, 73·06, was taken; and if 75·74 is divided by 73·06 the progressive error will make the result too small, for the salt solution did not stay at 73·06 while the reading 75·74 was being taken. Obviously the average of 75·76 and 75·74 must be divided by 73·06, or 75·74 must be divided by the average of 73·06 and 73·04, or some other combination used in which the *average* height of one column of liquid must have occurred at the same time as the average height of the other. This method of eliminating progressive errors is used in the process of weighing with a delicate balance and in many other processes of physical measurement.

What set of values near the end of the table can be used in the same way? Make five different calculations of density from successive parts of the table and see whether they show any evidence of *progressive* error.

238. Constant Errors.

It has already been stated that constant errors are more troublesome than accidental errors and that the latter give very little aid in determining the former. It is not the target (§ 156) that is found from individual measurements but only the centre of clustering, and *characteristic deviations show only how close determinations come to one another, not how close they come to the truth.*

Some constant errors are easily corrected with the aid of theoretical considerations; others may be very difficult to eliminate. Unfortunately there is no infallible rule for detecting them, and each experimental problem has its own special sources of error. The two beam-arms of a balance may be unequal, so that all weighings are proportionately erroneous; the end of a metre-stick may be worn, so that every setting of the zero-point is inaccurate; the neutral tint of litmus may be faultily judged, so that a chemical determination is biased. Consider such a simple process as the determination of atmospheric pressure with a mercurial barometer. The vacuum at the top is never perfect and there is often capillary action, both making the reading too low. If the barometer and its attached scale do not hang vertically every apparent reading will be too high. The scale itself is too long or too short except at a single temperature, and the mercury may have its accepted standard density only at a different temperature from the one that it has when the observation is made. Even if its density is standard the height of a column that will give a definite pressure will depend upon the strength of gravitational attraction and this varies with the latitude and altitude of the instrument. If an aneroid barometer is to be used instead of a mercurial one its mechanism introduces still more sources of error.

It is evident that the amount of constant error will generally be varied by changing observers, apparatus, methods, and times of observation; and the more radically different the sets of conditions are made the better, in all probability, will be the mutual neutralization of constant errors when the weighted average is taken. In practice, the values for most of the constants of nature have been obtained under such varying conditions. Atomic weights are obtained from various inter-relations of chemical compounds obtained from different sources and by different methods. The surface tension of water may be measured by the capillary ascent method, by the hanging drop method, by the capillary wave method, by the vibrating jet method, etc. The size of the molecules of a gas may be calculated from the rate at which heat is conducted through them, from the *covolume constant*, *b*, of Van der Waal's equation, from experimental determinations of the viscosity of the gas, from measurements of the maximum density obtainable by cooling and liquefying or solidifying it, etc. If various determinations agree closely in spite of the employment of essentially different methods it becomes more probable that constant errors have been satisfactorily removed, but it can never be certain that all of these methods have not some common source of error which would be eliminated only by using some entirely different method. Constant watchfulness, as stated in §§ 155, 156, and the exercise of good judgment are of the greatest importance in guarding against constant errors. If the student takes up further courses that involve accurate measurement he will usually find that various " sources of error " which have been found by previous experimenters will be explicitly stated. Many of them will be sources of constant error, and both his natural ability and his progress in learning will be put to the test in his management of them.

Exercises XX

1. Show that the use of a tangent galvanometer is most accurate when the angle of deflection is about 45°, *i.e.* an error of reading has less effect on the result when the deflection is 45°, than if greater or less than 45°.

2. Show that a Wheatstone bridge method of measuring a resistance is most accurate when the four arms of the bridge have equal resistance.*

3. Show that in measuring the focal length of a convex lens by observing object and image distances an error in observation has least effect when these distances are equal.

---

* For further discussion see S. W. Holman's " Precision of Measurements," and Stewart and Gee's " Elementary Practical Physics," vol. ii. pp. 448–450.

CHAPTER XXI

APPLICATIONS TO BIOLOGY

(A) FREQUENCY PROBLEMS

239. THE frequency curve obtained by the biologist is often unsymmetrical. The median mean and mode do not coincide, therefore the formula of the curve of frequency or errors and the theory of least squares are harder to apply. Consequently the biologist has to be content with an approximate application of the mathematical theory or follow Pearson and others who have sought to bring all the irregularities of the biological curves within the realm of mathematical argument. Again, the physicists or chemists do know that once they have got rid of their constant errors their observations, if numerous enough, must fit the normal curve; but the biologist or statistician generally does not know what selection has been going on amongst his variates, and some of his frequency curves are very hard to analyse.

The median and mode have proved more useful in biological study than the arithmetical mean. They are not affected so much as the last by extreme cases. Another advantage which they possess —from the biologist's point of view—is that they can be obtained with less arithmetic. If a large number of variates are arranged in order (of size, for example), the median can be selected by eye, and also the quartiles. The semi-interquartile range gives the dispersion (or probable error) reckoning from the median, and for much biological work this value is sufficiently accurate.

The mode also being the value possessed by the most common class is of great importance statistically, as, for example, the mode in coins issued by the mint, or postal orders stocked by a post office, or the stock sizes in clothes carried by the tailors and department stores.

240. The following three examples taken from biological text-books illustrate the difference in the type of problem.

(*a*) The veins of 2,600 beech leaves were counted with the following results :—

No. of veins	9	10	11	12	13	14	15	16	17	18	19	20	21	22	23
No. of leaves	1	4	7	40	125	320	480	600	520	305	140	40	16	2	0

A more exact value of the median in such a case is found as follows:

Divide the leaves into three classes, approximately equal in number. Let the number in the lower class be a, in the middle b, and in the upper c. In this case the middle class may be taken as the 16 class, and $\therefore b=600$, the lower class as everything below 16 and $\therefore a=977$, and the upper class everything above 16 and $\therefore c=1023$.

If instead of mere numbers (of veins) we had measurements, everything between $15\frac{1}{2}$ and $16\frac{1}{2}$ would be called 16. So that $15\frac{1}{2}$

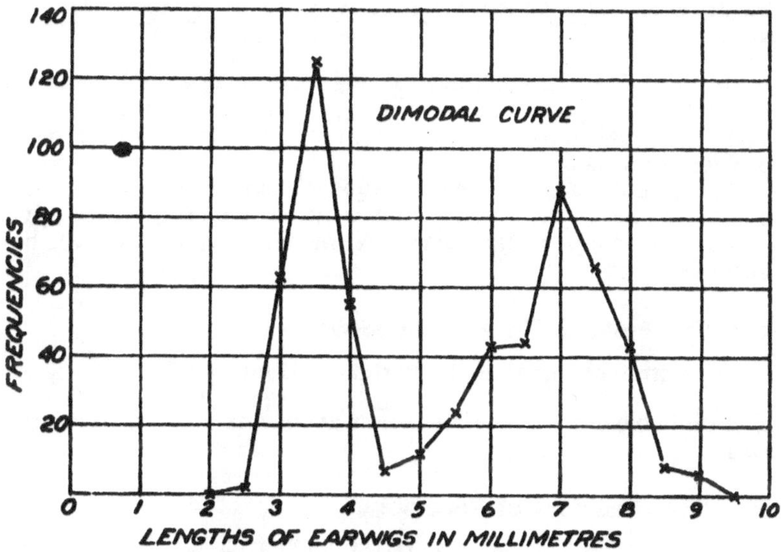

FIG. 77.

may be taken as the lower limit of the 16 class. Denote this by e. Denote the upper limit by f. The empirical rule now applied is

$$\frac{\frac{1}{2}n-a}{b}=\frac{x}{f-e}$$

where x=distance of the median above the lower limit of the middle class.

Applying this rule

$$\frac{1300-977}{600}=\frac{x}{16\frac{1}{2}-15\frac{1}{2}} \quad \therefore x=0{\cdot}54$$

Therefore the median$=15{\cdot}5+0{\cdot}54=16{\cdot}04$.
Straightforward arithmetic gives the A.M.$=16{\cdot}03$.

(*b*) The petals in the flowers of marsh marigolds were counted. None had less than 5 petals, none more than 8 petals.

No.	300	85	26	5
No. of petals	5	6	7	8

The mode is at the end of the frequency curve and it is evident that all our formulæ are useless in a problem like this. (Plot the frequency curve; it is J-shaped.)

(*c*) Bateson measured the lengths of the forceps of the earwigs on a certain island. The results were:

Lengths in mm.	2	2½	3	3½	4	4½	5	5½	6	6½	7	7½	8	8½	9	9½
No. of earwigs	0	1	63	125	55	7	12	24	43	44	88	66	43	8	6	0

In this table a length 3 mm. means between 2¾ and 3¼, and so on.

Plot the frequency curve (Fig. 77). It has two modes. Such variation is called ***dimodal***, and it gives the biologist an interesting problem to unravel.

In some dimodal biological frequency curves the modes are far apart (as in Fig. 77); sometimes they are closer together, giving merely two humps to the curve. Trimodal and higher curves have also been found.

241. The Probable Error of the Median.

$$\text{The probable error of the arithmetical mean} = \text{PE} = \frac{pe}{\sqrt{n}} \quad (\S\ 186).$$

The probable error of the median is about one and a quarter times this.

242. The Coefficient of Variability.

This term is used largely by the biometrician. It is a measure of the scattering of the variates, and therefore increases with the probable error or standard deviation. It is usually expressed as a percentage, and its value is given by the equation

$$\text{coefficient of variability} = \frac{\text{standard deviation}}{\text{arithmetical mean}} \times 100$$

It may be noted here that the statistician uses the standard deviation more than the dispersion or probable error, possibly because it can be calculated in nearly all cases and does not presume a "normal" frequency curve.

243. The Equation of the Normal Frequency Curve.

To fit the equation $y = be^{-(ax)^2}$ to the normal curve use the formula

$$\text{dispersion (or probable error)} \times a = 0{\cdot}4769$$

Therefore once the dispersion is found the a can be calculated. The b is of course the maximum height of the curve.

In the curve of § 202 the dispersion is 10, therefore

$$10 \times a = 0{\cdot}4769 \quad \therefore a = 0{\cdot}04769$$

Also the maximum height of the curve = 50, therefore the equation of the curve is

$$y = 50e^{-(0{\cdot}04769x)^2}$$

In the normal curve of § 248 the dispersion is approximately 1·65 inches, and the maximum height of the curve is 15·9 per cent.; therefore the equation of the curve is

$$y = 15{\cdot}9e^{-(0{\cdot}289x)^2}$$

244. Symmetry and Asymmetry.

If the mean is not equal to the median or mode the frequency curve is said to be asymmetrical. If the mean is greater than the mode, the curve is said to have positive skewness (as in Fig. 61). If the mean is less than the mode, the curve has negative skewness.

The asymmetry is measured approximately by the fraction

$$\frac{\text{Mean} - \text{Mode}}{\text{Standard Deviation}}$$

Professor K. Pearson has shown how to apply mathematics to the skew curves so common in biometrical measurements, and how to get mathematical equations to fit such skew curves.

245. Correlation.

Related variables may increase or decrease together. In this case the correlation is said to be *positive*. If one increases as the other diminishes the correlation is said to be *negative*. The greatest value of the correlation is unity, the least zero.

Examples of positive correlation are (1) ages of husbands and wives; (2) heights of fathers and sons; (3) yield of grain from one wheat plant and the number of culms per plant.

The problem of finding the coefficient of correlation may be treated graphically or algebraically.

(*a*) Let us take No. 2 in detail and try the geometrical treatment. Pearson measured the statures of fathers in classes of an inch range and the average heights of the sons of the fathers in each class.

The results are as follows, all measurements in inches:

Statures of fathers	61	62	63	64	65	66	67	68	69	70	71	72	73	74	75
Statures of sons	65½	64½	65¾	65¾	66¼	66½	66¾	68	68½	68¾	69½	70	71	70½	70¾

Draw two axes as shown (Fig. 78) and mark them with the heights of fathers and sons.

The average of all the fathers and all the sons happens to be 68 inches (see Exercise XXI, Question 12 for further details), so draw vertical and horizontal lines, respectively, at 68 on both axes. Plot the graph and we get the line AB.

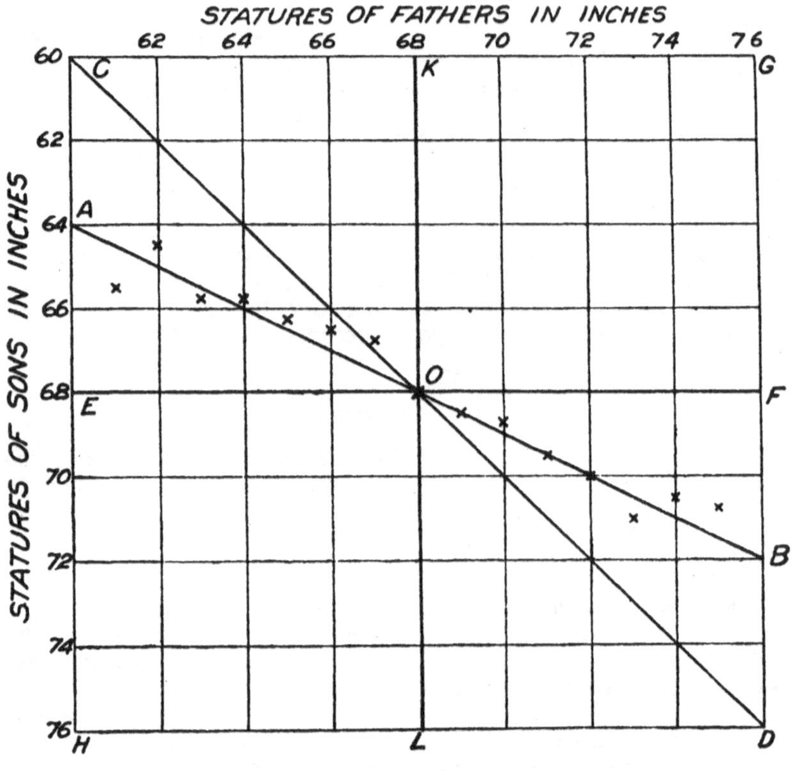

FIG. 78.

Now if fathers of height 62 had sons of height 62 and fathers of any height, H say, had sons of the same height, the graph would have been CD, a line at 45° to the horizontal. In this case the correlation would have been unity. If the sons of each class of father had an average height of 68 the graph would have been EF. In this case, there would be no relationship between the height of the fathers and the average heights of their sons, and the correlation would have been zero. As it happens, the line is AB and the tangent of the angle AOE is $\frac{1}{2}$. Therefore the coefficient of correlation is one-half.*

This example affords an instance of *regression.* For fathers away from the average of 68, their sons' heights on the average are half-way *towards* 68 ; that is, the sons' heights do not depart from the general mode as much as do their fathers'.

Question.—If the coefficient of correlation between father and son is $\frac{1}{2}$, what would you expect the value of the coefficient of correlation between mother and son to be ?

246. Consider the four quarters into which the diagram is divided. If the plotted sons' heights fall in $CKOE$ or $OFDL$ there is positive correlation, if they fall in $EOLH$ and $KGFO$, the correlation is negative. This gives a clue to the algebraical method of correlation.

A numerical example from Babcock and Clausen's "Genetics in Relation to Agriculture," and due to Love and Leighty, will make things clearer. A sample of 400 oat-plants were taken. The number of culms (*i.e.* stems) per plant were measured, also the yield per plant ; the results are set out in the table.

	Number of culms per plant.						No. of plants.
—	2	3	4	5	6	7	
Yield in grams per plant. 0–1 or $\frac{1}{2}$	3	—	—	—	—	—	3
1–2 or $1\frac{1}{2}$	28	19	3	—	—	—	50
2–3 or $2\frac{1}{2}$	18	66	20	1	—	1	106
3–4 or $3\frac{1}{2}$	1	42	58	7	1	—	109
4–5 or $4\frac{1}{2}$	—	7	59	11	3	—	80
5–6 or $5\frac{1}{2}$	—	—	26	14	2	—	42
6–7 or $6\frac{1}{2}$	—	—	—	4	3	—	7
7–8 or $7\frac{1}{2}$	—	—	1	1	—	—	2
8–9 or $8\frac{1}{2}$	—	—	—	—	1	—	1
No. of plants	50	134	167	38	10	1	400

Giving yields between 0 and 1 gram there are 3 plants and they each have 2 culms. Giving yields between 1 and 2 grams there are 50 plants. Of these 28 have 2 culms, 19 have 3, and 3 have 4 culms, and so on.

* If a line such as AB were not straight, the calculation of the coefficient of correlation would be more difficult.

The correlation is positive because the general trend of the bodv of figures is from top left to bottom right.

Writing now x's for number of culms and y's for yields, make a rectangle (Fig. 79) with a central vertical line drawn at A_x the average of all the x's and a central horizontal line at the average of all the y's. Let Δ_x be the distance of any x variate from the mean. Write it +ve if $A_x > a_x$ and −ve if $A_x < a_x$. Similarly for the y's. In the top left quarter the product of any Δ_x by any Δ_y is

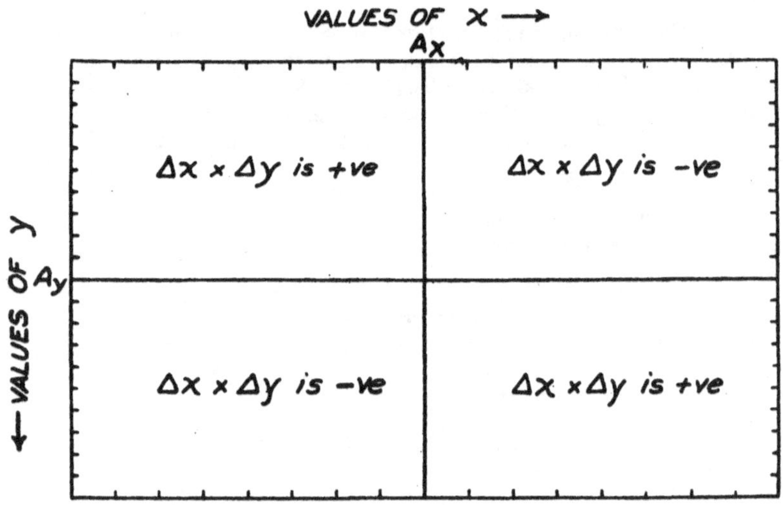

FIG. 79.

+ve, similarly in the bottom right quarter. In the other two quarters the product is −ve. The algebraical sum of all these products divided by the total number of variates is called the average *product-moment*, and this divided by $\sigma_x \times \sigma_y$, where σ_x, σ_y are the standard deviations in the x and y values, respectively, is called the coefficient of correlation.*

$$I.e.\text{ average product moment} = \frac{\Sigma \Delta_x \Delta_y}{n}$$

$$\text{coefficient of correlation } r = \frac{1}{\sigma_x \sigma_y}\left(\frac{\Sigma \Delta_x \Delta_y}{n}\right)$$

$$\text{and the probable error of } r = 0{\cdot}67\ \frac{1 - r^2}{\sqrt{n}}.$$

* This definition is due to Pearson. It makes the coefficient of correlation deduced by this method comparable with that deduced by the other methods.

Leaving the oat problem for the present, let us take the simpler hypothetical problem shown in the next table.

		x-values of variates.							$f(y)$
	—	1	2	3	4	5	6	7	
y-values of variates.	1	1	1	2	1	—	—	—	5
	2	2	2	3	2	1	—	—	10
	3	—	1	2	5	3	1	—	12
	4	—	—	1	3	2	2	2	10
	5	—	—	—	1	2	1	1	5
	$f(x)$	3	4	8	12	8	4	3	42

Find the mean value of x, also σ_x. Arrange the work thus:

f	x	fx	v_x	v_x^2	fv_x^2
3	1	3	3	9	27
4	2	8	2	4	16
8	3	24	1	1	8
12	4	48	0	0	0
8	5	40	1	1	8
4	6	24	2	4	16
3	7	21	3	9	27
42		42)168			102
		$A_x=4{\cdot}0$			$\sigma_x=\sqrt{\frac{102}{41}}=1{\cdot}58$

Similarly, find the mean value of y, also σ_y. They are:

$$A_y=3{\cdot}0, \quad \sigma_y=\sqrt{\tfrac{60}{41}}=1{\cdot}22$$

We now measure Δx right and left of the column of x headed 4 and Δy above and below the line of y numbered 3. Thus

for the reading in the top left corner	$\Delta_x\Delta_y=3\times2=6$
for the next reading to the right of this	$\Delta_x\Delta_y=2\times2=4$
for the next again	$\Delta_x\Delta_y=1\times2=2$
for the next again	$\Delta_x\Delta_y=0\times2=0$

Therefore the value of $\Sigma\Delta_x\Delta_y$ for the top row $=1\times6+1\times4+2\times2+1\times0=14$.

$\Sigma\Delta_x\Delta_y$ for the next row $=2(3\times1)+2(2\times1)+3(1\times1)+2(0\times1)+1(-1\times1)=14$.

$\Sigma\Delta_x\Delta_y$ for the middle row $=0$, for Δy is zero all along this line.

$\Sigma\Delta_x\Delta_y$ for the next row $=1(1\times-1)+3(0\times-1)+2(-1\times-1)+2(-2\times-1)+2(-3\times-1)=11$.

$\Sigma\Delta_x\Delta_y$ for the last row $=1(0\times-2)+2(-1\times-2)+1(-2\times-2)+1(-3\times-2)=12$.

$$\therefore \text{total } \Sigma\Delta_x\Delta_y=51$$

$$\therefore \text{average product moment}=\frac{51}{42}$$

$$\text{Coefficient of correlation}=r=\frac{1}{1{\cdot}58\times1{\cdot}22}\times\frac{51}{42}=0{\cdot}63$$

$$\text{and its probable error}=0{\cdot}67\frac{1{\cdot}63\times0{\cdot}37}{\sqrt{42}}=0{\cdot}06$$

Going back to the oat problem, regard the number of culms as x values and the yield as y values. The average number of culms is 3·567 and the average yield is 3·458 gm. It would take too long to calculate $\Sigma\Delta_x\Delta_y$ from these, so a compromise is made. Taking the mean number of culms as 4 and the average yield as $3\frac{1}{2}$ gms., we get integral values for the differences which we now call Δ'_x and Δ'_y, and $\Sigma\Delta'_x\Delta'_y=356$.

σ_x and σ_y calculated from the temporary means are 0·930 and 1·323.

Denoting the difference between the true mean x and the assumed mean x by w_x and the corresponding quantity for y w_y, the value of the coefficient of correlation is given by

$$r=\frac{1}{\sigma_x\sigma_y}\left(\frac{\Sigma\Delta'_x\Delta'_y}{n}-w_xw_y\right)$$

In the case of the oat problem

$$r=\frac{1}{0{\cdot}930\times1{\cdot}323}\left(\tfrac{356}{400}-0{\cdot}433\times0{\cdot}042\right)$$

$$=0{\cdot}71$$

$$\text{and its probable error}\quad=0{\cdot}67\frac{1-0{\cdot}5}{20}=0{\cdot}017$$

247. The Correlation Coefficient.

The *correlation coefficient* is of great importance to biometricians. The following rules guide us in its use:—

If $r<$ its probable error, there is no correlation;

$r>6$ times its probable error, there is correlation.

If the probable error is small the correlation is well marked if $r>0{\cdot}5$, and poor if $r<0{\cdot}3$.

Most of the work on correlation was initiated by Sir Francis

Galton. It has been used largely and its methods extended by Professor K. Pearson of the University of London. Galton used the graphical method, Pearson the algebraical. Between them and their followers it has been shown that not only can the coefficient of correlation be worked out for variations that can be measured accurately, but also for such variations as colour blindness, eye-colour, mental characteristics, and such-like which cannot be so estimated. Thus Pearson finds the correlation in conscientiousness between brother and sister to be 0·59, and between sister and sister 0·64.

248. Probability Paper.

This is a new kind of section paper invented by G. C. Whipple, of Harvard University, and A. Hazen, for use in statistics. To show how it is made and used let us take the following data quoted by Whipple. It is a table of heights of the soldiers in an army of 18,780 men.

Height in inches.	No. of soldiers.	$f \times (h - 60\frac{1}{2})$.	Percentage of soldiers in each class.		Percentage below the stated height.	
			Actual.	Calculated.	Height.	Percentage.
1	2	3	4	5	6	7
60–61	197	0	1·05	1·00	61	1·05
61–62	317	317	1·69	1·71	62	2·74
62–63	692	1,384	3·69	3·68	63	6·43
63–64	1,289	3,867	6·86	6·75	64	13·29
64–65	1,961	7,844	10·44	10·51	65	23·73
65–66	2,613	13,065	13·91	13·99	66	37·64
66–67	2,974	17,844	15·84	15·84	67	53·48
67–68	3,017	21,119	16·07	15·31	68	69·55
68–69	2,287	18,296	12·18	12·60	69	81·73
69–70	1,599	14,391	8·52	8·84	70	90·25
70–71	878	8,780	4·67	5·31	71	94·92
71–72	520	5,720	2·77	2·67	72	97·69
72–73	262	3,144	1·39	1·18	73	99·08
73–74	174	2,262	0·92+	0·61	74	100·00
	18,780	118,033	100·00	100·00		

We get the average height thus :—

$$A = \tfrac{1}{18780}(60\tfrac{1}{2} \times 197 + 61\tfrac{1}{2} \times 317 + 62\tfrac{1}{2} \times 692 + \ldots)$$
$$= \tfrac{1}{18780}\{60\tfrac{1}{2}(197 + 317 + 692) + (1 \times 317 + 2 \times 692 + 3 \times 1289 + \ldots)\}$$
$$= 60\tfrac{1}{2} + \tfrac{118033}{18780} = 60{\cdot}5 + 6{\cdot}285 = 66{\cdot}785 = 66{\cdot}8 \text{ inches nearly.}$$

By counting up to one half of 18,780, we find the median to be just below 67.

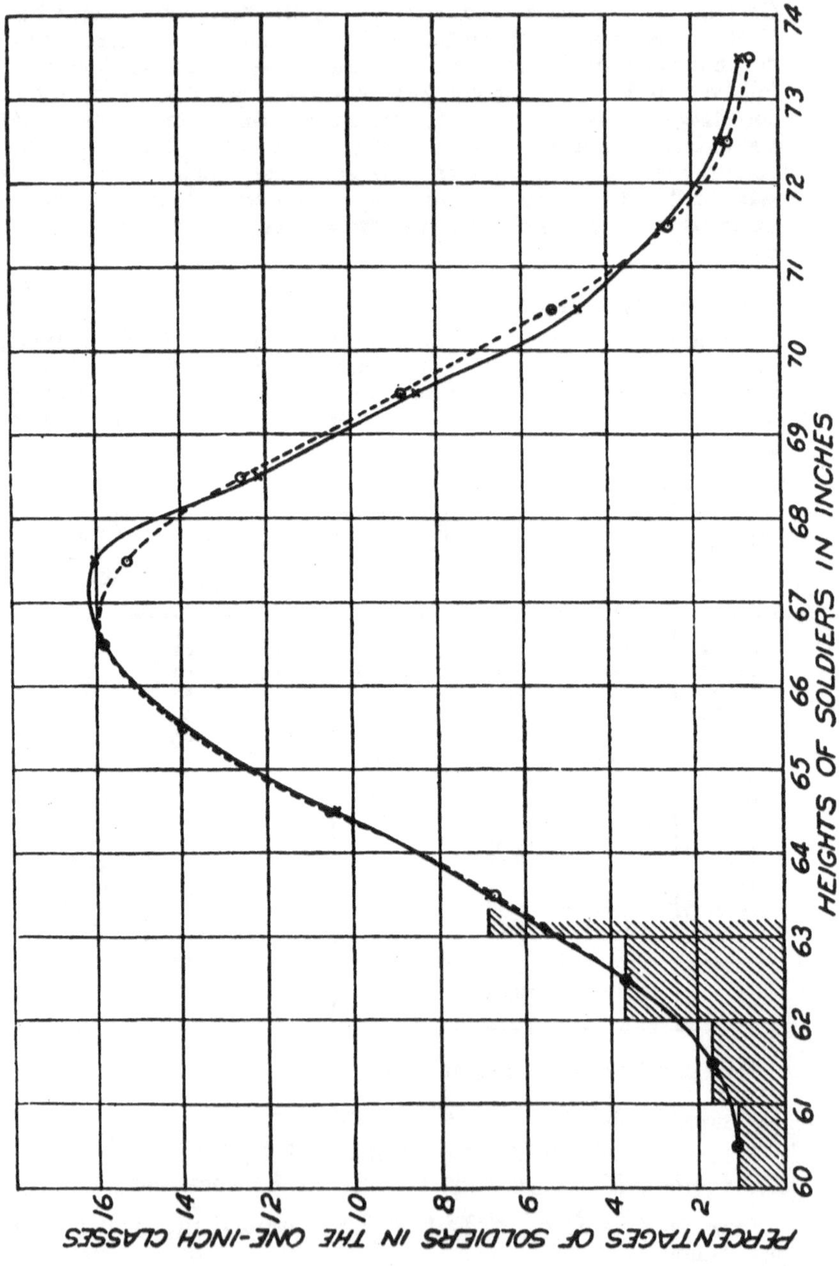
HEIGHTS OF SOLDIERS IN INCHES
PERCENTAGES OF SOLDIERS IN THE ONE-INCH CLASSES

Fig. 89

To get a better value apply the rule in § 240. Take the first five, the middle three and the last six. Then

$$a=4456,\ b=8604,\ c=5020, \text{ and } \frac{9390-4456}{8604}=\frac{x}{3} \quad \therefore\ x=1{\cdot}72$$

$\therefore$ Median $=66{\cdot}72=66{\cdot}7$ nearly.

A glance at the table shows that the mode is in the 3017 class, but a comparison of the numbers on the two sides of it shows that the 3017 class is unduly large. A little shifting of some of this class to the 2287 class would place the mode in the 2974 class or slightly below 67 inches.

The average, median, and mode are thus close together and the frequency curve (full line, Fig. 80) is fairly symmetrical. The best fitting normal curve (dotted line, Fig. 81) to the actual curve gives the figures in column 5; the discrepancy between columns 4 and 5 is very small.

Galton used another curve as well. Columns 6 and 7 give heights and percentages of soldiers below the stated heights. The curve is plotted in Fig. 81. It is ogee in shape. From it Galton readily deduced the median M (50 per cent.). It is 66·8 in our case. Galton also used centiles. These are the 1 per cent. dividing lines. The median is the fiftieth centile. The tenth centile or decile D is at 63·6 inches; *i.e.* 10 per cent. are below 63·6 inches, 90 per cent. are above. The twentieth centile is at 64·7 inches and so on.

The lower quartile, Q_1, is the twenty-fifth centile. For the soldiers it is 65·1 inches. The upper quartile, Q_3, is 68·4 inches. The two quartiles are practically the same distance from the average. The semi-interquartile range is $\frac{1}{2}$ of 3·3, so that the dispersion or probable error is 1·65 inches. Galton made up tables like the one here shown, and it is fairly obvious that we can draw conclusions

Centiles . . .	10	20	25	30	40	50	60	70	75	80	90
Values in above problem . .	63·6	64·7	65·1	65·5	66·2	66·8	67·4	68·0	68·4	68·8	70·0
Deviations	−3·2	−2·1	−1·7	−1·3	−0·6	0	0·6	1·2	1·6	2·0	3·2
Deviations made symmetrical	−3·2	−2·05	−1·65	−1·25	−0·6	0	0·6	1·25	1·65	2·05	3·2
Deviations in terms of the probable error	−1·94	−1·24	1·00	−0·76	−0·37	0	0·37	0·76	1·00	1·24	1·94
Correct values for normal curve	−1·90	−1·25	1·00	−0·78	−0·38	0	0·38	0·78	1·00	1·25	1·90

such as the following: "20 per cent. of the variates have negative deviations of magnitude greater than 1·25 times the probable error, or neglecting signs 40 per cent. of the variates have deviations greater than 1·25 *p.e.*, or 60 per cent. have deviations less than 1·25 *p.e.* We may use the above table to supplement the table of § 190.

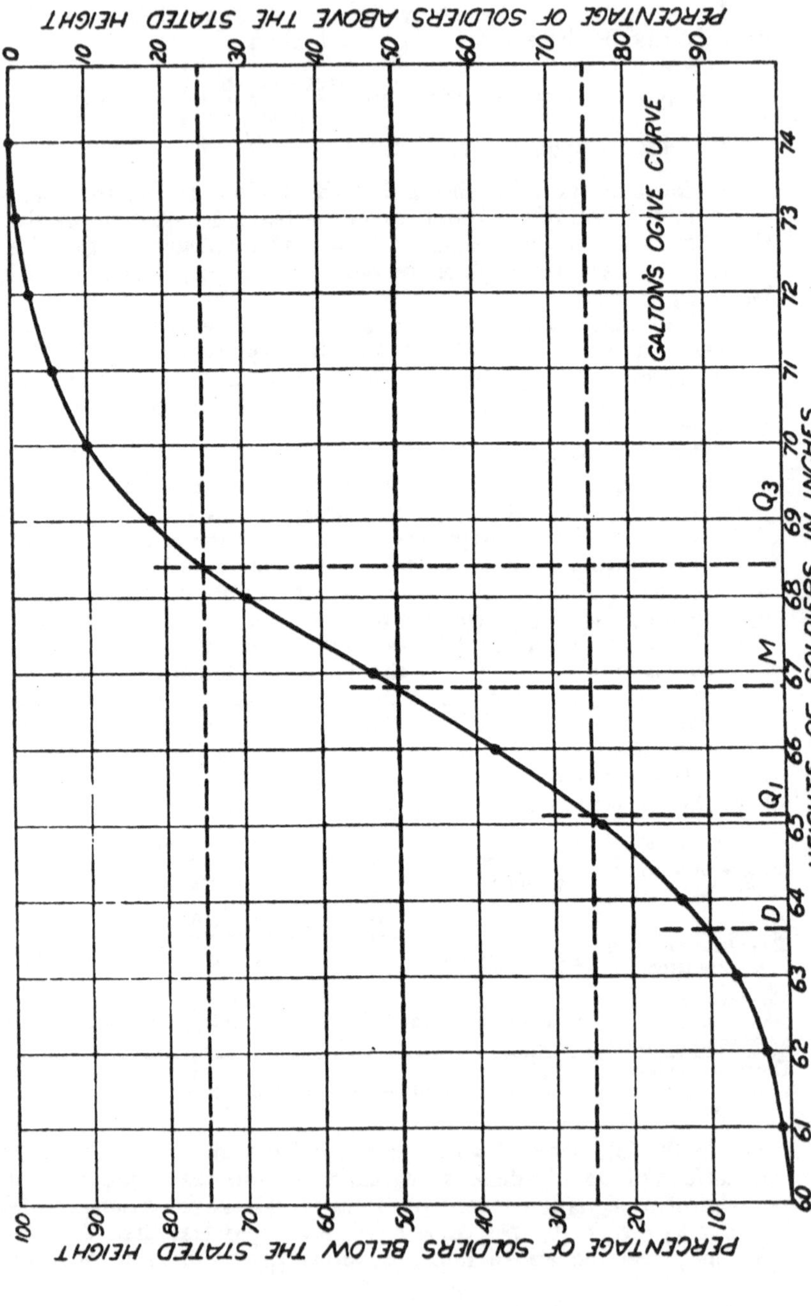

PERCENTAGE OF SOLDIERS ABOVE THE STATED HEIGHT
0
10
20
30
40
50
60
70
80
90
GALTON'S OGIVE CURVE
74
73
72
71
70
69
68
67
66
65
64
63
62
61
60
Q3
M
Q1
D
HEIGHTS OF SOLDIERS IN INCHES
100
90
80
70
60
50
40
30
20
10
PERCENTAGE OF SOLDIERS BELOW THE STATED HEIGHT

Fig. 81

Thus denoting a deviation by v, and the probable error by $p.e.$, we get the following tables, one being the converse of the other:—

Ratio $\frac{v}{p.e.}$	Probability of a variate having a deviation less than v	Probability of a variate having a deviation less than v.	Ratio $\frac{v}{p.e.}$
0·2	0·107	0·05	0·09
0·4	0·217	0·1	0·19
0·6	0·314	0·2	0·38
0·8	0·411	0·3	0·58
1·0	0·500	0·4	0·77
1·5	0·688	0·5	1·00
2·0	0·823	0·6	1·25
2·5	0·908	0·7	1·54
3·0	0·957	0·8	1·90
5·0	0·999	0·9	2·44
∞	1·000	0·95	2·91
		1·00	∞

Galton's ogee curve can be drawn freehand so as to go through the plotted points even when the variates do not conform to the normal law of errors, so that a good curve is not necessarily evidence of normal fitting. In order that a better test should be available, Dr. Hazen devised probability paper.* On this paper the percentage columns of Galton's curve are extended in such a way from the central 50 per cent. line that the ogee curve finally becomes straight. It will be seen from Fig. 81 that Galton's curve is nearly straight from 30 to 70 per cent., so that little alteration need be made for percentages within 20 per cent. of the mean. Beyond that the greater the difference from 50 per cent. the greater must be the extension.

The paper is symmetrical about the 50 per cent. line and we give here the values deduced by Whipple from probability considerations for the relative distances of the lines of some of the other percentages.

Per cent. line.	Relative distance.
50	0·000
40	0·376
30	0·777
25	1·000
20	1·248
15	1·537
10	1·906
5	2·439
1	3·450
0·1	4·585
0·01	5·550

* Arithmetic Probability Paper. Designed by Hazen, Whipple and Fuller, Codex Book Co., Inc., New York, N.Y., U.S.A.

This table is practically identical with the first and last rows of Galton's table (p. 265), as it must be from geometrical considerations. The paper is shown in Fig. 82 with the data from the soldiers' table plotted on it. The line is quite straight, showing that the soldiers'

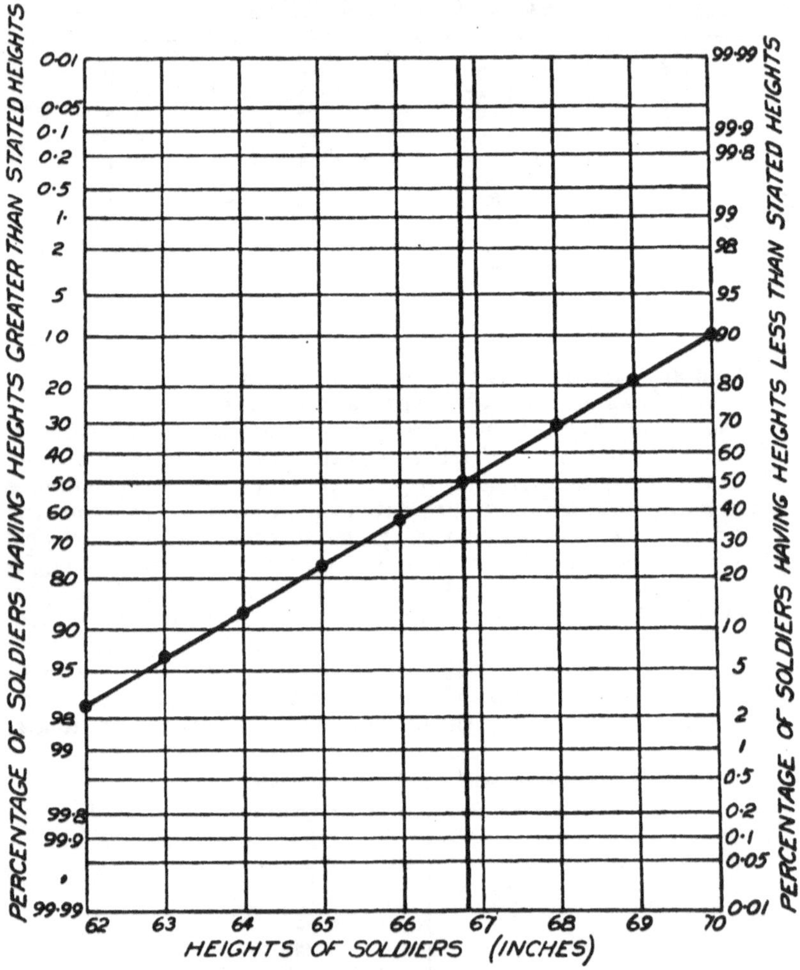

FIG. 82.

frequency curve is a good " error " curve. Notice particularly the legend at the sides of the figure.

Fig. 83 shows another way of showing Galton's ogee curve. If all the men were placed in line, Fig. 83 shows the curve that could be drawn touching the tops of all their heads.

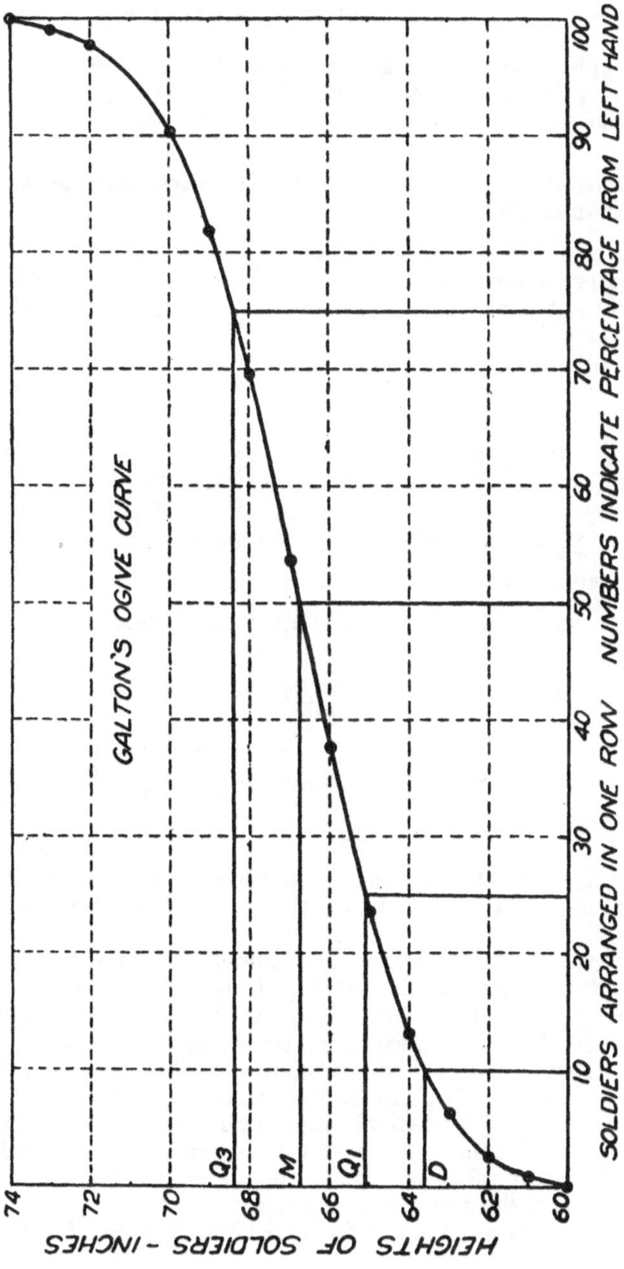
GALTON'S OGIVE CURVE
SOLDIERS ARRANGED IN ONE ROW NUMBERS INDICATE PERCENTAGE FROM LEFT HAND
10
20
30
40
50
60
70
80
90
100
HEIGHTS OF SOLDIERS - INCHES
74
72
70
68
66
64
62
60
Q3
M
Q1
D

FIG. 83.

Practise plotting some of your other frequency data on probability paper.

Whipple, Hazen and Fuller have also designed logarithmic-probability paper in which a logarithmic scale replaces the uniform horizontal scale of the arithmetic probability paper.

249. Probable Error in the Probability or Expectancy of an Event. Bernoulli's Formula.

If in n trials an event happens np times, p is said to be the probability of the event happening. Let q be the probability of the event not happening. Then, since the event is bound either to happen or not to happen, $p+q=1$.

It can be shown by mathematics that the probable error in np

$$=0{\cdot}6745\sqrt{npq}$$

Note that the greater the value of n the smaller is the percentage probable error in the number of events happening.

Example 1.—Suppose in a certain town of population 100,000 the general annual death rate is 12·1 per 1,000. The number of deaths per year $=\frac{12{\cdot}1}{1000}\times 100{,}000=1210$.

What variations may be expected from year to year in this number?

$$p=\frac{12{\cdot}1}{1000},\quad q=\frac{987{\cdot}9}{1000}.\quad \text{Also } n=100{,}000.$$

$$\therefore \text{probable error}=0{\cdot}6745\sqrt{100{,}000\times\frac{12{\cdot}1}{1000}\times\frac{987{\cdot}9}{1000}}$$

$$=23{\cdot}3 \text{ deaths.}$$

It is an even chance that the number of deaths lie within the range 1210 ± 23, *i.e.* that the death rate is within 0·23 of 12·1 per thousand.

Even twice or three times this amount might be put down to chance, but if the death rate per thousand increased from the average by four or five times 0·23 it would be a serious matter.

Example 2.—Of interest to social workers. Suppose that in a certain town of population 100,000, a certain epidemic took off 240 of the people. The disease was traced to a certain commodity, say some article of food which was supplied by only two companies, say *K* and *L*, *K* supplying two-thirds of the community and *L* one-third. What number of deaths should be found amongst their customers and their probable errors?

We should expect 160 and 80 deaths among *K* and *L*'s customers, respectively.

$$\text{Probable error in the } 160 = \tfrac{2}{3}\sqrt{67{,}000 \times \frac{240}{100{,}000} \times \frac{99{,}760}{100{,}000}} = 8 \cdot 5$$

$$\text{Probable error in the } 80 = \tfrac{2}{3}\sqrt{33{,}000 \times \frac{240}{100{,}000} \times \frac{99{,}760}{100{,}000}} = 6 \cdot 0$$

Thus it is an even chance that the number of deaths among *L*'s customers lies between 74 and 86 or outside this. If over 90 deaths occurred among *L*'s customers we should suspect *L*'s supplies. Similarly for K

250. Mendelism.

This is another interesting branch of biology in which the correct analysis of the numerical data helps to indicate the biological processes involved. The subject is too large for treatment in this book. The elementary student may refer with advantage to Doncaster's "Heredity" and Davenport's "Statistical Methods."

(B) Illustrations of the Use of Graphic Analysis and Mathematical Equations in Biological Investigations

251. Mathematical Theory of Wounds.

In the *Journal of Experimental Medicine* of 1916, cases are taken where the area of a surface wound was measured by a planimeter at successive intervals of two or four days and a curve plotted between area of wound and the time. The curve was found amenable to mathematical treatment, and seems to have an equation of the form

$$S = S_0 \epsilon^{-\lambda t}$$

Having got this equation from the earlier readings it is now possible for the doctor to forecast the behaviour of the wound, and if for any reason the wound does not heal up as forecasted, the existence of complications is deduced and these are sought for.

Example.—Drs. Tuffier and Desmaries observed the healing of a surface wound on the right leg of a patient of age thirty-one years (*J. Exp. Med.*, February 1, 1918). Observations—areas in square cms.

Day	0	4	8	12	16	20	24	28	32	36	40	44
Area	26	20	11	9	8	6	5	2	$2\frac{1}{2}$	2	1	$\frac{1}{2}$

Plot area against time. The curve looks like an exponential curve. Now plot log (area) against time. A fairly good straight line results.

Confirm that equation of straight line is

$$\log(\text{area}) = 1 \cdot 41 - 0 \cdot 033t$$

and therefore that the equation of original curve is nearly

$$\text{area} = 25\tfrac{1}{2}\ 10^{-0 \cdot 033t}$$

252. The Law of Fatigue.

Dr. Macalister and Mr. Haughton (*Proc.* Roy. Soc., London) describe some experiments in which a pair of 10-lb. dumb-bells, held in one hand, were raised simultaneously from the vertical to the horizontal position and again lowered at a rate regulated by a metronome made for the purpose. No rest was allowed at the beginning or end of the motion, which took place under the following conditions :—

(1) To keep time with the metronome.
(2) To raise the weights in the transverse plane.
(3) To supinate the hands.
(4) To abstain from all bending of the knees or spinal column.
(5) The experimenter not to count the lifts.

The experiments were made at intervals of not less than 24 hours so as to avoid "training." Each experiment was performed until it was impossible to do another lift without violating (4). One of Dr. Macalister's records was as follows :—

Time of lift $=t$ secs.	$\frac{1}{2}$	$\frac{2}{3}$	1	$1\frac{1}{2}$	2	3	4	6
No. of lifts $=n$.	19·0	24·6	26·2	23·6	20·0	14·2	12·4	8·5

What is the relation between t and n ?

Plotting n against t gives a hump-backed curve. The investigators had built up a theory which indicated that an equation

$$\frac{n}{t}(1+\beta t^2)=\text{a constant}$$

should be fulfilled. Plotting t^2 against $\frac{t}{n}$ should therefore give a straight line. This they did with the result that they found that

$$t^2=50{\cdot}7\frac{t}{n}-1$$

or

$$\frac{1+t^2}{t}=\frac{50{\cdot}7}{n} \quad \text{or} \quad \frac{n}{t}(1+t^2)=50{\cdot}7$$

Confirm this for yourself and having got an equation $\left(\frac{1+t^2}{t}=\frac{52}{n}\right.$ seems to be as good as the one found), calculate values of n for the given values of t and compare the experimental and calculated values.

253. Racing Records.

Professor John Perry, in "Elementary Practical Mathematics," quotes the following : "If t seconds is the *record* time of a trotting (in harness) race of m miles, we having the following published records.

m	1	2	3	4	5	10	20	30	50	100
t	119	257	416	598	751	1575	3505	6479	14,141	32,153

Find the law connecting m and t.

To do this plot, first, m against t and see if a straight line results. If not try various functions. Eventually you will find that log t plotted against log m yields a straight line whose equation is

$$\log t = \log 118 + 1{\cdot}183 \log m$$

whence

$$t = 118 m^{1{\cdot}183} \text{ *}$$

Queries: (1) What is the reason for his particular law and fatigue? (2) Does the same form of law occur with other kinds of races? (See questions at the end of the chapter.)

254. Growth and Temperature.

Van 't Hoff's law for chemical reactions is that: "As the temperature increases in arithmetical progression the velocity of reaction increases in geometrical progression," *i.e.*

Velocity at temperature $(t+n)^\circ$ = Velocity at $t^\circ \times x^n$.

Does this law hold for vegetable and animal growths? Miss Leitch ("Annals of Botany," 1916), quoted by D'Arcy Thompson in "Growth and Form," obtained the following results for the growth of rootlets of peas (*Pisum sativum*) at different temperatures:

Temperature.	Growth in mm. per hour.
°C	
4	0·12
8	0·29
12	0·47
16	0·74
20	1·01
24	1·52
28	2·04

Plot growth (g) against temperature (t). The result is a curve getting steeper as t increases. This suggests that we plot log g against t. Do so. The graph is nearly straight and its equation is

$$\log g = -0{\cdot}77 + 0{\cdot}0385t$$

or

$$g = \frac{1}{10^{0{\cdot}77}} 10^{0{\cdot}0385t}$$

$$= 0{\cdot}170 \times 1{\cdot}092^t$$

so that we see that whenever the temperature goes up 1° C. the growth increases in the ratio 1 : 1·092. Show that if we consider 10° steps the growth increases in the ratio 1 : 2·42.

* Or between this and $t = 115m^{1{\cdot}19}$.

255. The Law of Sensations—Weber and Fechner's Law.

Weber and Fechner stated that it would be found that if x = the constant intensity of a stimulus and x_0 = its threshold value, the intensity of the sensation s is given by the relation

$$s=\log\frac{x}{x_0} \quad \text{or} \quad x=x_0 10^s$$

i.e. to make the sensation increase in arithmetical progression the stimulus must increase in geometrical progression.

An experiment on vision gave the following results.* The stimulus in each case was a rotating disc made up of white and black sectors. The relative brightness of the white and black paper measured photometrically was as 50 to 1.

The relative brightness of the disc as stimuli may be expressed in degrees of either white or black sectors. It is more convenient arithmetically to select the black. Thus, if the angles are white sector 60°, black sector 300°, the photometric value may be expressed as $60\times 50+300=3{,}300$.

To carry out the experiment, 4 such discs were used—*A*, *B*, *C*, *D*. *A* was set at any particular value and while they were all in rotation the sector angles of *B*, *C*, *D* were altered by a mechanism so that *B* was just noticeably brighter than *A*, *C* than *B*, and *D* than *C*. The sensation value differences were equal—in this case the differences are said to be *liminal* because just noticeable. The discs were then stopped and their stimuli values computed. It was found that the ratio of *A* to *B* was the same as that of *B* to *C* and of *C* to *D*. Another way of putting it is to say that when the sensation values change by equal steps the logarithms of the stimuli change by equal steps. The following gives the figures obtained in one experiment:

A series of grey sensations with liminal differences.	Corresponding stimuli made up of sectors of angles.		Objective brightness.	Logarithm of objective brightness.	Differences between successive logarithms.	Ratio of the stimuli.	Values of stimuli expressed in terms of *A*.
Sensation	white	black					
A	60°	300°	3300·0	3·5185			*A*
B	63	297	3447·0	3·5375	0·0190	$\frac{B}{A}=1·045$	$A\times 1·045$
C	66·2	293·8	3603·8	3·5568	0·0193	$\frac{C}{B}=1·046$	$A\times 1·045^2$
D	69·5	290·5	3765·5	3·5758	0·0190	$\frac{D}{C}=1·045$	$A\times 1·045^3$

The value of the ratio is immaterial. It will depend on the conditions. With any other equal sensation steps we should still get a constant ratio, the size of the ratio depending, of course, upon the size of the step in our sensations.

* For this example we are indebted to Professor Bott of the Department of Psychology in the University of Toronto.

Weber's law may also be written, "Rate of increase of sensation with increase of stimulus is inversely proportional to the magnitude of the stimulus." The data illustrating Weber's law may be very conveniently plotted on semi-logarithmic paper using the uniform scale for the sensation values and the logarithmic scale for the stimuli values.

Another example of Weber's law occurs in the measurement of stellar magnitudes. Stars which are visible to the eye are divided into six magnitudes, the first being the brightest. The average first magnitude star is 100 times as bright as the average sixth magnitude star and the relative intensities are: (1) 100, (2) 39·8, (3) 15·9, (4) 6·3, (5) 2·5, (6) 1, the common ratio being 2·51.

256. Magnitude and Similitude.

The areas of similar figures are to one another as the squares of their linear dimensions.

The volumes of similar figures are to one another as the cubes of their linear dimensions.

If a brick B is similar to a brick A, but of double the length, it will have 4 times the area and 8 times the volume. If we take a sequence of similar bodies increasing in size as we go up the series, the areas increase at a greater rate than the lengths and the volumes increase at a greater rate than the areas. If we take a series of similar bodies diminishing in size as we go down the series, the volumes diminish at a greater rate than the areas.

For bodies similarly constituted weight is proportional to volume, therefore for similar bodies (*i.e.* similar in shape and constitution) their weights are proportional to the cubes of their lengths.

For example, if we have a number of animals of the same kind, say salmon, and the graph between weight and cube of length is straight, we should say there was no alteration in form.

The strength of a beam is proportional to its cross-section. If we have two similar bridges, A and B, their strengths are proportional to the squares of their linear dimensions, their weights to the cubes of their linear dimensions. Therefore, if the smaller bridge is just strong enough to carry its weight, the larger bridge will not be strong enough. It will collapse. An animal loses moisture largely through its skin. Suppose there is a natural balance for animals of one size. What will happen with animals of similar shape and similar skins as the size diminishes? The water to be evaporated will be proportional to the volumes, the evaporating area to the area of the surface. As size diminishes volumes decrease faster than areas, therefore the very small animals would soon dry up. Nature gives many of them a horny skin to cut down the evaporation.

Consider a fruit on a stem and the same with double the linear dimensions. The weight of the fruit increases to 8 times, the strength of the stem to 4 times, therefore the strength of the stem is insufficient and the plant changes its habit and trails instead of standing up.

In the case of large animals the neck has to increase to an ungainly size to support the huge head, *e.g.* the elephant. Of course, if an animal takes to the water, the water helps to support its weight.

For similar fishes (or ships) the power developable is proportional to the volumes (of the muscles or furnaces), that is to the cubes of their lengths. Friction with the water is a surface effect and hence proportional to the squares of the lengths. Therefore, as the size increases the speed increases; in fact, it may be proved that the normal velocities are proportional to the square roots of their lengths.

Much the same argument applies to birds and aeroplanes. The bigger they are the faster they must go to keep up, and this means they must be given the necessary power. Hence, if a large bird can fly it can fly fast, but many large birds cannot fly at all unless they have the runway necessary to get up to the flying speed.

For a further study on these lines, see D'Arcy Thompson: "Growth and Form."

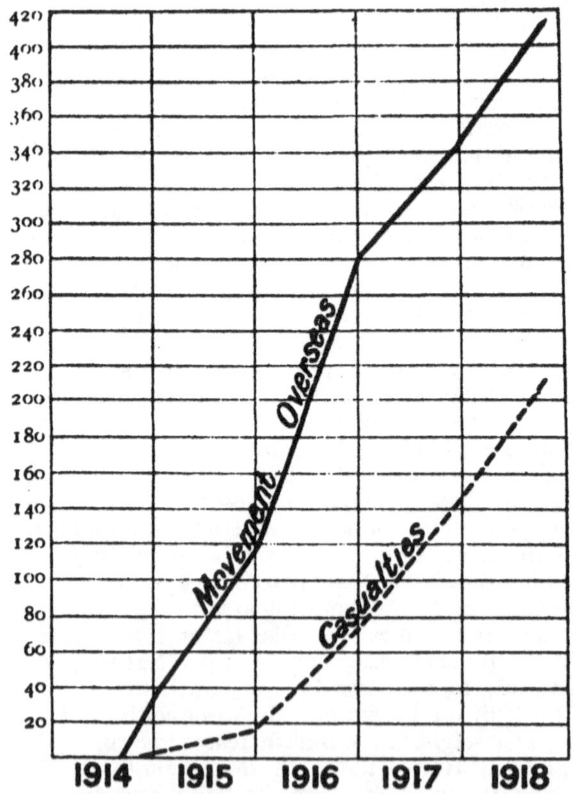

FIG. 84.—MOVEMENT OVERSEAS AND CASUALTIES OF THE CANADIAN EXPEDITIONARY FORCE IN THE WAR OF 1914–1918.

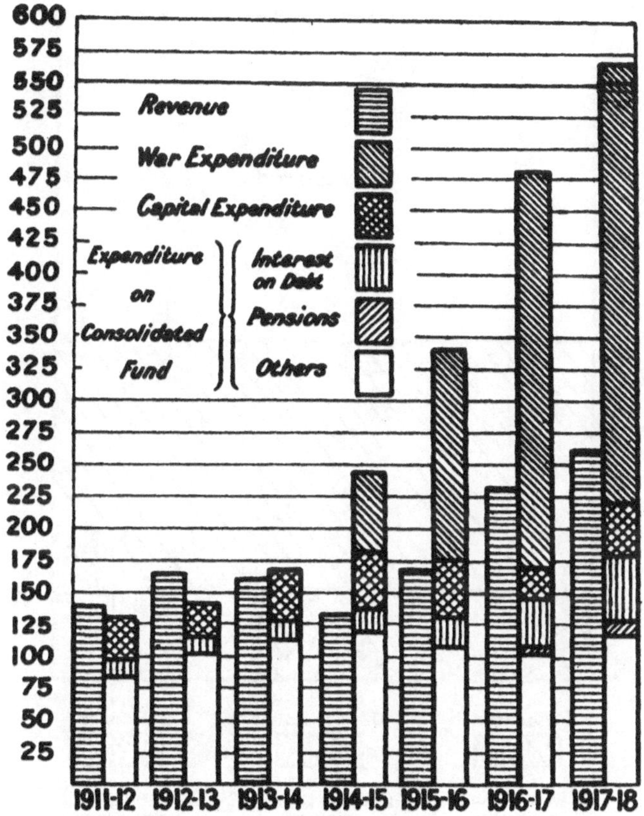

FIG. 85.—REVENUE AND PRINCIPAL EXPENDITURE OF CANADA. (In millions of Dollars.)

257. The Representation of Statistics.

Statistics of varying quantities are best represented by curves (Fig. 84).* Other statistics are often represented more or less pictorially by lines, and areas. The best way is to use straight lines or areas of uniform breadth (Fig. 85), the length in each case

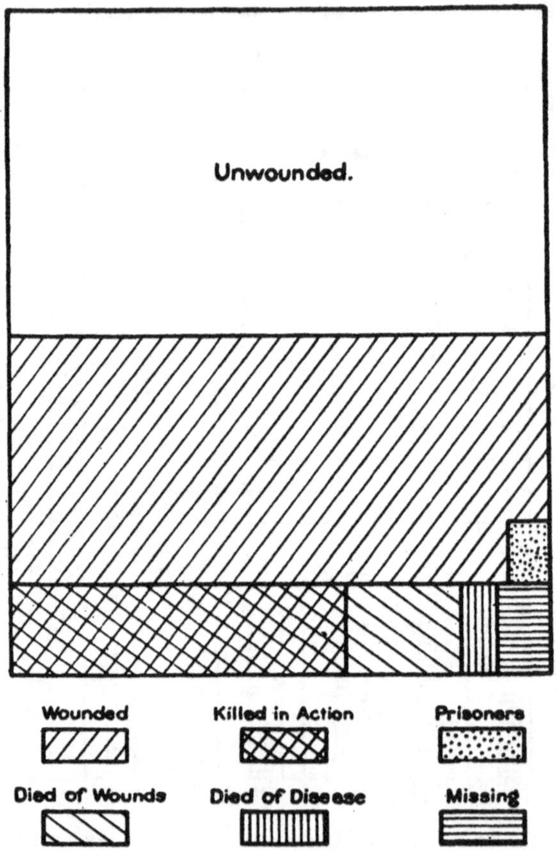

FIG. 86.—CASUALTIES IN THE CANADIAN EXPEDITIONARY FORCE.

being proportional to the quantity to be represented, or areas of any convenient shape may be employed (Fig. 86).

Another popular method is to take a circle and to divide it up into sectors, whose angles (and therefore areas) are proportional to the quantities under comparison (Fig. 87). The worst method

* Figs. 84, 85, 86 were given in a pamphlet issued by the Department of Public Information, Ottawa, entitled "Canada's Part in the Great War."

is the pictorial method where, for example, an armed soldier is made to represent a nation's military strength, or a ship its navy, or a bag of corn a country's corn production. One instinctively judges a bag of corn by its weight, so that its volume should represent the country's production, and invariably the artist has made its linear dimension proportional to the production. These diagrams therefore create a false impression. As an example, consider Fig. 88, which is a reproduction from a pamphlet issued by the Canadian Food Board in 1917. The three fishes are supposed to represent 94, 29, and 52 pounds, respectively. Their lengths are proportional to 94, $28\frac{1}{2}$, 57, so that the artist has made the linear dimensions

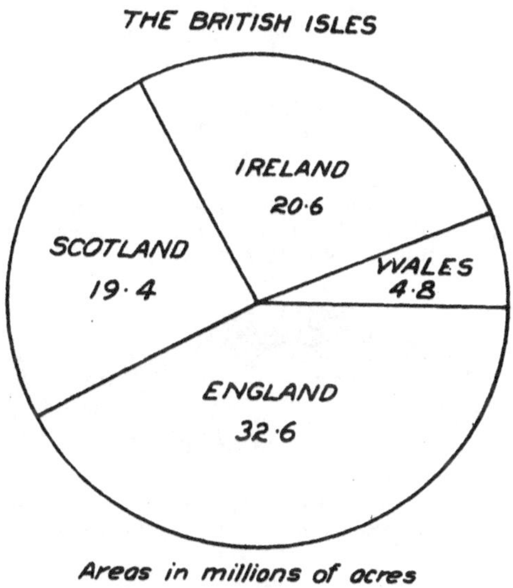

FIG. 87.

approximately proportional to the weights. One judges, however these fish by their imagined weights. The fish being similar, these weights are in the ratio $(94)^3 : (28\frac{1}{2})^3 : (57)^3$, *i.e.* as $94 : 2\frac{1}{2} : 18$, so that instead of getting the impression that 20 lbs. of fish are consumed in Canada, we imagine a consumption of only $2\frac{1}{2}$ lbs., which is very wide of the mark. The ratios of the first two weights being 94 : 29, the ratio of the lengths should have been $\sqrt[3]{94} : \sqrt[3]{29}$, *i.e.* as 94 : 63.

A convenient pamphlet on graphical representation for the use of non-mathematical writers has been issued by the American Statistical Society. A copy may be seen in Pearl's "Medical Biometry and Statistics" (see Bibliography at the end of the book).

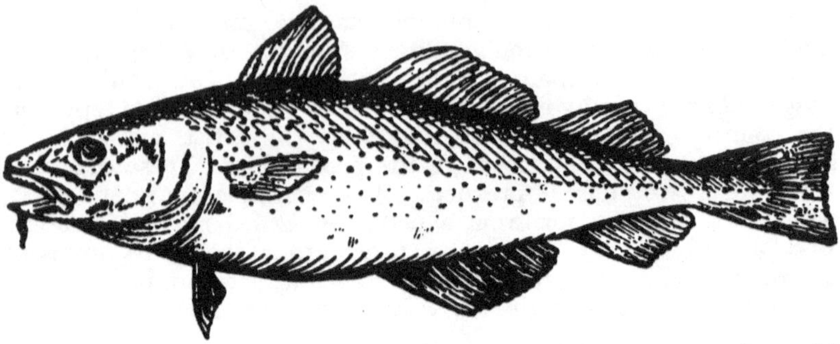

This cut represents the total Canadian catch of fish in 1917. It would equal 94 pounds a head a year for each man, woman, and child.

This cut represents the fish actually eaten in Canada It is equal to only 29 pounds a head. Where did the remaining 65 pounds go? Exported, *i.e.* sent out of the Dominion for others to eat.

This cut represents what the Canadian Food Board sets as a standard fish diet—52 pounds per head a year, or 1 pound a week for every Canadian during war and afterwards.

FIG. 88.

EXERCISES XXI

1. The lengths of a thousand ears of Indian corn were measured and arranged in half-inch groups as follows:

Length	3	3½	4	4½	5	5½	6	6½	7	7½	8	8½	9	9½	10	10½	11	11½	12
No. in each group	5	6	13	17	18	55	61	73	80	98	113	134	142	100	53	26	5	1	0

Plot. Find the characteristics. Comment.

2. The table shows month by month for the first year the deaths per 10,000 babies born.

Month	1	2	3	4	5	6	7	8	9	10	11	12
Deaths	438	113	94	80	71	64	58	53	49	46	42	39

Plot. Comment. Find also for each month the rate of mortality per 1,000 of those living at the beginning of the month.

3. The Anthropometrical Committee of the British Association measured and weighed men of the British Isles. Their results are as follows :—

Heights, in socks.	Number.
ins.	
57–8*	2
58–	4
59–	14
60–	41
61–	83
62–	169
63–	394
64–	669
65–	990
66–	1223
67–	1329
68–	1230
69–	1063
70–	646
71–	392
72–	202
73–	79
74–	32
75–	16
76–	5
77–78	2
	Total 8585

Weights.	Number.
pounds	
90–†	2
100–	34
110–	159
120–	390
130–	867
140–	1623
150–	1559
160–	1326
170–	787
180–	476
190–	263
200–	107
210–	85
220–	41
230–	16
240–	11
250–	8
260–	1
270–	0
280–290	1
	Total 7749

Plot the frequency curves on ordinary paper and on probability paper. Comment. Find the characteristics.

4. The following observations are made.

Frequency.	Value.
2	5·1
5	5·2
16	5·3
10	5·4
7	5·5

5·2 means between 5·15 and 5·25.
Apply the rule of § 240 to find the median.

5. 3,837 families in America containing deaf and dumb persons were studied and the number of children per family noted.

No. of children	1	2	3	4	5	6	7	8	9	10	11	12	13	14	15	16	17	18	19	20	21	22	23	24
No. of families	110	320	460	550	510	470	380	310	260	170	110	75	50	15	13	10	8	6	4	2	2	1	1	0

Plot the frequency curve. Get the characteristic values—by calculation or by the curve. Calculate its asymmetry.

6. Wood and Stratton quote (*Journ. Agric. Science*, vol. iii. Pt. 2) the following analyses of 160 mangel roots for dry matter :

Number	1	8	12	34	44	32	17	9	2	1
Percentage of dry matter	10+	11+	12+	13+	14+	15+	16+	17+	18+	19+

Plot the frequency curve. Read 10+ as 10·5 per cent.
Find the mean, and the other characteristic values.

* 57–8 signifies from $56\frac{15}{16}$–$57\frac{15}{16}$. † 90 lbs. signifies here from $89\frac{1}{2}$ to $99\frac{1}{2}$ lbs

7. The glumes were measured on 595 wheat plants, which were a second generation from a cross between Polish wheat (glume length, 28 mm.) and Rivet wheat (glume length, 9 mm.). In the table a length 7 signifies 7 to 8.

Length, mm.	No. of plants.	Length, mm.	No. of plants.	Length, mm.	No. of plants.
7	1	17	90	27	13
8	2	18	68	28	21
9	24	19	31	29	20
10	65	20	11	30	27
11	56	21	8	31	12
12	5	22	9	32	9
13	3	23	3	33	6
14	6	24	5	34	2
15	20	25	11	35	1
16	60	26	11	36	1

Plot and comment on the grouping.

8. The distribution of stature in 1,000 women (mothers) and 1,000 fathers, measured by Professor Pearson, is as follows:

Height.	Number of mothers.	Number of fathers.
50 inches and below	0	
Between 51 and 52 inches	1	
,, 52 ,, 53 ,,	1	
,, 53 ,, 54 ,,	2	
,, 54 ,, 55 ,,	2	
,, 55 ,, 56 ,,	3	
,, 56 ,, 57 ,,	5	0
,, 57 ,, 58 ,,	15	1
,, 58 ,, 59 ,,	31	1
,, 59 ,, 61 ,,	75	2
,, 60 ,, 60 ,,	130	2
,, 61 ,, 62 ,,	153	3
,, 62 ,, 63 ,,	173	6
,, 63 ,, 64 ,,	155	25
,, 64 ,, 65 ,,	110	44
,, 65 ,, 66 ,,	74	70
,, 66 ,, 67 ,,	38	110
,, 67 ,, 68 ,,	14	149
,, 68 ,, 69 ,,	8	149
,, 69 ,, 70 ,,	6	129
,, 70 ,, 71 ,,	3	110
,, 71 ,, 72 ,,	1	80
,, 72 ,, 73 ,,	0	50
,, 73 ,, 74 ,,		33
,, 74 ,, 75 ,,		19
,, 75 ,, 76 ,,		7
,, 76 ,, 77 ,,		4
,, 77 ,, 78 ,,		3
,, 78 ,, 79 ,,		2
,, 79 ,, 80 ,,		1
,, 80 and above		0
	1000	1000

Plot the frequency curves.

Find the mode, median, arithmetical mean, average deviation of an individual, and the dispersion or probable deviation of an individual.

From the diagrams find the semi-interquartile ranges and show that these correspond to the dispersions.

9. Love and Leighty measured up 400 plants of the sixty-day oats (the same as given in § 246) and measured the number of culms on each plant and the height, with the following results :—

Height of plants in cms.	Number of culms per plant.						
	2	3	4	5	6	7	
45–50	—	1	1	—	—	—	2
50–55	2	3	3	—	—	1	9
55–60	4	9	4	2	2	—	21
60–65	6	10	13	4	1	—	34
65–70	8	40	42	6	1	—	97
70–75	15	41	53	12	2	—	123
75–80	10	22	43	11	3	—	89
80–85	4	8	8	3	1	—	24
85–90	—	—	—	—	—	—	0
90–95	1	—	—	—	—	—	1
	50	134	167	38	10	1	400

Find the coefficient of correlation between number of culms and height of plant, and its probable error.

10. In the case of the oat statistics given in § 246, find the average yield per plant, the mean square error, the probable error and the coefficient of variability.

11. The following table gives the ages in years of husbands and wives for all pairs in England and Wales who were residing together on the night of the census of 1901. Table based on 5,317,520 pairs, but condensed by expressing the numbers to the nearest thousand.

Ages of husbands.	Ages of wives.														
	15+	20+	25+	30+	35+	40+	45+	50+	55+	60+	65+	70+	75+	80+	85+
15+	2	2	—	—	—	—	—	—	—	—	—	—	—	—	—
20+	16	173	46	4	1	—	—	—	—	—	—	—	—	—	—
25+	4	185	402	84	10	2	1	—	—	—	—	—	—	—	—
30+	1	41	265	411	84	12	2	1	—	—	—	—	—	—	—
35+	—	9	69	251	369	80	12	2	1	—	—	—	—	—	—
40+	—	3	17	71	219	309	66	12	2	1	—	—	—	—	—
45+	—	1	6	20	66	178	252	59	10	2	1	—	—	—	—
50+	—	—	2	8	19	57	146	195	44	10	2	—	—	—	—
55+	—	—	1	3	8	18	46	110	141	35	6	1	—	—	—
60+	—	—	—	1	3	18	16	39	81	101	23	4	1	—	—
65+	—	—	—	1	1	3	6	11	26	53	58	13	2	1	—
70+	—	—	—	—	1	1	2	5	8	18	31	31	6	1	—
75+	—	—	—	—	—	1	1	2	3	5	10	14	12	2	
80+	—	—	—	—	—	—	—	1	1	1	2	4	5	3	1
85+	—	—	—	—	—	—	—	—	—	—	1	1	1	1	—

Discuss the correlation between the ages of husbands and wives.

12. Correlation between statures of fathers and sons (one or two sons only of each father: 65 inches includes from $64\frac{1}{2}$ to $65\frac{1}{2}$, and so on). Data due to Pearson and Lee.

Statures of fathers.

Statures of sons.	59	60	61	62	63	64	65	66	67	68	69	70	71	72	73	74	75	
60	—	—	—	—	2	2	4	—	—	—	—	—	—	—	—	—	—	8
61	—	—	—	—	2	—	—	—	4	—	—	—	—	—	—	—	—	6
62	—	1	1	—	2	4	1	1	2	2	—	—	—	—	—	—	—	14
63	—	1	1	9	9	8	16	20	11	5	—	1	1	—	—	—	—	82
64	4	—	6	15	12	17	32	37	12	5	6	3	5	—	—	—	—	154
65	8	4	2	8	13	38	54	43	30	22	14	10	—	—	—	—	—	246
66	—	2	4	9	21	38	40	67	70	64	21	8	10	4	—	—	—	358
67	—	6	8	19	14	55	79	106	103	78	50	55	13	2	4	—	—	592
68	—	—	6	8	30	40	41	97	126	94	118	53	34	38	0	—	—	694
69	—	—	4	—	21	20	51	73	64	96	116	86	40	14	9	—	4	598
70	—	—	—	—	4	10	23	75	47	78	90	78	58	25	14	6	4	512
71	—	—	—	—	—	13	20	35	43	76	59	83	43	32	20	4	4	432
72	—	—	—	—	—	1	12	5	28	31	43	45	40	34	11	2	—	252
73	—	—	—	—	—	—	3	3	10	30	26	24	30	25	13	2	2	168
74	—	—	—	—	4	—	6	6	—	21	9	10	26	13	13	—	8	116
75	—	—	—	—	—	—	—	—	—	4	8	—	10	3	7	2	—	34
76	—	—	—	—	—	—	—	—	—	5	1	—	2	4	4	—	—	16
77	—	—	—	—	—	—	—	—	—	5	1	4	—	—	6	—	—	16
78	—	—	—	—	—	—	—	—	—	—	4	4	—	1	3	—	—	12
79	—	—	—	—	—	—	—	—	—	—	—	—	—	1	1	—	—	2
	12	14	32	68	134	246	382	568	550	616	566	464	312	196	114	16	22	4312

Find the correlation coefficient.

13. There are two variables a and b, and corresponding values are listed in the same column in the following table:

a	7	5	6	3	9
b	4	2	5	1	8

Explain Galton's method which is used to find the degree of correlation between two variables and find the value of the degree of correlation in the case quoted.

14. In the correlation problem of § 246, since $\sigma_x = \sqrt{\frac{\Sigma(\Delta_x)^2}{n-1}}$ and $\sigma_y = \sqrt{\frac{\Sigma(\Delta_y)^2}{n-1}}$ it follows that the coefficient of correlation is very nearly equal to

$$\frac{\Sigma(\Delta_x\Delta_y)}{\sqrt{(\Delta_x)^2(\Delta_y)^2}}$$

Show that this expression is equal to unity if the correlation is perfect, and that the geometrical method of § 245 is thus in agreement with the algebraical method of § 246.

15. Weldon in a paper * on the variations of a certain decapod crustacea—*Crangon vulgaris*—measured the length of its carapace and expressed it as a thousandth part of the whole length of the animal. Four hundred were examined at Plymouth. The numbers in the first column give the percentage which had fractional lengths greater than the numbers in the second column.

Grade.	Length.
0	
5	261·50
10	258·95
20	256·05
25	255·07
30	254·10
40	252·27
50	250·52
60	249·09
70	247·29
75	246·00
80	244·74
90	241·39
95	238·60

State the median, the quartiles, and the semi-interquartile range.
Plot a curve between the above numbers, also the frequency curve.

16. In Denmark between the years 1860 and 1879 (inclusive) the average number of births was 57,583; the average percentage of boys born being 51·43. The deviations from the average percentage was year by year as follows: +0·23, −0·26, +0·07, +0·19, +0·25, −0·05, +0·44, −0·23, −0·17, −0·41, +0·17, +0·06, +0·01, −0·06, +0·14, +0·01, +0·02, −0·45, +0·05, +0·02.

Find the probable error in the number of births of boys and the probable percentage error. How many of the actual percentages were above this; how many below?

17. Take any one frequency distribution, *e.g.* the number of soldiers and their height of § 248, and plot a curve, taking the order of the individual as abscissa and the height as ordinate. This is Galton's ogee curve, and is the same as that of § 81 turned over on its side and looked at in a mirror. The curve is, of course, the same as that which we should get if we arranged all the men in order along a straight line and drew a smooth curved line just touching the tops of their heads (see Fig. 83).

18. Drs. Tuffier and Desmaries measured the area of a wound on the right knee of a French soldier. Results in square centimetres:

Day	0	6	11	14	18	21	25	28	32	35	38	40
Area	11	6	$4\frac{1}{2}$	$2\frac{1}{4}$	$3\frac{3}{4}$	$1\frac{1}{2}$	$1\frac{1}{4}$	$1\frac{1}{2}$	1	$\frac{1}{2}$	$\frac{1}{3}$	$\frac{1}{10}$

Find the equation of the curve.
What do you imagine happened just before the eighteenth and twenty-eight days?

19. The following data were obtained on the growth of a certain plant.

Temperature (°C.)	18	23·5	26·6	28·5	30·2
Growth in 48 hours (mm.)	1·1	10·8	29·6	26·5	64·6

Find the relationship between growth and temperature.

* *Proc.* Roy. Soc. London, vol. 47, 1890, p. 445.

20. The Amateur Swimming Records (*Whitaker's Almanack*) :

Yards, s	100	150	220	300	440	500	880	1,000	1,700
Seconds, t	$54\frac{3}{5}$	$92\frac{2}{5}$	$145\frac{2}{5}$	$206\frac{2}{5}$	323	$362\frac{1}{5}$	$685\frac{3}{5}$	$814\frac{1}{5}$	$1,414\frac{1}{2}$

Find the relation between s and t.

21. The Skating Records (*Whitaker's Almanack*) :

Miles, s	1	2	3	4	5	10	20
Minutes, t	2·6	5·7	8·58	12·0	14·6	31·2	66·6

Find the relation between s and t.

22. The Amateur Walking Records (*Whitaker's Almanack*) :

Miles, s	1	2	3	4	5	7	10	15	20	25	50	100
Minutes, t	6·4	13·2	20·4	27·2	36·0	50·7	75·0	119·2	166·4	217·1	472·5	1084·2

Find the relation between s and t.

23. The Amateur Cycling Records (*Whitaker's Almanack*) :

Kilometres, s	1	5	10	60	70	80	90
Minutes, t	0·9	3·6	6·2	35·4	41·3	47·2	53·1

Find the relation between s and t.

24. The Amateur Racing Records (*Whitaker's Almanack*) :

Miles, s	$\frac{1}{4}$	$\frac{1}{2}$	$\frac{3}{4}$	1	2	3	4	5	10
Seconds, t	48	112	189	256	550	857	1104	1473	3041

Find the relation between s and t.

25. An experiment was performed similar to that described in § 255 The differences were in this case supra-liminal, but still judged on observation to be equal. The readings were :

Sensations.	Composition of Rotating Discs.
A	60° white 300° black
B	99° ,, 261° ,,
C	160° ,, 200° ,,
D	256° ,, 104° ,,

Show that sensations which are judged to differ by an equal increment correspond to stimuli which differ by a given ratio. Also plot on semi-logarithmic paper.

26. In the operation of gill nets on a certain lake in Ontario the results with white fish were as follows :

Caught in the $4\frac{1}{2}''$ net	75 averaging $16\frac{1}{2}''$ in length
Caught in the $3''$ net	175 ,, $13\frac{1}{2}''$,,
Caught in the $2''$ net	300 ,, $12''$,,

Plot the curve. Could it be extrapolated to give the number of fish of 1, 2, 3, . . . etc., inches long which might be present in the lake ? Are the data sufficient ?

27. Find the equation between the x and the y of the following table. The x represents the weight of an embryo and the y the height.

x, gms.	10	30	50	70	90	110	130	150	170	190	210	230	250	270
y, mm.	58·8	76·4	91·1	99·0	108·1	115·1	122·7	129·5	135·0	141·1	144·0	150·0	152·8	155·6

x, gms.	290	310	330	350	370	390
y, mm.	158·6	161·3	160·5	171·0	169·5	173·6

You may save time by first drawing a smooth curve between x and y and then selecting four or five representative points having simple numbers for co-ordinates.

28. Data on the size of plaice.

Length, cms.	24	26	28	30	32	34	36	38	40	42	44
Weight, gms.	128	173	221	271	328	384	454	529	614	679	800

Do the plaice remain of similar shape as they grow up? (Hint: Plot w against l^3).

29. The table gives the average number of children per wife in America for the periods named. Compiled from twenty-two genealogical records of America.

Before 1700	7·37
1700–1749	6·83
1750–1799	6·43
1800–1849	4·94
1850–1869	3·47
1870–1879	2·47

Plot and comment.

30. Small drops of water fall through the air, but the minute drops of mists and fogs remain suspended. Why is this?

31. Bodies are attracted to the sun by the gravitational pull but repelled by the solar-radiation pressure. Show that very large bodies move in to the sun while microscopic dust is repelled.

32. Two similar tuning-forks are made of the same steel. The second is twice as long as the first. What will be their relative pitches.

33. Prove Froude's law of the correspondence of speeds, *i.e.* that the normal speeds of similar ships are proportional to the square roots of their lengths.

34. *Physical measurements on public school boys* (*Nature*, 1899). 14,000 observations made. The table applies only to boys of 15 years of age, and is to be read in this way: 5 per cent. of the boys are less than 68·3 inches high, and 5 per cent. weigh less than 132 pounds, and so on.

Per cent.	Inches.	Pounds.	Per cent.	Inches.	Pounds.
5	68·3	132	55	62·6	100½
10	67·3	126	60	62·1	98½
15	66·5	121	65	61·7	96½
20	65·8	117½	70	61·2	94½
25	65·1	115	75	60·7	92
30	64·7	112	80	60·0	89½
35	64·2	109½	85	59·5	86½
40	63·9	107	90	58·8	84
45	63·4	105	95	57·7	79½
50	63·0	102½			

Plot these on probability paper as well as on ordinary squared paper. (If no probability paper is available place tracing paper over Fig. 82.) Comment.

CHAPTER XXII

ON AREAS

258. THE measurement of areas is important. We give here a number of useful rules. Many of the areas which have to be computed have a straight base and vertical sides (Fig. 89). If an area

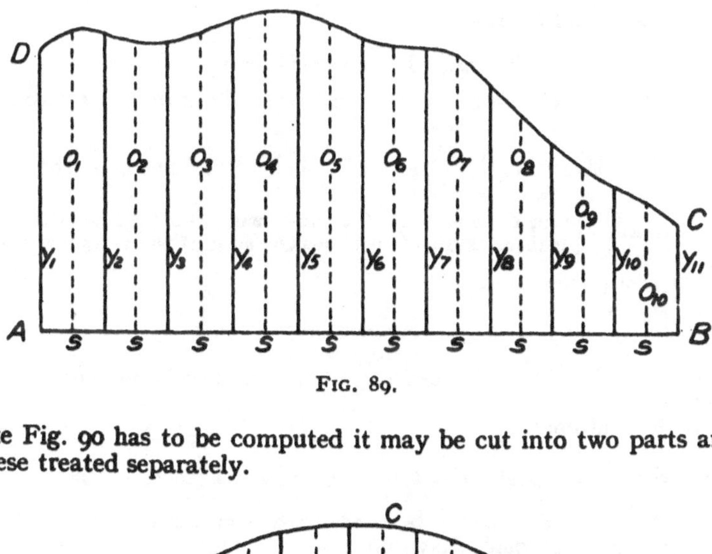

FIG. 89.

like Fig. 90 has to be computed it may be cut into two parts and these treated separately.

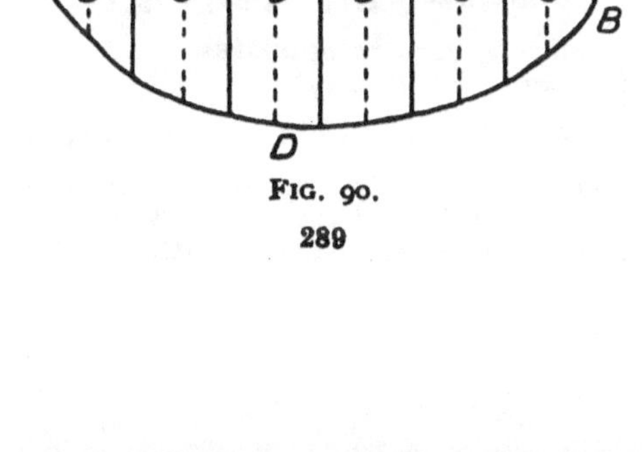

FIG. 90.

Rules for Finding Areas.

General Expression.—If the equation between x and y is of the form $y=f(x)$, the area $S=\sum y dx=\sum f(x)dx$ between the limits of x. The figure is supposed to be enclosed between the outline curve and the axis of x. The intercept on the x axis is divided into n equal parts and ordinates are erected at the points of division, giving $n+1$ ordinates in all. (Some of these may be of zero length.) The ordinates are $y_1, y_2, y_3, \ldots y_{n+1}$. The distance between successive ordinates is denoted by s.

1. *Mid-ordinate Rule.*—Draw the mid-ordinate of each compartment. (See dotted lines in Figs. 89 and 90.) Then

$$S=\text{base of figure}\times\text{average length of mid-ordinate}$$
$$=\frac{AB}{n}(o_1+o_2+o_3+\ldots+o_n)=s\Sigma o.$$

2. *Trapezoidal Rule.*—

$$S=s\{\tfrac{1}{2}(y_1+y_{n+1})+y_2+y_3+y_4+\ldots+y_n\}.$$

3. *Simpson's One-third Rule.*—Used when the number of ordinates is odd:

$$S=\frac{s}{3}\{(y_1+y_{n+1})+2(y_3+y_5+\ldots)+4(y_2+y_4+\ldots)\}$$

$$=\frac{s}{3}\times\begin{cases}\text{sum of first and last plus twice the sum of other odd}\\ \text{ordinates plus four times the sum of the even ordinates.}\end{cases}$$

With only three ordinates

$$S=\frac{s}{3}(y_1+4y_2+y_3).$$

4. *Simpson's Three-eighths Rule.*—Number of ordinates 7 or 10 or 13 or 16, etc.

With 10 ordinates

$$S=\frac{3s}{8}\{(y_1+y_{10})+2(y_4+y_7)+3(y_2+y_3+y_5+y_6+y_8+y_9)\}.$$

5*a*. *Weddle's Rule.*—To be used with 7 ordinates—usually more accurate than Simpson's first rule:

$$S=\frac{3s}{10}\{y_1+y_3+y_4+y_5+y_7+5(y_2+y_4+y_6)\}.$$

5*b*. *Dufton's Rule.*—11 ordinates

$$S=\text{base}\times\tfrac{1}{4}(y_2+y_5+y_7+y_{10}).$$

5*c*. *Lowry's Rule.*—4, 7, or 10 ordinates.

With 10 ordinates

$$S=s\{2(y_2+y_6+y_{10})+(y_3+y_5+y_7+y_9)\}.$$

6. *Counting Squares.*—Use the small squares by preference. Neglect those that have less than half of the square within the figure. Count as in, those which are more than halfway within.

7. *Weighing Methods.*—Draw figure on (uniformly) thin material, *e.g.* paper or metal foil. Cut out the lamina. Compare weight of lamina with weight of lamina of known area cut out of same material.

8. *Planimeter.*—Hatchet (rough) (§ 260) or Amsler's (accurate).

9. *Ordinary Mensuration Rules.*—*E.g.* (1) circle $= \pi r^2$, where r = radius; (2) triangle $= \frac{1}{2}bh = \frac{1}{2}ab \sin C = \sqrt{\{s(s-a)(s-b)(s-c)\}}$, where b = base, h = height, a and c are the other sides; s = half the sum of the sides; C is the angle between a and b.

10. If the equation of the curve is known find the value of $\int y dx$, between the limits of x, by the Integral Calculus.

Mean value of $f(x)$ from x_1 to $x_2 = \int_{x_1}^{x_2} f(x)dx$, divided by $(x_2 - x_1)$

= area of the figure divided by the length of the base.

Distance of the centre of the area from the y-axis

$$=\bar{x}=\frac{\text{moment of area}}{\text{area}}=\frac{\int f(x).xdx}{\int f(x)dx}=\frac{1}{2}s\frac{o_1+3o_2+5o_3+\ldots}{o_1+o_2+o_3+\ldots}$$

Distance of the centre of the area from the x-axis

$$\bar{y}=\frac{\text{moment of area}}{\text{area}}=\frac{\frac{1}{2}\int (f(x))^2.dx}{\int f(x)dx}=\frac{1}{2}\frac{\Sigma o^2}{\Sigma o}.$$

Moment of inertia of the area about the

$$y \text{ axis} = \frac{s^3}{4}(o_1+9o_2+25o_3+\ldots).$$

Moment of inertia of the area about the

$$x \text{ axis} = \frac{s}{3}(o_1^3+o_2^3+o_3^3+\ldots).$$

259. Note on Counting Squares.

For ordinary frequency polygon and histogram work the method of counting squares is often as good as any of the more elaborate methods. For other work, such as finding an area between a curve whose equation is $y=f(x)$, attention must be paid to the scales of x and y. If one division corresponds to unit increase of x and also to unit increase of y, the actual number of unit squares will then correspond with the values obtained by the mensuration rules applied to the x's and y's. If there is not this correspondence the x, y, value of a single square must be calculated and this figure multiplied by the total number of squares within the boundary to give the x-y value of the area. The same procedure must be applied to all the approximate methods where the ordinates and base lines are measured in terms of the divisions of the squared paper and not in terms of x's and y's.

260. The Hatchet Planimeter.

This instrument is very useful for finding the areas of regular and irregular figures. Its use should be more widely known.

A hatchet planimeter in its simplest form may consist of a metal rod about a quarter of an inch in diameter, bent twice at right angles to form a long arm about 10 inches long and two short arms about 3 inches long. One arm (Fig. 91) is sharpened to a point

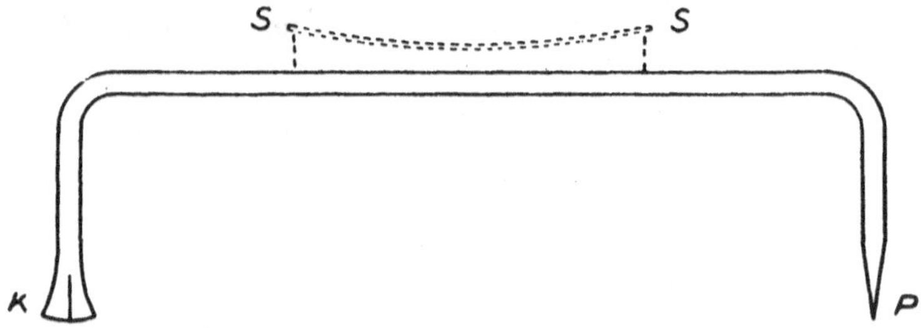

FIG. 91.

P, and the other to a convex knife-edge or axe-edge *K* in line with *P*. The distance from the point to the middle of the knife-edge is usually made, for convenience, a definite distance, *e.g.* 10 inches or 25 cms.

The rod must be stout enough to maintain this distance constant.*

To determine the area of a figure proceed as follows:

(*a*) Estimate approximately the centre of area *G* (see Fig. 92),

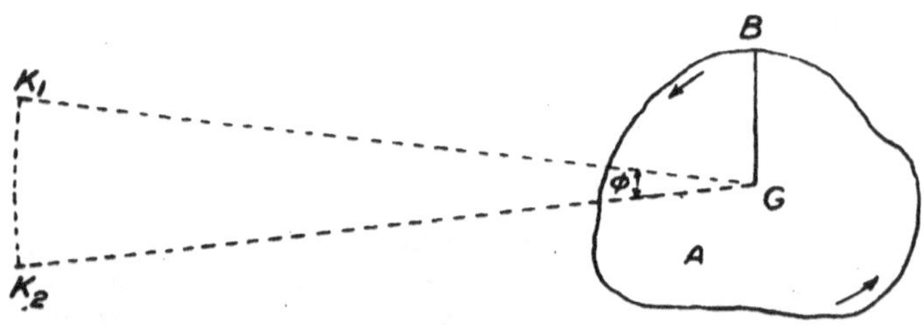

FIG. 92.

and through this point draw a straight line *GB* across the figure to the periphery.

(*b*) Stand the instrument on its short legs so that its long arm is roughly at right angles to *GB*, with the point *P* at the centre of gravity *G*. When in this position a dent K_1 is made on the paper by lightly pressing on the knife-edge *K*. Holding the instrument upright and lightly grasping the *P*-arm between finger and thumb,

* Hatchet planimeters may be bought from Messrs. Harling, Finsbury Pavement, London, E.C. Home-made instruments will give good results.

make the point P pass from the centre along the straight line GB to the point B in the periphery of the figure, and then to trace once round the outline of the figure until the point B is again reached, thence to the centre G again. The knife edge will now be at a place K_2. Press it lightly down to make a dent K_2. The distance between the two marks K_1K_2 is measured and the product of this length and the constant length PK gives approximately the area of the figure. Measure the arc K_1K_2 with the scale SS (Fig. 91) on the planimeter, the concave side of the scale being placed towards the area. This scale is bent to a curve of radius equal to PK.*

To obtain the result more accurately, it is advisable when the point P (after tracing the outline of the figure) arrives at the centre to take the tracer point up to the periphery as before, and then around in the opposite direction. This should, with care, bring the edge K back either to the first mark K_1 or near to it. Call this last position K_1'. The nearness of these marks depends to some extent on the accuracy with which the centre of area has been estimated.

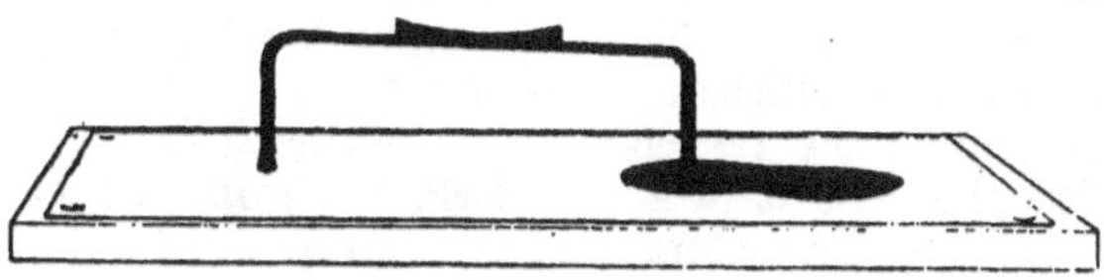

FIG. 93.

Measure both K_1K_2, $K_1'K_2$. Or, you may draw the line GB downwards instead of upwards and retrace. Or, having gone around once and got K_2 you may go around again and get K_3 and again to get K_4, and find the average of K_1K_2, K_2K_3, K_3K_4, etc.

Find how nearly the separate values agree and so give some indication of the accidental error to which the instrument is liable.

If the scale of the planimeter is in inches you may work wholly in inches; if not work in centimetres.

Fig. 93 represents a planimeter standing on a drawing board ready for use on the area of the metal template indicated in black.

Experiments.—Make your areas under test in the bottom right-hand corner of the paper so that the axe-edge of the planimeter does not run off the left-hand edge. Measure PK with a good millimetre scale.

(1) Make a rectangle on drawing paper or on squared paper 15 cms. long by 10 cms. broad. Find its area by the planimeter. Do it twice. Find the mean. Check by arithmetic and find the percentage error.

* For laboratory use it is convenient to have the scale always at hand. It is, of course, not necessary to mount the scale on the planimeter. Some prefer to measure the distance K_1K_2 with a pair of dividers and a diagonal scale.

(2) Make a circle of 7 cms. radius. Find its area. Repeat. Proceed as above. Divide the mean value of the area by the square of the radius and deduce the value of π in the formula

$$\text{Area} = \pi r^2$$

Find the percentage error.

From (1) and (2) deduce the percentage error of the instrument.*

(3) Find what error is made by a serious departure from the centre of gravity. Repeating (2) taking your starting point 1 cm. to the right and 1 cm. up from the centre of the circle, and find what error is made.

(4) Mark the outline of a template on your paper. The area will of course be too large unless your pencil is sharp and kept close to the edge of the template. Find the area—do it three times. Quote each. Find the mean.

You may keep the template on the paper and run the point around the edge. This makes easier running.

(5) Check the area of the template by tracing the periphery on squared paper and counting the squares.

(6) Also check the template by weighing it and comparing the weight with the weight of a rectangle cut from the same sheet of metal. (Weigh to an accuracy of 0·01 gm.)

(7) With a loop of thread, two pins, and a sharp pencil draw an ellipse about 6 inches long by 4 inches wide. Find its area.

Also calculate the area from $\frac{\pi}{4}Mm$, where M and m are the lengths of the major and minor axes. Find the percentage error.

NOTE.—Mathematics shows that the correct formula to be used is

$$c^2\phi \text{ or } c \times \text{arc } K_1K_2 = A\left(1 + \frac{\rho^2}{2c^2}\right)\dagger$$

where c = length of the arm of the planimeter ;

ϕ = angle between first and last positions of arm of planimeter (see Fig. 92) ;

A = area of figure ;

$A\rho^2$ = moment of inertia of the area about an axis drawn through the centroid at right angles to the area.

If it is not possible to pick out the centroid as the starting point, obtain K_1K_2 as before, then turn the figure through two right angles and repeat the process, taking the tracer point downwards

* For a 25 cms. long planimeter and an area about 12 by 12 cms. the planimeter overestimates the area by about 1½ per cent. Knowing this it is possible to make a correction to get the right value.

† See *Journal* of the Royal Astronomical Society of Canada, July, 1921, for full details of the theory and several practical examples.

and around the figure. Obtain a new value for K_1K_2, and the formula now is

$$c \times \text{mean of arcs} = A\left(1+\frac{k^2}{2c^2}\right)$$

where Ak^2=moment of inertia of the area about an axis drawn through the starting point at right angles to the area.

In the case of a circle $\rho^2=\frac{1}{2}r^2$

$$\therefore A=(c\times \text{arc } K_1K_2)\div\left(1+\frac{1}{4}\frac{r^2}{c^2}\right).$$

EXERCISES XXII

1. Find by mensuration or counting squares the area between the lines $y=5x+3$, $y=2$, $y=8$, $x=0$.

2. Plot the curve $y^2=4x$ from $x=0$ up to $x=16$. Erect ordinates at the integral values of x and find the area between the curve, the axis of x, and the ordinate at $x=9$ by one of the ordinate rules. Check your result by counting squares.

3. Find also for the curve $y^2=4x$ the area between the curve, the axis of x, and the ordinates at $x=10$ and $x=20$.

4. Find the area between the parabola $y^2=4x$, the hyperbola $xy=10$, and the two straight lines whose equations are $x=1$ and $x=3$.

5. Draw on squared paper a quadrant of a circle of radius 6 cms. (or 6 inches). Draw ordinates at every centimetre (or inch). Draw also the mid-ordinates. Find the area of the quadrant by the approximate rules, and counting squares. Compare with the true answer and find the percentage errors. Find the position of the centre of gravity of the quadrant.

6. Show that a loop of string lying on a plane surface encloses the greatest area when its shape is circular.

7. With two pins, a loop of thread, and a sharp pencil describe an ellipse about 7 inches long and 5 inches wide.

Find its area by the approximate rules and compare the result with the result obtained by the formula

$$\text{Area} = \pi \text{ (semi-major axis) (semi-minor axis)}$$

8. Plot the curve $y=\frac{1}{10}x^2$ from $x=0$ to $x=10$. What is the name of this curve? Draw ordinates at every centimetre division and find the area between the curve, the axis of x and the line $x=10$. Find the average value of y over this range.

What error is made by using Simpson's one-third rule with only three co-ordinates?

9. Plot the curve $y=\frac{1}{1+x}$ from $x=2$ to $x=10$.

Find the area between the curve, the end ordinates, and the axis of x. Discuss the relative accuracies of the methods you adopt.

10. Plot the curve

$$y=\log_{10}\frac{1}{x}$$

for values of x given by 0·1, 0·2, 0·3, etc., up to 1·0. Find in two ways the area enclosed between the curve, the axis of x, and the ordinate erected at $x=0·1$.

11. Plot the curve $xy=1$ from $x=0·1$ to $x=1·0$. Find the area between the curve, the axis of x and the ordinates at $x=0·1$ and $x=1·0$. Show that the area is numerically equal to the value of μ in the equation

$$\epsilon^{\mu}=10$$

Plot also log y against log x and find the equation of the graph obtained.

12. Plot the curve $y=x^2-3x+20$ between $x=0$ and $x=10$. Find the area between the curve, the axis of x and the ordinates at $x=0$ and $x=10$.

13. Plot the curve $y=\log_{10}x$ for values of x from 1 to 7. Find the area included between the curve, the axis of x, and the ordinate at $x=7$. Count squares and use the approximate rules.

14. Plot the curve $y=\log_e x$ from $x=1$ to $x=9$. Find the area included between the curve, the axis of x and the line $x=9$.

15. Plot the curve $y=\log_e x$ from about 10 values of y from $y=0$ to $y=3·0$. Draw an ordinate at $x=21$. Find the area included between the curve, the axis of x, and this ordinate. Count squares and use the approximate rules.

16. Plot the curve $y=\epsilon^{-x}$ for values of x from $x=0$ to $x=4·0$. Find the area between the curve, the axis of x, and the ordinates $x=1$, $x=4$.

17. Plot the curve $y=\sin\theta$ from $\theta=0$ to $\theta=\pi$, taking intervals of $\frac{\pi}{12}$. Find the area between the curve and the axis of θ. Deduce the average value of $\sin\theta$ from $\theta=0$ to $\theta=\frac{\pi}{2}$.

18. Plot the curve $y=\sin^2\theta$ from $\theta=0$ to $\theta=\pi$. Proceed as in Question 17.

19. Plot the curve

$$y=\cos\theta$$

between $\theta=0°$ and $\theta=60°$. Find in any way you please the average value of $\cos\theta$ over this range.

20. A quantity of air whose volume is 120 cubic feet at a pressure of 40 pounds per square inch is compressed under isothermal conditions (pv=a constant) to a volume of 15 cubic feet Plot the $p.v.$ curve. Measure the area between it and the axis of volume. Find the work done in the compression.

21. On a base of exactly 2 inches construct a regular pentagon. Find its area.

22. A reservoir is at its deepest 42 feet deep. Its cross-sectional areas have been measured at different depths with the following results :

Depth in feet	0	7	14	21	28	35	42
Area in square feet	52,000	37,300	23,950	15,850	11,650	4,600	0

Find the volume of water.

23. In some experiments on traction the pull on a car and the time were measured at equidistant points with the results shown :

E, pounds	325	310	290	255	230	225	258
s, feet	0	200	400	600	800	1000	1200
t, time, secs.	0	17½	30	42½	52½	65	76½

Plot the pounds-feet curve and the pounds-second curve. Find the average value of the pull in both cases. How is it that the space-average pull may differ from the time-average pull ?

24. The number of gas molecules N_V, having speeds between V and $V+dV$ given by

$$N_V=\frac{4N}{a^3\sqrt{\pi}}\epsilon^{-\frac{V^2}{a^2}}V^2dV$$

where N=total number of molecules.

Taking $a=1$ plot the curve $y=\epsilon^{-V^2}V^2$ by convenient steps from $V=0$ to $V=3$, and find the modal velocity, the median velocity and the mean velocity. What percentage of the molecules has a velocity less than the mode ? What fraction of the molecules has a velocity within 5 per cent. of the mode ?

25. Describe the Hatchet planimeter. Show how it is used. Indicate, if you can, the proof of the formula upon which it is based. Give some idea of the percentage error we should get if we used a hatchet planimeter of arm 10 inches long on an approximately circular area of average breadth 5 inches.

26. Try the hatchet planimeter on some of the areas in the exercises above. Also map out the outline of your hand on paper. Find its area by the planimeter. Mark out a rectangle on the same paper. Cut out both areas and weight them. Get the area of the diagram of the hand by comparison of the weights.

27. Use the planimeter and a good atlas to compare the areas of countries, lakes, counties, states. Why must not maps on Mercator's projection be used ? What method of projections correctly reproduces the area ?

28. Show that the volume of a prismoid is given by

$$V=\frac{H}{6}(T+B+4M)$$

where H is the height and T, B, M are the areas of the top, bottom, and mid-sections respectively.

29. Show that the volume of a barrel is given by

$$V=\frac{\pi H}{36}(5D^2+4d^2)$$

where H is the height, and D and d the bung and head diameters respectively.

MISCELLANEOUS EXERCISES

Work out as quickly and accurately as possible the following calculations, which are examples of common occurrence in the departments of mechanics, properties of matter, and heat in an elementary physical laboratory. All measurements are in C.G.S. units unless otherwise stated. Use either logarithms or slide-rule or abbreviated arithmetic.

1. Calculation of the atmospheric pressure from a barometric height

$$H=75.91\times 13.596\,(1-0.000182\times 20)\times 980.5$$

dynes per square centimetre.

2. Calculation of a thermal conductivity from Searle's bar experiment

$$\frac{500\times(30.2-14.7)}{127}=k\,.\,\frac{\pi}{4}(3.88)^2\times\frac{86.3-54.7}{10.0}$$

Find k.

3. Calculation of the density of a metal from observations of the dimensions in centimetres of a rectangular block and its mass in grams

$$7.82\times 5.46\times 2.73\times D=991$$

4. Calculation of the latent heat of a vaporization of a liquid

$$\begin{aligned}&14.3L+14.3(100.3-35.6)\\&=(327+168\times 0.095)(35.6-10.4).\end{aligned}$$

5. The number of dynes per square centimetre in 479 pounds per square foot.

6. The modulus of rigidity of brass

$$n=\frac{4\times 26.1\times(2.44)^3\times 7.50\times 980}{(0.0725)^4}\cdot\frac{\text{dynes}}{\text{cm.}^2}$$

7. The coefficient of viscosity by flow through a capillary tube

$$=\frac{\pi\times 12.5\times 980\times 45\times(8.465\times 10^{-4})^2}{8\times 6.5\times 1.91}$$

8. Young's Modulus from the up and down vibrations of a weight hanging from a wire :

$$Y=\frac{8\pi\times 4536\times 354}{0.216^2\times(0.0208)^2}\text{ dynes per sq. cm.}$$

9. Determination of g from an experiment with a spiral spring :

$$1.494=2\pi\sqrt{\frac{(1093+\frac{161}{3})\times 0.0493}{g}}.\quad\text{Find } g.$$

10. Force exerted by a heated iron bar:

$$F=3{\cdot}0\times10^{7}\times(\tfrac{1}{2})^{2}\frac{\pi}{4}\times0{\cdot}000012\times600 \text{ pounds.}$$

11. Temperature correction of a barometer reading:

$$h=763{\cdot}27\times\left(\frac{1+17\times0{\cdot}000018}{1+17\times0{\cdot}000182}\right)\text{ mm.}$$

12. Calculation of a moment of inertia:

$$I=665\left(\frac{15{\cdot}4^{2}}{12}+\frac{1{\cdot}25^{2}}{4}\right)\text{ gm. cm.}^{2}$$

13. Calculation of a Specific Heat:

$$S=\frac{(550+160\times0{\cdot}095)\times(21{\cdot}2-15{\cdot}8)}{662\times(100{\cdot}7-21{\cdot}2)}$$

14. Calculation of a vapour density:

$$D=2\times\frac{0{\cdot}1066}{54{\cdot}30}\times\frac{374}{273}\times\frac{760}{298}\times\frac{1}{0{\cdot}00009001}$$

15. The volume of a rectangular block whose dimensions as obtained with an ordinary millimetre scale are: length 14·2 cms., breadth 6·5 cms., depth 3·7 cms.

16. A pressure in dynes per sq. cm.

$$P=45{\cdot}2\times981\times13{\cdot}6$$

17. The same pressure as (16) in pounds per sq. in.

18. Convert 44,100 foot-pounds into kilowatt-hours.

19. A density, corrected for air-buoyancy:

$$D=\frac{7{\cdot}643\times0{\cdot}9982}{7{\cdot}643+0{\cdot}128}+\frac{0{\cdot}128\times0{\cdot}001293\times\frac{273}{293}\times\frac{751}{760}}{7{\cdot}643+0{\cdot}128}$$

20. Relation between Y, n, and k:

$$\frac{1}{1{\cdot}91\times10^{12}}=\frac{1}{9k}+\frac{1}{3\times7{\cdot}8\times10^{11}}.\quad\text{Find } k.$$

21. Y, by bending

$$=\frac{1020\times980\times76{\cdot}3^{3}}{4\times0{\cdot}46\times1{\cdot}86\times1{\cdot}24^{3}}\text{ dynes per sq. cm.}$$

22. g, in Toronto

$$=(980{\cdot}62-1{\cdot}9462\cos 87^{\circ}\,20'-0{\cdot}00022\times105)\frac{\text{cms.}}{\text{sec.}^{2}}$$

23. Calculation of a specific heat from a cooling experiment:

$$\frac{124{\cdot}1+45\times0{\cdot}095}{410}=\frac{95{\cdot}6s+45\times0{\cdot}095}{195}$$

24. Calculation of Young's Modulus from a bending of a beam:

$$Y=\frac{4\times1000\times980\times83{\cdot}5^{3}}{126\times(1{\cdot}67)^{3}\times0{\cdot}24}\text{ dynes per cm.}^{2}$$

25. Calculation of the pressure coefficient of a gas :

$$H_t = 112{\cdot}5\left\{\frac{133{\cdot}47(1+100\times 0{\cdot}000025)+0{\cdot}77\left(\dfrac{1+\frac{1}{273}\times 100}{1+\frac{1}{273}\times 15}\right)}{133{\cdot}47+0{\cdot}77\left(\dfrac{1}{1+\frac{1}{273}\times 15}\right)}\right\}\text{cms.}$$

26. The height of a hill from barometric readings at base and summit.

$$\text{Height in cms.} = 1{\cdot}840\times 10^6\left(1+\frac{15}{273}\right)\log_{10}\frac{29{\cdot}92}{28{\cdot}88}$$

27. The density of a substance from the following observations :

Weight of body in air (of density 0·0012 gm./c.c.) = 5·643 gm.
Weight of body in water (of density 0·9983 gm./c.c.) = 4·218 gm. Use formula

$$D = \frac{m_1}{m_1-m_2}\Delta - \frac{m_2}{m_1-m_2}a$$

28. The coefficient of viscosity of a liquid by flow through a capillary tube :

$$n = \frac{42{\cdot}6\times 980\times \pi\times(0{\cdot}0514)^4\times 60{\cdot}0}{8\times 56{\cdot}5\times 14{\cdot}16}$$

29. The surface tension of a liquid from measurements of a drop :

$$0{\cdot}363 = 2\cos 26^\circ\, 20'\sqrt{\frac{T}{980\times 13{\cdot}6}}$$

Find T.

30. The mass of a litre of moist air :

$$M = \frac{22{\cdot}40\times 1{\cdot}293\times 273}{293\times 760{\cdot}0}(759{\cdot}7-\tfrac{3}{8}\times 17{\cdot}1)$$

Find M.

31. On latent heat and calorimetry. Steam is passed into a worm immersed in a liquid which is brought to the boiling point and then partially boiled away.

$$m\times 539+m(99{\cdot}9-78{\cdot}3) = (2040\times 0{\cdot}621+269\times 0{\cdot}094)(78{\cdot}3-15{\cdot}7)+2040\times 202.$$

Find m.

32. The density of a body (see Question 27) :

Weight in air (of density 0·0012) = 17·2625 ;
Weight in water (of density 0·9982) = 10·3874.

33. The latent heat of vaporization of steam from the readings shown in the formula

$$16{\cdot}92\{L+(99{\cdot}6-39.5)\} = (355.30+165{\cdot}1\times 0{\cdot}094)(39{\cdot}5-12{\cdot}1).$$

34. The value of g from an experiment with a Borda's pendulum :

$$g = 4\pi^2\frac{\left(99{\cdot}914+\frac{2}{5}\cdot\frac{3{\cdot}92^2}{99{\cdot}91}\right)\left(1+\frac{20}{32(99{\cdot}91)^2}\right)}{\left(\frac{314}{313}\times 2\times\frac{86400}{86409}\right)^2}$$

35. The radius of a thermometer bulb from the equation

$$\frac{4}{3}\pi r^3 \times 90 \times 0{\cdot}00045 = 28 \times 0{\cdot}039.$$

36. The density of a liquid with a specific gravity bottle :

$$D = \frac{16{\cdot}3142 - 11{\cdot}0062}{17{\cdot}5714 - 11{\cdot}0062}(0{\cdot}9982 - 0{\cdot}0012) + 0{\cdot}0012$$

37. A platinum resistance thermometer has at 0·0° C., 100·0° C., 443·8° C. resistances of 257·4, 356·9, 673·0 units, respectively. The equation connecting resistance and temperature is

$$R_t = R_0(1 + at + bt^2)$$

where R_0, R_t, are the resistances at 0° C. and t° C., respectively, and a and b are constants.

(*a*) Find a and b.

(*b*) Find at what temperature according to the above equation the resistance will be zero.

38. The specific heat of bronze from the following data :

Weight of bronze	21·30 gm.
Weight of copper calorimeter	52·87 gm.
Weight of water in calorimeter	105·73 gm.
Initial temperature of bronze	99·5° C.
Initial temperature of water and calorimeter	17·0° C.
Final temperature of mixture	18·5° C.
Specific heat of copper	0·094

39. The calculation of vapour density from an experiment with Victor Meyer's method:

$$D = \frac{0{\cdot}054}{0{\cdot}0000900 \times 18{\cdot}0 \times \dfrac{273}{294} \times \dfrac{765 - \dfrac{140}{13{\cdot}6} - 18}{760}}$$

40. Density of ice by weighing:

$$\delta = \frac{7{\cdot}9317 \times 0{\cdot}99987 \times 13{\cdot}596}{7{\cdot}9317 \times 13{\cdot}596 + 10{\cdot}2956 \times 0{\cdot}99987}$$

41. Coefficient of viscosity of air :

$$\eta = \frac{\pi \times 981 \times 14{\cdot}1 \times (0{\cdot}051)^4}{8 \times 20{\cdot}5 \times 8{\cdot}35}$$

42. Specific heat of quartz by a steam calorimeter :

$$s = \frac{(0{\cdot}7420 - 0{\cdot}002872 \times 85{\cdot}25)539}{16{\cdot}730 \times 85{\cdot}25}$$

43. Mechanical equivalent of heat by Callendar's apparatus :

$$J = \frac{3768 \times 552 \times 47{\cdot}65 \times 981{\cdot}2}{388{\cdot}9 \times 5{\cdot}94}$$

44. Comparison of thermal conductivities of brass and steel by bar method :

$$\frac{K_b}{K_s} = \left\{\frac{\log(1{\cdot}45 + \sqrt{1{\cdot}45^2 - 1})}{\log(1{\cdot}28 + \sqrt{1{\cdot}28^2 - 1})}\right\}^2$$

45. Determine the latent heat of vaporization of water from the following observations:

	gm.
Mass of copper calorimeter	105·88
Mass of calorimeter plus water	384·28
Mass of calorimeter plus water plus condensed steam .	394·24
Temperature of steam	100·0° C.
Temperature of cold water at the beginning of the experiment	14·4° C.
Temperature of water finally	35·1° C.

46. Find the number of
 (*a*) calories in a British thermal unit;
 (*b*) dynes in the weight of a pound;
 (*c*) gms. per sq. cm. in a pound per sq. in.;
 (*d*) "dynes per sq. cm." in one "poundal per sq. ft.";
 (*e*) foot-pounds in one calorie;
 (*f*) litres in one cubic foot;
 (*g*) "c.c. per sec." in one "cub. ft. per min.";
 (*h*) "dynes per sq. cm." in one "pound-wt. per sq. in."

47. The coefficient of linear expansion of cast iron is $1{\cdot}06 \times 10^{-6}$ in centigrade units. What is it in Fahrenheit units?

48. The thermal conductivity of cast iron is 0·108 in C.G.S. centigrade units. What is it in terms of the pound, foot, minute, and degree Fahrenheit?

49. The earth receives from the overhead sun 1·925 calories per sq. cm. per minute. Find the horse-power equivalent of the solar power received per acre.

ANSWERS

Exercises I

3·1416)180·000(572957
157080
22920
21991
929
628
301
283*
18
16
2

At the mark * notice that 3·6 comes nearer to being "4 to carry" than "3 to carry," since it is more than three and a half.

The sixth figure which is given in the quotient is intended to represent the value of the 2/3 remaining after the last subtraction. It could also have been obtained by continuing the regular process of abridged division: cancel the 3, leaving 0·3 for the divisor; then $2 \div 0{\cdot}3 = 7$; multiplying, 7×3 gives 2 to carry, $7 \times 0 + 2 = 2$.

2. 28·60741. After the figure 4 of the quotient has been written down the next step is to multiply 236453 by 4. Ordinarily it is sufficient to take the nearest cancelled figure and say $4 \times 6 = 24$, giving 2 to carry; but as the product comes close to 25, which is on the boundary between 2 to carry and 3 to carry, it is well to investigate one more cancelled figure, saying $4 \times 4 = 16$, giving 2 to carry toward 4×6; the latter then becomes 26, giving 3 to carry towards the written product instead of 2.

8. When they are equal.
9. When they are equal.
11. When the number is unity.

Exercises II

5. (*a*) equal; (*b*) sp. gr. < density.
6. $u\rho = 1$.
9. 352, $70{\cdot}3 \times x$, $44{\cdot}7 \times y$, $0{\cdot}0160 \times z$.

Exercises III

2. $\frac{3\pi}{4}$, 341°.
7. 57, 57, −57.
8. $\tan A = 1/\tan G$.
14. 206265.
15. 12/206265.

Exercises IV

2. Five, four.
5. Six.
7. Between 77·158 and 77·162.
   Between 15·431 and 15·433.

9. 33·0 cm.
16. 0·04 per cent., 1/150000 per cent., 0·6 per cent., 1·3 per cent., $1\frac{1}{3}$ per cent.
17. 8 per cent., 1·9 per cent., 0·4 per cent.
18. 0·003903, 20·59°.
19. a=0·003064, b=−0·00000051.

Exercises V

1. 2·594, 2·704, 2·717.
2. 2·718.
3. If x=natural logarithm of π, $x=2{\cdot}3026 \log_{10}\pi=1{\cdot}147$.
7. $\frac{1}{269}=0{\cdot}0372$, $\frac{1}{\pi}=0{\cdot}3183$, $\frac{1}{e}=0{\cdot}3679$.
10. Approximately 2,000 years, $\lambda T=\log_e 2=0{\cdot}693$.
11. 3·85 days.

Exercises VI

1. Expression$=\frac{400}{800-3}=\frac{1}{2(1-\frac{3}{800})}=0{\cdot}5018$.
2. $(500+4)(500-2)=251000$, ∴ ans.=0·251.
3. 0·095 is 5 per cent. less than 0·1. One-tenth of 264·3=26·43. 5 per cent. of this is one-twentieth of it, *i.e.* 1·32. Taking this away we get 25·1.
4. $1+m\delta_1+n\delta_2+p\delta_3+\ \ldots$
5. 0·15 per cent.
6. 9·87.
7. (*a*) 7·071, (*b*) 0·99672, (*c*) 20·0075.
8. (*a*) 2·00018, (*b*) 6·0030.
9. 0·33 per cent. less.
10. 751·8 mm.
11. (*a*) 761·8, (*b*) 762·3, (*c*) 30·142, (*d*) 759·18, (*e*) 200·65, (*f*) c=0·00043, (*g*) c=0·000026.
12. 17·2705.
13. 8·903.
14. The other side.
15. 2π, $2\pi r$, $4\pi r^2$.
16. $8x+3$, $-\frac{1}{2\sqrt{x}}$, $-\sin x$, $\frac{5}{(x+4)^2}$
17. $128-32t$. At the expiration of 4 secs.
18. 0·53 radian, 0·38 radian.
19. 0·999868, 0·999730, 0·99821, 0·99531.
20. 1/1863, 0·4343/1863.

Exercises VII

6. (*a*) $\frac{22}{7}$, $\frac{44}{14}$, $\frac{66}{21}$, $\frac{540}{172}$; (*b*) $\frac{10}{7}$, $\frac{87}{32}$; (*c*) $\frac{17}{12}$, $\frac{58}{30}$, $\frac{9}{4}$, $\frac{48}{20}$, $\frac{8}{3}$.
7. (*a*) 225, 50·2, 17,400, 0·0064; (*b*) 3,370, 358, 2,300,000, 0·000512. (*c*) 28·0, 8·85, 2·8; (*d*) 7, 2·9, 1·34+, 0·7; (*e*) 0·238, 0·318, 273; (*f*) 1·06.
8. 19·5, 4·06, 0·15, 25·5, 0·88.
9. K 31·9 per cent., Cl 28·9 per cent., O 39·2 per cent.
10. 30·6 sq. yds.
11. 29·5 lbs.
12. (*a*) 29·8, (*b*) 153, (*c*) 37, (*d*) $4{\cdot}21\times10^7$, (*e*) 0·000216, (*f*) 2·00.
13. (*a*) 12, 81, 21, 78, 28, 45, 71, 87, 93, 4, 81, 74, 77, 56, 91.
(*b*) 91, 48, 55, 80, 21, 87, 6, 62, 74, 98, 15, 33.
(*c*) 63, 41, 76, 94, 35, 37.
14. 4·0 inches.

Exercises VIII

10. $r=\frac{1}{2}+6\theta/8\pi$.
11. $r=\frac{1}{2}\times(1{\cdot}08)^{8\theta/\pi}$.
12. $x/4{\cdot}3+y/3{\cdot}4=1$, or $y=-0{\cdot}79x+3{\cdot}4$.
16. $dy/dx=2x$.
17. $dy/dx=-1/x^2$.
18. $dy/dx=-1/(2\sqrt{x})$.
19. $dy/dx=e^x$.
20. $dy/dx=-e^{-x}$
21. $dy/dx=1/x$.
22. $dy/dx=0{\cdot}4343/x$.
23. $dy/dx=\cos x$.
24. $dy/dx=-\sin x$.
25. $48{\cdot}4^\circ$ F.
27. 0·86 inch, 2·42 inches per day.
28. 0·83, 1·83, 3·27 inches; 0·570, 0·854 inch per day.
29. **Fastest Bristol to Taunton, slowest Newton to Plymouth**; 63·4 miles per hour.
30. The equation of the original curve is $y=\frac{1}{10}e^x$. Derived curve has same shape.
31. Original curve is $y=e^x$. Derived curve has same shape.
32. Time of ascent = 50 secs.; $s=1600t=32t^2$.
33. $v=32t$ ft. per sec.; $a=32$ ft. per sec. per sec.
34. $a=4$, $v=5+4t$, $s=6+5t+2t^2$.
35. 80, 16, 0, -16, -128 feet per sec.
36. 0·4050, 0·000274.
37. $2\pi r$, 28·27.
38. $D=0{\cdot}00074C+0{\cdot}998$, $C=1340(D-0{\cdot}998)$.
39. The company gave away 40 cubic feet for every 176 cubic feet paid for by the customer; $18\frac{1}{2}$ per cent. loss.
40. 0·999868, 0·999730, 0·99821, 0·99531.
0·0000138, 0·000152, 0·000290.
41. Simple harmonic motion, for the acceleration is proportional to the distance from the centre.
42. $x=1$, $y=2$.
43. $y=30\frac{1}{4}$, $x=5\frac{1}{2}$.
44. $x=2{\cdot}24$.

Exercises IX

1. $y=7{\cdot}7-4{\cdot}7x$.
2. $a=18{\cdot}84$, $b=-0{\cdot}0185$, $c=5{\cdot}76\times10^{-6}$.
4. $a=3{\cdot}9\times10^{-3}$, $b=-8\times10^{-7}$.
7. $1/D^2=0{\cdot}000025N$.
8. $1/D^2=0{\cdot}0003+0{\cdot}000318N$.
9. $y=0{\cdot}043x-1{\cdot}80$.
10. $E=0{\cdot}277R+4{\cdot}75$.
11. $E=0{\cdot}148R+3{\cdot}44$.
12. $E=0{\cdot}118R+1{\cdot}75$.
13. $E=0{\cdot}164R+2{\cdot}6$, efficiency $=0{\cdot}8$.
14. $pv=40{,}000$.
15. $pv=3083$.
16. $pv^{1{\cdot}068}=$ a constant.
17. $s=490t^2$.
18. $\log E=1{\cdot}1340\log t+0{\cdot}7827$.
19. $\log\theta=1{\cdot}61-0{\cdot}044t$ or $\theta=40{\cdot}5\times10^{-0{\cdot}044t}$.
20. $\lambda=0{\cdot}0131$.
21. 6·9, 14·4, 27·3.
22. Approximately, $\log p=7{\cdot}96-2265/\theta$, from $\theta=273$ to 373. Graph gives 8·9 cms. Correct = 9·2 cms.
23. Approximately, $\log p=7{\cdot}455-2075/\theta$, from $\theta=373$ to 473. Errors are $+2{\cdot}0$, $+0{\cdot}7$, $-1{\cdot}1$, $-2{\cdot}7$, $-1{\cdot}5$, $+7{\cdot}3$ cms.
24. $y=3x/(1+4x)$.
25. $y=100x/(30+x)$ and $y=100(1-e^{-x/30})$.
26. $\log\eta=1{\cdot}7(1-t/43)$ or $\eta=50\times10^{-t/26}$.
27. $pv^{1{\cdot}0646}=479$.
28. $I=0{\cdot}0512V^{1{\cdot}525}$.
29. 0·98.
30. $A=6{\cdot}1007$, $B=1518$, $C=122{,}500$.
31. $y=(18x+120)/x$.
32. $pv^{1{\cdot}33}=750$.
33. Expansion $pv^{1{\cdot}24}=655$, compression $pv^{1{\cdot}24}=233$.
34. $n=2{\cdot}2\log N/M$.
35. (a) $y=2(x+3)^2$. (b) $y+16=4(x+2)^2$.

36. $y=12-2x-0.1x^2$.
37. $\eta=(0.01793)/(1+0.023t)^{1.54}$.
38. Molecular Wt. × Depression = 18, nearly.
39. $V=kR^2$.
40. $T=75.2+50.4(d-1.0000)$.
41. $N^2=0.0257+0.0037n$. $H=0.16$ C.G.S. unit.
42. $\tan\theta=0.23n$, $H=0.167$ C.G.S. units.
43. $y_1=0.38$ radian, $y_2=0.53$ radian.
44. $xy'=0.4343$.
45. 2 nearly, 2.4° and 1.2° per cm. respectively.
46. From −60° to 100° C. $\log\eta=0.800\log\theta-5.713$.
From +60 to 200° C. $\log\eta=0.775\log\theta-5.651$.
47. $N=14{,}260e^{-8970d}$
48. For first five readings, $W=1.70x^{0.835}$.
For the rest $W=8.30x^{0.844}$.
49. $V=128e^{-0.139t}$.
50. $\phi=A+B/\lambda^2$.
51. $I=I_0e^{-t/76}$, t in seconds. Time for half value 53 secs.
52. Fd^2 = a constant.
53. $C=0.775$, $k=0.147$.
54. $k=121$.
55. $\lambda=0.000144$, $E_\lambda=4.02\times10^{12}$.
56. $k=0.00132$.
57. $\log(A-x)=\log A-0.002177t$ or $x=14.11(1-\epsilon^{-0.00501t})$.
58. $C_1/C_2^2=0.50$.

Exercises X

3. Yes.
4. Taking the square root of the product.
5. (*a*) 1.43 inches, (*b*) −39.9°, (*c*) 2.26 inches.
6. In London, lat. 51.5° N. the figures are 25.4°, 38.5°, 62.0°, 53.0°.
7. 4049, 4167, 4272, 4398, 4593.
8. +0.145, +0.030 c.c.
9. −0.09, −0.45.
10. 209°, 1700′; 203° 4900′; 27.6″, 2300′; 22.5″, 7800′; 31.7″, 215°: 26.5″, 206°.
11. No.
12. (*a*) $x=3.41$, $y=0.97$. (*b*) $x=2.66$ and -0.66.
(*c*) 1.60, 0.46, −2.06. (*d*) 0.214, −0.934.
13. −0.88925, 0.22567, 0.66359.
14. 2.36 and 4.43 inches.
15. $\theta=1.297603$.
16. (*a*) 12.7 mm.; (*b*) 657.3, 681.6, 706.8, 732.9, 788.0; (*c*) 814.5, 873.0, 935.6, 1000. The correct are 816.9, 875.1, 937.9, 1004.
17. (*b*) 44, 75, 253 gms. of salt per kilogram of water. (*c*) 1.034, 1.065, 1.094, 1.148 gm. per c.c. (*d*) $C=1340(D-0.998)$. (*e*) 1.055, 1.169.
18. 1.0595, 1.0980.
19. 25.8 gms., 85.5 gms., 204.0 gms.
20. 28.5 gms., 55.0 gms.
21. 13.4 gms., 45 gms.
22. −2.32°, −5.12°, −8.46°.
23. The Great War started in 1914, 730,000, 900,000, 1937.
24. $\log N=0.0503(Y-1881)+3.07$. (*a*) Between 117,000 and 128,000. (*b*) Between 166,000 and 186,000.
25. (*a*) 16.61 millions, (*b*) 39 millions, (*c*) 0.0151, (*d*) $\log P=0.00521Y-2.40$ over whole range. $P=14.1\times10^{0.0052(Y-1851)}$ from 1851 to 1901.
26. (*a*) 10.5 millions, (*b*) 5.8 millions, (*c*) $\log P=0.00753Y-13.54$.
The Pearl-Reed formula $1/P=A+B\epsilon^{-nt}$ gives $1/P=-0.031+0.430\epsilon^{-0.0155t}$, t being measured in years from 1851.
27. (*a*) 1.0937. (*b*) 2.789. (*c*) 58.14 mm. (*d*) 58.14 mm.
29. 23.4, 2.0, 24.1, 0.5.

Exercises XI

1. a, b, c are the intercepts with the axes.

2. $x^2+y^2-\dfrac{z^2}{9}=0$.

3. A point.

4. A vertical cliff

6. 10·04 cms., 28·6°; 8·85, 7·24, 8·50 cms.

7. 7·8 cms., 24·6°, 31·0°.

8. 18·5°.

9. (*a*) 18·5°; (*b*) $1/3\sqrt{2}$, 13·2°; (*c*) 0°; (*d*) 64·6° N. or S. of E.; (*e*) 1131′.

10. Dip 31¾° Strike N. 66° E.

11. Dip 16·1°. Strike is E. and W.

12. Ht=1·635″, angle 70½°.

13. (*a*) 2·83″, (*b*) 109°.

15. (*a*) 94·2°, (*b*) 70°, 70°, 40°.

Exercises XII

3. Sixteen- or eighteen-hundredths of a unit.

4. 24·44 inches.

5. $10\,m_1+20e_2$.

7. (*a*) 169, (*b*) 170·1, (*c*) 167·4, (*d*) 169±1 cm.^{2}.

8. (*a*) 168·7, (*b*) 168·86, (*c*) 168·59, (*d*) 168·7±1 inches^2.

9. 429·7±2·1 cm.^{2}

10. 83·1±0·3 cm.^{2}

11. 161·9±0·2 cm.

12. 4·25±0·04.

Exercises XIII

4. 17/18, 18 to 17.

8. Thirteenths.

12. 2·0117 seconds. About 0·00005 sec.

14. 985 cm. per sec.^{2}.

15. The length of the vernier=39 scale divisions and it is divided into 40 equal parts.

Exercises XIV

3.

Sample.	1	2	3
Average	0·93	2·00	0·99
Mode	0·97	2·02	0·975
Median	0·94	2·03	0·98
Quartiles	0·867 and 1·020	1·87 and 2·13	0·92 and 1·08
S.I.Q.R.	0·076	0·14	0·08
Average deviation	0·08	0·16	0·09

4. 23·1, 22·7, 21·7, 2·5.

6. No objection to the 3½ plan.

12. The quotient obtained by dividing the distance from the median to the mode, divided by the semi-interquartile range.

17. U-shaped type.

18. 27·34 mm., average deviation=0·03 mm.=0·11 per cent. of the average length.

19. 5·17, 5·15, 5·0, 4·55, 5·86, 0·65.

Exercises XV

6. Median=31, S.I.Q.R.=2½. Probable error of median=0·7.

7. Mode 21·7, Median 22·7, Arithmetical Mean 23·1, *ad* 2·5.

8. A.M. 20·154, a.d. 0·080, A.D. 0·025, p.e. 0·076, P.E. 0·024.

9. A.M. 11·79, a.d. 0·05, A.D. 0·02, p.e. 0·05, P.E. 0·02.

10. A.M. 2·755, Me. 2·75, Mo. 2·75, p.e. 0·014, P.E. 0·002.

11. A.M. 36·94, a.d. 0·12, A.D. 0·05, p.e. 0·10, P.E. 0·045.

12. A.M. 69·0, Me. 69, Mo. 69·2, Q_1 67·3, Q_3 70·6, S.I.Q.R. 1·64, p.e. 1·65.

14. A.M. 7·79, a.d. 0·54, A.D. 0·017, p.e. 0·049, P.E. 0·016.

15. A.M. 1.164, a.d. 0·008, A.D. 0·003, p.e. 0·007, P.E. 0·0022.
16. Bimodal.
19. A.M. 1·0545, p.e. 0·010, P.E. 0·0032.
20. A.M. 1·200, P.E. 0·003, $T=28\cdot58$ dynes per cm.
21. A.M. 7·1400, P.E. 0·00006.
22. A.M. 340·5, P.E. 0·9.
23. A.M. 209·020, P.E. 0·0013.
24. A.M. 0·230, a d 0·016, A.D. 0·0045, P.E. 0·0038.

Exercises XVI

1. Fine platinum wire is best; the finest obtainable.
2. Weighting according to the average deviation would give practically the same result.
4. 1021. 5. 1·23±0·03. 6. 58·48±0·03.
7. 11·3436±0·0005. Yes; for the observer's ranges do not overlap.
8. 1·42890±0·00007. 1·42892. 9. 426·9.
10. Yes. 256·42±0·06.
11. W.A. 299,868, P.E. 14. Ranges nearly overlap.

Exercises XVII

4. Initially the A.M., a.d., and p.e. are 20·154, 0·08, 0·075. The reading 19·9 is rejected. Final values are A.M. 20·18, a.d. 0·062, A.D. 0·021, p.e. 0·052, P.E. 0·017.
5. Reject 1·278. Final values are A.M. 1·2649, a.d. 0·0025, A.D. 0·0008, p.e. 0·0023, P.E. 0·0007.
6. Reject 511. Final values, A.M. 538·1, a.d. 6·3, A.D. 2·1, p.e. 5·0, P.E. 1·7.
7. No rejections. A.M. 74·2, a.d. 0·42, A.D. 0·11, p.e. 0·38, P.E. 0·094.
8. Reject in succession 7·62 and 7·70. Final values A.M. 7·82, a.d. 0·02, A.D. 0·007, p.e. 0·019, P.E. 0·006.
9. The first mean is the oft-quoted value 29,002·3 feet. Reject 29,026·1. Final values A.M. 28,997·5, a.d. 5·2, A.D. 2·3, p.e. 4·3, P.E. 1·9.
10. Reject 1·087. Final values A.M. 1·051, a.d. 0·008, A.D. 0·003, p.e. 0·007, P.E. 0·002.
11. Reject the reading between 4·8 and 4·9. Final mean 5·47 with a p.e. of 0·14.

Exercises XVIII

1. Better allow 4½ cm. over and above the 57·0 cm.
2. 200·00±0·45 sq. cms.
3. 1·9 per cent.
4. Let $A=\frac{1}{n}(a_1+a_2+a_3+\ \ldots\ +a_n)$. Writing the probable errors in $a_1, a_2, a_3 \ldots$ as $p_1, p_2, p_3 \ldots$ and in A as P, we get $P^2=p_1^2+p_2^2+p_3^2+\ldots p_n^2$. But $p_1=p_2=p_3=\ldots$, therefore $P^2=np^2$ or $P=p/\sqrt{n}$.
5. 32,740±590 c.c.
6. ($2\cdot08\times10^{12}$±1·8 per cent.) dynes per sq. cm.
7. 60,590±60 gm. cm.².
8. 979·9±1·8 cm. per sec. per sec.

Exercises XIX

2. By least squares $y=9\cdot85-0\cdot56x$ or $x/17\cdot6+y/9\cdot85=1$. By graph $y=10\cdot2-0\cdot61x$ is a good fit.
3. $E=2\cdot8+0\cdot16R$. 5. 0·83. 6. 163·5, 163·5.
7. (1) $x=5\cdot847$, $y=4\cdot406$. (2) $x=5\cdot842$, $y=4\cdot368$.
8. $x=6\frac{1}{8}$ feet, $y=10\frac{7}{8}$ feet, $z=3\frac{1}{2}$ feet.
9. (1) $y=1\cdot638+0\cdot4138x$; (2) $y=1\cdot636+0\cdot4142x$; (3) $y=1\cdot625+0\cdot4167x$.
10. (a) $y=-0\cdot035+1\cdot348x$; (b) $y=1\cdot2+66x$ nearly.
11. $y=0\cdot544x$. 12. $y=2\cdot66+0\cdot422x$.

13. $x=6.21$, $y=-0.468$, $z=-0.769$.
15. $h=0.55$ cm., $m=453$.
17. 4.293 sec.
18. $v=3.19513+0.44253x-0.07653x^2$.
19. $\log p=21.07449-2903.39/T-4.71734 \log T$.
20. $\log p=6.1007-1518/T-122500/T^2$.

Exercises XX

1. Let $C=k \tan a$, and an error δa in a produce an error δC in C. Then, by § 220, $\delta C=k \sec.^2 a \;\; \delta a$ and $\frac{\delta C}{C}=\frac{\sec^2 a}{\tan a}\delta a=\frac{2\delta a}{\sin 2a}$. Therefore, for a given error, in a the ratio of the error in C to C itself is least when $\sin 2a$ is greatest, *i.e.* when $2a=90°$ or $a=45°$.

2. Let R_1 and R_2 be the two resistances in the bridge and a and b the corresponding lengths of bridge wire. $a+b=$whole length of wire$=$a constant, L, say. We have

$$\frac{R_1}{R_2}=\frac{a}{b}$$

Regarding R_2 as fixed we have to show that the error in R_1 bears to R_1 itself the smallest ratio when the point of balance is at the centre of the bridge wire.

We have $R_1=\frac{a}{L-a}R_2$, and by § 220 $\delta R_1=\frac{(L-a)-a(-1)}{(L-a)^2}\delta a=\frac{L\delta a}{(L-a)^2}$

$$\therefore \quad \frac{\delta R_1}{R_1}=\frac{L\delta a}{a(L-a)}$$

For a given δa, $\frac{\delta R_1}{R}$ is least when $a(L-a)$ is a maximum, and since $a(L-a)$ is the product of two terms whose sum is a constant, it is a maximum when $a=L-a$, *i.e.* when $a=\frac{L}{2}$.

3. Let $L=$distance between image and object and v the image distance.

$$\frac{1}{f}=\frac{1}{v}+\frac{1}{L-v} \qquad \therefore f=\frac{v(L-v)}{L}=\frac{1}{L}(Lv-v^2)$$

Therefore by § 220, $$\delta f=\frac{L-2v}{L}.\delta v.$$

Therefore $$\frac{\delta f}{f}=\frac{L-2v}{L}.\frac{L}{v(L-v)}\delta v=\frac{L-2v}{v(L-v)}\delta v$$

Now $\frac{L-2v}{v(L-v)}$ is zero when $v=\frac{L}{2}$. Therefore, for a given error in v the fractional error in f is least when $v=\frac{L}{2}$.

Exercises XXI

4. 5.33.
6. 14.5, p.e. 1.1 per cent, P.E. 0.09 per cent.
8. Mothers: Mo. 62½, Med. 62½, A.M. 62.6, Q_1 60½, Q_3 64½, S.I.Q.R. 2.0.
   Fathers: Mo. 68, Med. 68½, A.M. 68¼, Q_1 66½, Q_3 70½, S.I.Q.R. 2.0.
9. 0.042 ± 0.034.
10. 3.458, 1.323, 0.891, 38.3 per cent.
11. $r=+0.91$. 12. $r=+0.51$. 13. $r=+0.94$.
16. 0.16 per cent.; 10 above, 10 below.
18. $\log A=1.01-0.0347t$.
19. $\log G=0.138t-2.34$ or $G=\frac{1}{210}(1.38)^t$.
20. $t=0.840s-44$ over the region covered. 21. $t=2.63s^{1.076}$

22. $t=5{\cdot}9s^{1\cdot125}$, after 3 miles.
23. $t=0{\cdot}59s$ for the long-distance races.
24. $s=0{\cdot}1+0{\cdot}0033t$ for the first 5 miles.
$t=265s^{1\cdot23}$ from $\frac{1}{4}$ mile to $\frac{3}{4}$ mile.
$t=255s^{1\cdot08}$ from $\frac{3}{4}$ mile to 10 miles.
25. Ratio$=1{\cdot}575$.
26. Log $n=7{\cdot}1-4\frac{2}{7}\log l$ or $\log n=4{\cdot}19-\frac{4}{7}l$ covers the data well. The Length classes are, however, too indefinite.
27. $y=54{\cdot}347+1{\cdot}555x+68{\cdot}549\log x$(see Pearl, "Medical Biometry").
28. No.
32. The smaller fork gives an octave above the larger fork.

Exercises XXII

1. 168.
2. $85\frac{1}{3}$.
3. 77·1.
4. 5139.
5. 28·27, $\bar{x}=\bar{y}=2{\cdot}54$.
8. $33\frac{1}{3}$, $3\frac{1}{3}$. None.
9. 1·300.
10. 0·292.
11. 2·30, $\log x+\log y=0$.
12. $383\frac{1}{3}$.
13. 3·31.
14. 11·8.
15. 43·95.
16. 0·3496.
17. 2, $2/\pi$.
18. $\pi/2$, $\frac{1}{2}$.
19. $3\sqrt{3}/2\pi$ or 0·83.
20. 143,800 foot-pounds.
21. 6·88 sq. ins.
22. 832,900 cubic feet.
23. Space average 261 pounds, time average 271 pounds.
24. Mode 1·000, Median 1·085, Mean 1·128, 43 per cent., $\frac{1}{12}$.
25. About $1\frac{1}{2}$ per cent.

Miscellaneous Exercises

1. $1{\cdot}008\times10^{8}$ dynes per sq. cm.
2. 1·63 calories per sq. cm. per unit temperature gradient.
3. 8·50 gms. per c.c.
4. 540 calories per gm.
5. 229,400.
6. $4{\cdot}04\times10^{11}$ dynes per sq. cm.
7. 0·0125 dynes per sq. cm. per unit velocity gradient.
8. $2{\cdot}00\times10^{12}$ dynes per sq. cm.
9. 981 cms. per sec. per sec.
10. 42,400 pounds.
11. 761·14 mm.
12. 13,400 gm. cm.^{2}.
13. 0·058 calories per gm. per degree.
14. 153.
15. 342 c.c.
16. 603,000 dynes per sq. cm.
17. 8·74 pounds per sq. inch.
18. 0·0166 k.w.h.
19. 0·9817 gm. per c.c.
20. $1{\cdot}15\times10^{12}$ dynes per sq. cm.
21. $6{\cdot}80\times10^{10}$ dynes per sq. cm.
22. 980·51 cms./sec.^{2}.
23. 0·59 calories per gm. per degree.
24. $1{\cdot}62\times10^{10}$ dynes per sq. cm.
25. $H_t=113{\cdot}0$ cms.
26. 29,900 cms.
27. 3·919 gms. per c.c.
28. 0·0086 dynes per sq. cm. per unit velocity gradient.
29. 547 dynes per cm.
30. 26·75 gm.
31. 880 gm.
32. 2·503 gm./c.c.
33. 540 calories per gm.
34. 980·78 cm. per sec.^{2}.
35. 1·85 cms.
36. 0·8075 gm. per c.c.
37. $a=0{\cdot}00394$, $b=-7\times10^{-7}$, temperature$=-244^\circ$ C.
38. 0·096 calorie per gm. per degree.
39. 37.
40. 0·913 gm. per c.c.
41. 0·00022 dynes per sq. cm. per unit velocity gradient.
42. 0·187 calorie per gm. per degree.
43. $4{\cdot}21\times10^{7}$ ergs. per calorie.
44. 1·6.
45. 532 calories per gm.
46. (*a*) 252, (*b*) $4{\cdot}45\times10^{5}$, (*c*) 70·3, (*d*) 14·89, (*e*) 3·09, (*f*) 28·32, (*g*) 472, (*h*) $6{\cdot}9\times10^{4}$.
47. $5{\cdot}9\times10^{-6}$.
48. 0·94.
49. 7270.

APPENDIX

TABLES

TABLES

Explanatory Notes

Equivalents . . . . page 315

Constants. Characteristic Deviations . . . . page 316

Density of Water . . . . page 317

Notice that the density (mass per volume) of water (under atmospheric pressure) is never as great as unity.* A parallel column gives the specific gravity with reference to water at the temperature of maximum density.

Inverse Tangents and Circular Measure . . . . page 317

Squares and Square Roots . . . . pages 318, 319

The squares and square roots of all numbers are obtained with four-figure accuracy by using this table like any logarithm table. Complete five-figure and six-figure squares of all three-figure number are obtained as follows: the last two figures of the required square (N^2) are printed in italic opposite the last two figures of N in the margin. These can then be used without ambiguity to correct the four-figure value obtained in the ordinary manner: *e.g.* required the square of 417. The table gives the approximate value (pointed off by inspection) as 173900. Opposite 17 are found the italic figures *89*. The square is accordingly 173889; not 173989, for the latter would round off to 174000 instead of 173900.

Squares and Probable Error Reduction Table . . . . . page 322

If either $\Sigma(v^2)/n$ or $\Sigma(v^2)/n(n-1)$ is located between two consecutive numbers in the *third* column, $(n\pm 1/2)^2/(0{\cdot}67449)^2$, of the same table, then the value of $0{\cdot}67449\sqrt{\Sigma(v^2)/n}$ or $0{\cdot}67449\sqrt{\Sigma(v^2)/n(n-1)}$, as the case may be, will be found opposite it in the *first* column. A very rough mental calculation will prevent taking a value which is $\sqrt{10}$ times too large or small.

Constants . . . . page 322

The characteristics $\bar{1}$ and $\bar{2}$ have been replaced by 9 and 8 respectively.

* The terms *density* and *specific gravity* are often confused, even in text-books. Sometimes an arbitrary unit of volume is substituted for the cubic centimetre in order that the maximum density of water may appear to be unity, and a note is added to the effect that the density is stated in "grams per millilitre." (A *litre* is defined as the volume of 1000 gms. of water at the temperature of maximum density.)

Circular Functions . . . . page 322

In the table of circular functions the "radian value," natural sine, cosine, tangent, and cotangent are given for every degree of the quadrant (above 45° use the *lower* and *right-hand* margin), also the logarithmic sine and cosine. By subtracting the two latter from each other and from zero any of the six logarithmic functions may be obtained from the table by inspection.* Sines and cosines of any intermediate values can safely be obtained by interpolation, and tangents up to tan 70°. For the sine, tangent, and numerical measure of a small angle the equations at the corners of the table should be used as factors: *e.g.* sin 3′ $=3\times 0{\cdot}000290888=0{\cdot}000872664$.

For inverse "radians" and tangents see p. 317.

Four-Place Logarithms . . . . pages 320, 321

To find the logarithm of a given number: For the integral part ("characteristic") of the logarithm, count to the left from units place to the first figure (other than zero) of the number, thus:

$$\overset{7}{9}\,\overset{6}{3}\,\overset{5}{0}\,\overset{4}{0}\,\overset{3}{0}\,\overset{2}{0}\,\overset{1}{0}\,\overset{0}{0}{\cdot} \qquad \text{or} \qquad \overset{0}{0}{\cdot}\overset{-1}{0}\,\overset{-2}{0}\,\overset{-3}{0}\,\overset{-4}{3}\,0\,5$$

log 93000000 = 7· + . . . ; log 0·000305 = −4· + . . . For the decimal part ("mantissa"), find the first (two) figures of the given number in the left-hand column (N) of the table and the third figure at the top of another column. The required mantissa will be found in line with the first two figures and in the column headed by the third. Consider the second or third figure, if lacking, to be zero: *e.g.* log 7 = log 70 = log 700 = 0·8451 ; log 0·000023 = log 0·0000230 = −5· + 0·3617. The mantissa is always kept positive, and a logarithm like the last is abbreviated $\bar{5}{\cdot}3617$ to save space. If the number has four significant figures find the logarithms of the next smaller and next larger marginal numbers and assume that logarithmic differences are proportional to the corresponding numerical differences. Thus, 1·873 would be located (on a scale) 3-tenths of the way from 1·87 to 1·88 ; therefore log 1·873 is likewise 3/10 of the way from 2718 to 2742, namely 2725. (Three-tenths of the *tabular difference*, 24, will be found from the small marginal tables to be 7, and

* When two logarithms are to be added or subtracted it will be found more convenient, after a little experience, to work from left to right than the reverse. This is especially easy in finding reciprocals by subtracting from zero, as in Exercises V, Question 6: beginning at the left subtract each figure from 9, except the last one, which is to be subtracted from 10. For example,

$$\begin{array}{rl} \log 1 = 0 = & 1{\cdot}\ 9\ 9\ 9\ 9\ 10 \\ \log \pi = & 0{\cdot}\ 4\ 9\ 7\ 1\ \ 5 \\ \hline \therefore\ \log 1/\pi = & \bar{1}{\cdot}\ 5\ 0\ 2\ 8\ \ 5 \end{array}$$

Try working the following exercises from left to right. Before each addition note whether the next pair of figures will add up to more than nine and so give "one to carry." If they add up to exactly nine, look a step farther to the right, and so on.

$$\begin{array}{r} 4\,1\,5\,6\,5\,2\,8\,4\,3\,6\,4\,6\,4\,4\,3\,6\,1\,4\,6 \\ \pm 3\,2\,2\,8\,3\,2\,8\,9\,4\,3\,2\,3\,8\,4\,2\,3\,8\,5\,6 \\ \hline = \qquad\qquad\qquad\qquad\qquad \end{array}$$

In subtracting from left to right, before setting down each partial difference notice whether it will need to be decreased by unity on account of the figures that follow.

2718+7=2725. The approximate tabular difference, D, is given for each line, so that only the final digits need be subtracted.) If the given number has 5 or more significant figures a table in which the logarithms are stated to 5 or more places must be used.

To find the number ("antilogarithm") that corresponds to a given logarithm: If the given logarithm does not occur in the body of the table determine its position in respect to the next higher and lower tabular logarithms and use proportional parts as before: *e.g.* 0·1345 is found to be 10/32 of the way from 0·1335 to 0·1367, hence its antilogarithm will be $1{\cdot}36\frac{10}{32}$, or 1·363. Notice that 10/32 can be reduced to tenths and 3/10 to twenty-fourths, mentally, by using the small multiplication tables (PP) in the margin.

Reciprocals are easily determined mentally by using a table of logarithms: *e.g.* $1/e$=0·3678. (Footnote on page 314.)

Exponentials . . . . page 323

A subscript number takes the place of a decimal point followed by the corresponding number of ciphers: *e.g.* 0_44343 means 0·00004343, $_31234$ means 0·0001234, etc.

Notice that $e^{x+y}=e^xe^y$; for example, $e^{2.5}$ is 7·3891×1·6487, and $\log_{10}e^{2.5}$ is 1·0857.

Napierian or Natural Logarithms . . . . page 324

The table extends from 0·00 to 10·00 by steps of 0·05 and from 10 to 100 by jumps of 5·0. If table is not fine enough multiply the common logarithm of the number by 2·3036.

The Probability Integral . . . . page 325

The tabular value gives the area under the curve $y=e^{-x^2}$ between the ordinates 0 and x in terms of the total area between $x=0$ and $x=\infty$. In the small table of y and x/0·47694 the value of l for Chauvenet's criterion will be found in the first column opposite the value of $1-1/2n$ in the second column.

EQUIVALENTS

1 inch	=2·5400 cms.	1 cm.	=0·3937 inch
1 foot	=30·48 cms.	1 cm.	=0·3281 foot
1 sq. in.	=6·45 sq. cms.	1 sq. cm.	=0·1550 sq. in.
1 sq. ft.	=929·0 sq. cms.	1 cubic metre	=35·32 cubic ft.
1 cub. in.	=16·39 cub. cms.	1 metre	=3·281 ft.
1 cub. ft.	=28·32 litres		
1 pound	=453·6 gms.	1 kilogm.	=2·205 pds.
1 ounce	=28·35 gms.		
1 grain	=0·0648 gm.		
1 mile	=1609·3 metres	1 kilometre	=0·621 mile
1 quart (Imp.)	=1137 cub. cms.	1 litre	=0·880 imp. qt.
1 quart (U.S.)	=946 cub. cms.		=1·057 U.S. qts.
1 gallon (Imp.)	=4·536 litres		
1 long ton	=1016·0 kgms.	1 metric ton (1000 km.)=2204 pds.	
1 short ton	=907·2 kgms.		
1 pound per cub. ft.	=0·01602 gm per c.c.	1 pound wt.=445,000 dynes	
1 mile per hour	= 44·70 cm. per min.	1 pound-wt. per sq. in.=68,970 dynes per sq. cm.	
1 cub. ft. per min.	=472 cub. cm. per sec.	1 pound-wt. per sq. in.=70·31 gm. wt. per sq. cm.	
1 British Thermal Unit=252 calories		1 calorie= 3·08 ft.-pds.	
1 Horse-power=746 watts		1 kilowatt=1·34 horse-power.	

Approximate Equivalents

25 mm.	=1 inch	30 gm.	=1 oz.
10 cm.	=4 inches	30 c.c.	=1 fl. oz.
40 inches	=1 metre	11 pds.	=5 kgm.
8 kilometre	=5 miles	15 lb. per sq. in.	=1 kgm. per sq. cm.
1000 sq. cm.	=1 sq. ft.		

Thermometry

$F=\frac{9}{5}C+32$ $\qquad$ $C=5(F-32)/9$

Logarithms

To convert common logarithms into hyperbolic logarithms multiply by 2·3026.

To convert hyperbolic logarithms into common logarithms multiply by 0·4343.

$\text{Log}_{10}\, 2{\cdot}3026=0{\cdot}3622$. $\text{Log}_{10}\, 0{\cdot}4343=\bar{1}{\cdot}6378$.

Constants

$\pi=3{\cdot}14159$
$\log_{10}\pi=0{\cdot}49715$
$\pi^2=9{\cdot}8696$
$1/\pi=0{\cdot}3183$
$\sqrt{\pi}=1{\cdot}7725$

$\sqrt{2}=1{\cdot}4142$
$\sqrt{3}=1{\cdot}7320$
$\sqrt{5}=2{\cdot}2361$
$\sqrt{6}=2{\cdot}4491$
$\sqrt{7}=2{\cdot}6458$
$\sqrt{8}=2{\cdot}8284$
$\sqrt{10}=3{\cdot}1623$
$\sqrt{11}=3{\cdot}3166$
$\sqrt{12}=3{\cdot}4641$

1 radian $=57{\cdot}2958°$
$=206265''$
1 degree $=0{\cdot}01745$ radian.

Characteristic Deviations

$p/m=0{\cdot}4769$	$p/s=0{\cdot}6745$	$s/p=1{\cdot}1482$
$a\ m=0{\cdot}5642$	$p/a=0{\cdot}8753$	$a/p=1{\cdot}1829$
$s\ m=0{\cdot}7071$	$a/s=0{\cdot}7979$	$s/a=1{\cdot}2533$

m = the modulus in the equation $y=\epsilon^{-\frac{x^2}{m^2}}$.

Mensuration

triangle : base, b ; altitude, a ; area, $ab/2$.
parallelogram : base, b ; altitude, a ; area, ab.
circle : radius, r ; circumference, $2\pi r$; area, πr^2.
ellipse : major axis, $2a$; minor axis, $2b$; area, πab.
cylinder : radius, r ; length, l ; surface, $\pi r^2+2\pi rl+\pi r^2$; volume, $\pi r^2 l$.
cone : radius, r ; height, h ; surface, $\pi r^2+\pi r\sqrt{r^2+h^2}$; volume, $\pi r^2 h/3$.
pyramid : area of base, a ; height, h ; perimeter of base, p ; slant height, s ; surface, $ps/2$; volume, $ah/3$.
sphere : radius, r ; surface, $4\pi r^2$; volume, $4\pi r^3/3$.

DENSITY OF WATER

temp.	density (gm. per cm.³)	spec. grav. (:H_2O at 4° C.)
0°C	.999841	.99987
2	.999941	.99997
4	.999973	1.00000
6	.999941	.99997
8	.999849	.99988
10	.999700	.99973
12	.999498	.99952
14	.999244	.99927
16	.998943	.99897
18	.998595	.99862
20	.998203	.99823
22	.997770	.99780
24	.997296	.99732
26	.996783	.99681
28	.996232	.99626
30	.995646	.99567
32	.99502	.99505
34	.99437	.99440
36	.99768	.99371
38	.99296	.99299
40	.99221	.99224
42	.99144	.99147
44	.99063	.99066
46	.98979	.98982
48	.98893	.98896
50	.98804	.98807
52	.98712	.98715
54	.98618	.98621
56	.98522	.98525
58	.98422	.98425
60	.98321	.98324
62	.98217	.98220
64	.98110	.98113
66	.98002	.98005
68	.97891	.97894
70	.97778	.97781
72	.97663	.97666
74	.97545	.97548
76	.97426	.97429
78	.97304	.97307
80	.97180	.97183
82	.97054	.97057
84	.96927	.96930
86	.96797	.96800
88	.96665	.96668
90	.96531	.96534
92	.96396	.96399
94	.96258	.96261
96	.96119	.96122
98	.95978	.95981
100	.95835	.95838

The temperatures are according to the constant-volume hydrogen scale.

INVERSE TANGENTS AND CIRCULAR MEASURE

n	*is the circular measure and tangent of*
.0001	0° 00′ 20″.6 = .000100
.0002	41 .3 = .000200
.0003	1 01 .9 = .000300
.0004	1 22 .5 = .000400
.0005	1 43 .1 = .000500
.0006	2 03 .8 = .000600
.0007	2 24 .4 = .000700
.0008	2 45 .0 = .000800
.0009	3 05 .6 = .000900
.001	0° 03′ 26″.3 = .001000
.002	6 52 .5 = .002000
.003	10 18 .8 = .003000
.004	13 45 .1 = .004000
.005	17 11 .3 = .005000
.006	20 37 .6 = .006000
.007	24 03 .9 = .007000
.008	27 30 .1 = .008000
.009	30 56 .4 = .009000

n	*is the circular measure of*	*is the tangent of*
.01	0° 34′ 22″.6	0° 34′ 22″.6
.02	1 08 45 .3	1 08 44 .7
.03	1 43 07 .9	1 43 06 .1
.04	2 17 30 .6	2 17 26 .2
.05	2 51 53 .2	2 51 44 .7
.06	3 26 15 .9	3 26 01 .1
.07	4 00 38 .5	4 00 15 .0
.08	4 35 01 .2	4 34 26 .1
.09	5 09 23 .8	5 08 34 .0
.10	5° 43′ 46″.5	5° 42′ 38″.1
.11	6 18 09 .1	6 16 38 .3
.12	6 52 31 .8	6 50 34 .0
.13	7 26 54 .4	7 24 24 .9
.14	8 01 17 1	7 58 10 .6
.15	8 35 39 .7	8 31 50 .8
.16	9 10 02 .4	9 05 25 .0
.17	9 44 25 .0	9 38 53 .0
.18	10 18 47 .7	10 12 14 .3
.19	10 53 10 .3	10 45 28 .7
.20	11° 27′ 33″.0	11° 18′ 35″.8
.21	12 01 55 .6	11 51 35 .2
.22	12 36 18 .2	12 24 26 .7
.23	13 10 40 .9	12 57 09 .9
.24	13 45 03 .5	13 29 44 .6
.25	14 19 26 .2	14 02 10 .5
.26	14 53 48 .8	14 34 27 .2
.27	15 28 11 .5	15 06 34 .5
.28	16 02 34 .1	15 38 32 .1
.29	16 36 56 .8	16 10 19 .8
.30	17° 11′ 19″.4	16° 41′ 57″.3
.31	17 45 42 .1	17 13 24 .4
.32	18 20 04 .7	17 44 40 .8
.33	18 54 27 .4	18 15 46 .4
.34	19 28 50 .0	18 46 40 .9
.35	20 03 12 .7	19 17 24 .2
.36	20 37 35 .3	19 47 56 .0
.37	21 11 58 .0	20 18 16 .1
.38	21 46 20 .6	20 48 24 .4
.39	22 20 43 .3	21 18 20 .8

n	*is the circular measure of*	*is the tangent of*
.40	22° 55′ 05″.9	21° 48′ 05″.1
.41	23 29 28 .6	22 17 37 .1
.42	24 03 51 .2	22 46 56 .7
.43	24 38 13 .9	23 16 03 .7
.44	25 12 36 .5	23 44 58 .2
.45	25 46 59 .2	24 12 39 .9
.46	26 21 21 .8	24 42 08 .7
.47	26 55 44 .5	25 10 24 .7
.48	27 30 07 .1	25 38 27 .6
.49	28 04 29 .7	26 06 17 .5
.50	28° 38′ 52″.4	26° 33′ 54″.2
.51	29 13 15 .1	27 01 17 .7
.52	29 47 37 .7	27 28 27 .9
.53	30 22 00 .4	27 55 24 .9
.54	30 56 23 .0	28 22 08 .6
.55	31 30 45 .6	28 48 38 .8
.56	32 05 08 .3	29 14 55 .8
.57	32 39 30 .9	29 40 59 .3
.58	33 13 53 .6	30 06 49 .4
.59	33 48 16 .2	30 32 26 .2
.60	34° 22′ 38″.9	30° 57′ 49″.5
.61	34 57 01 .5	31 22 59 .5
.62	35 31 24 .2	31 47 56 .1
.63	36 05 46 .8	32 12 39 .3
.64	36 40 09 .5	32 37 09 .3
.65	37 14 32 .1	33 01 25 .7
.66	37 48 54 .8	33 25 29 .3
.67	38 23 17 .4	33 49 19 .5
.68	38 57 40 .1	34 12 56 .5
.69	39 32 02 .7	34 36 20 .4
.70	40° 06′ 25″.4	34° 59′ 31″.3
.71	40 40 48 .0	35 22 29 .1
.72	41 15 10 .7	35 45 14 .0
.73	41 49 33 .3	36 07 46 .0
.74	42 23 56 .0	36 30 05 .2
.75	42 58 18 .6	36 52 11 .6
.76	43 32 41 .3	37 14 05 .4
.77	44 07 03 .9	37 35 46 .6
.78	44 41 26 .5	37 57 15 .2
.79	45 15 49 .2	38 18 31 .5
.80	45° 50′ 11″.8	38° 39′ 35″.3
.81	46 24 34 .5	39 00 26 .8
.82	46 58 57 .1	39 21 06 .3
.83	47 33 19 .8	39 41 10 .3
.84	48 07 47 .8	40 01 48 .9
.85	48 42 05 .1	40 21 52 .3
.86	49 16 27 .7	40 41 43 .9
.87	49 50 50 .4	41 01 23 .8
.88	50 25 13 .0	41 20 52 .0
.89	50 59 35 .7	41 40 08 .7
.90	51° 33′ 58″.3	41° 59′ 14″.0
.91	52 08 20 .9	42 18 07 .9
.92	52 42 43 .6	42 36 50 .6
.93	53 17 06 .3	42 55 22 .2
.94	53 51 28 .9	43 13 42 .7
.95	54 25 51 .6	43 31 52 .3
.96	55 00 14 .2	43 49 51 .1
.97	55 34 36 .9	44 07 39 .1
.98	56 08 59 .5	44 25 16 .6
.99	56 43 22 .2	44 42 43 .5

COMPLETE SQUARES UP TO 999². 4-FIG. SQUARES AND ROOTS

N	0	1	2	3	4	5	6	7	8	9	*D*
0	*00*	*01*	*04*	*09*	*16*	*25*	*36*	*49*	*64*	*81*	
10	1000	1020	1040	1061	1082	1102	1124	1145	1166	1188	22
11	1210	1232	1254	1277	1300	1322	1346	1369	1392	1416	24
12	1440	1464	1488	1513	1538	1562	1588	1613	1638	1664	26
13	1690	1716	1742	1769	1796	1822	1850	1877	1904	1932	28
14	1960	1988	2016	2045	2074	2102	2132	2161	2190	2220	30
15	2250	2280	2310	2341	2372	2402	2434	2465	2496	2528	32
16	2560	2592	2624	2657	2690	2722	2756	2789	2822	2856	34
17	2890	2924	2958	2993	3028	3062	3098	3133	3168	3204	36
18	3240	3276	3312	3349	3386	3422	3460	3497	3534	3572	38
19	3610	3648	3686	3725	3764	3802	3842	3881	3920	3960	40
20	4000	4040	4080	4121	4162	4202	4244	4285	4326	4368	42
21	4410	4452	4494	4537	4580	4622	4666	4709	4752	4796	44
22	4840	4884	4928	4973	5018	5062	5108	5153	5198	5244	46
23	5290	5336	5382	5429	5476	5522	5570	5617	5664	5712	48
24	5760	5808	5856	5905	5954	6002	6052	6101	6150	6200	50
25	6250	6300	6350	6401	6452	6502	6554	6605	6656	6708	52
26	6760	6812	6864	6917	6970	7022	7076	7129	7182	7236	54
27	7290	7344	7398	7453	7508	7562	7618	7673	7728	7784	56
28	7840	7896	7952	8009	8066	8122	8180	8237	8294	8352	58
29	8410	8468	8526	8585	8644	8702	8762	8821	8880	8940	60
30	9000	9060	9120	9181	9242	9302	9364	9425	9486	9548	62
31	9610	9672	9734	9797	9860	9922	9986	1005	1011	1018	6 4
32	1024	1030	1037	1043	1050	1056	1063	1069	1076	1082	7
33	1089	1096	1102	1109	1116	1122	1129	1136	1142	1149	7
34	1156	1163	1170	1176	1183	1190	1197	1204	1211	1218	7
35	1225	1232	1239	1246	1253	1260	1267	1274	1282	1289	7
36	1296	1303	1310	1318	1325	1332	1340	1347	1354	1362	7
37	1369	1376	1384	1391	1399	1406	1414	1421	1429	1436	8
38	1444	1452	1459	1467	1475	1482	1490	1498	1505	1513	8
39	1521	1529	1537	1544	1552	1560	1568	1576	1584	1592	8
40	1600	1608	1616	1624	1632	1640	1648	1656	1665	1673	8
41	1681	1689	1697	1706	1714	1722	1731	1739	1747	1756	8
42	1764	1772	1781	1789	1798	1806	1815	1823	1832	1840	9
43	1849	1858	1866	1875	1884	1892	1901	1910	1918	1927	9
44	1936	1945	1954	1962	1971	1980	1989	1998	2007	2016	9
45	2025	2034	2043	2052	2061	2070	2079	2088	2098	2107	9
46	2116	2125	2134	2144	2153	2162	2172	2181	2190	2200	9
47	2209	2218	2228	2237	2247	2256	2266	2275	2285	2294	10
48	2304	2314	2323	2333	2343	2352	2362	2372	2381	2391	10
49	2401	2411	2421	2430	2440	2450	2460	2470	2480	2490	10
50	2500	2510	2520	2530	2540	2550	2560	2570	2581	2591	10
51	2601	2611	2621	2632	2642	2652	2663	2673	2683	2694	10
52	2704	2714	2725	2735	2746	2756	2767	2777	2788	2798	11
53	2809	2820	2830	2841	2852	2862	2873	2884	2894	2905	11
54	2916	2927	2938	2948	2959	2970	2981	2992	3003	3014	11

PP

	64	63	62	61	60
1	6.4	6.3	6.2	6.1	6.0
2	12.8	12.6	12.4	12.2	12.0
3	19.2	18.9	18.6	18.3	18.0
4	25.6	25.2	24.8	24.4	24.0
5	32.0	31.5	31.0	30.5	30.0
6	38.4	37.8	37.2	36.6	36.0
7	44.8	44.1	43.4	42.7	42.0
8	51.2	50.4	49.6	48.8	48.0
9	57.6	56.7	55.8	54.9	54.0

	59	58	57	56	55	54
1	5.9	5.8	5.7	5.6	5.5	5.4
2	11.8	11.6	11.4	11.2	11.0	10.8
3	17.7	17.4	17.1	16.8	16.5	16.2
4	23.6	23.2	22.8	22.4	22.0	21.6
5	29.5	29.0	28.5	28.0	27.5	27.0
6	35.4	34.8	34.2	33.6	33.0	32.4
7	41.3	40.6	39.9	39.2	38.5	37.8
8	47.2	46.4	45.6	44.8	44.0	43.2
9	53.1	52.2	51.3	50.4	49.5	48.6

	53	52	51	50	49	48
1	5.3	5.2	5.1	5.0	4.9	4.8
2	10.6	10.4	10.2	10.0	9.8	9.6
3	15.9	15.6	15.3	15.0	14.7	14.4
4	21.2	20.8	20.4	20.0	19.6	19.2
5	26.5	26.0	25.5	25.0	24.5	24.0
6	31.8	31.2	30.6	30.0	29.4	28.8
7	37.1	36.4	35.7	35.0	34.3	33.6
8	42.4	41.6	40.8	40.0	39.2	38.4
9	47.7	46.8	45.9	45.0	44.1	43.2

	47	46	45	44	43	42
1	4.7	4.6	4.5	4.4	4.3	4.2
2	9.4	9.2	9.0	8.8	8.6	8.4
3	14.1	13.8	13.5	13.2	12.9	12.6
4	18.8	18.4	18.0	17.6	17.2	16.8
5	23.5	23.0	22.5	22.0	21.5	21.0
6	28.2	27.6	27.0	26.4	25.8	25.2
7	32.9	32.2	31.5	30.8	30.1	29.4
8	37.6	36.8	36.0	35.2	34.4	33.6
9	42.3	41.4	40.5	39.6	38.7	37.8

	41	40	39	38	37	36
1	4.1	4.0	3.9	3.8	3.7	3.6
2	8.2	8.0	7.8	7.6	7.4	7.2
3	12.3	12.0	11.7	11.4	11.1	10.8
4	16.4	16.0	15.6	15.2	14.8	14.4
5	20.5	20.0	19.5	19.0	18.5	18.0
6	24.6	24.0	23.4	22.8	22.2	21.6
7	28.7	28.0	27.3	26.6	25.9	25.2
8	32.8	32.0	31.2	30.4	29.6	28.8
9	36.9	36.0	35.1	34.2	33.3	32.4

For four-figure squares and square roots use this table in the same way as a logarithm table. For five- and six-figure (exact) squares of three-figure numbers: (first) obtain the four-figure value from the table *and point it off correctly;* for example, $619^2 = 383200$; (second) the last two figures of the perfect square will be found in *italics* opposite the last two figures of the given

COMPLETE SQUARES UP TO 999². 4-FIG. SQUARES AND ROOTS

N	0	1	2	3	4	5	6	7	8	9	*D*
55	30*25*	3036	3047	3058	3069	3080	3091	3102	3114	3125	11
56	31*36*	3147	3158	3170	3181	3192	3204	3215	3226	3238	11
57	32*49*	3260	3272	3283	3295	3306	3318	3329	3341	3352	12
58	33*64*	3376	3387	3399	3411	3422	3434	3446	3457	3469	12
59	34*81*	3493	3505	3516	3528	3540	3552	3564	3576	3588	12
60	36*00*	3612	3624	3636	3648	3660	3672	3684	3697	3709	12
61	37*21*	3733	3745	3758	3770	3782	3795	3807	3819	3832	12
62	38*44*	3856	3869	3881	3894	3906	3919	3931	3944	3956	13
63	39*69*	3982	3994	4007	4020	4032	4045	4058	4070	4083	13
64	40*96*	4109	4122	4134	4147	4160	4173	4186	4199	4212	13
65	42*25*	4238	4251	4264	4277	4290	4303	4316	4330	4343	13
66	43*56*	4369	4382	4396	4409	4422	4436	4449	4462	4476	13
67	44*89*	4502	4516	4529	4543	4556	4570	4583	4597	4610	14
68	46*24*	4638	4651	4665	4679	4692	4706	4720	4733	4747	14
69	47*61*	4775	4789	4802	4816	4830	4844	4858	4872	4886	14
70	49*00*	4914	4928	4942	4956	4970	4984	4998	5013	5027	14
71	50*41*	5055	5069	5084	5098	5112	5127	5141	5155	5170	14
72	51*84*	5198	5213	5227	5242	5256	5271	5285	5300	5314	15
73	53*29*	5344	5358	5373	5388	5402	5417	5432	5446	5461	15
74	54*76*	5491	5506	5520	5535	5550	5565	5580	5595	5610	15
75	56*25*	5640	5655	5670	5685	5700	5715	5730	5746	5761	15
76	57*76*	5791	5806	5822	5837	5852	5868	5883	5898	5914	15
77	59*29*	5944	5960	5975	5991	6006	6022	6037	6053	6068	16
78	60*84*	6100	6115	6131	6147	6162	6178	6194	6209	6225	16
79	62*41*	6257	6273	6288	6304	6320	6336	6352	6368	6384	16
80	64*00*	6416	6432	6448	6464	6480	6496	6512	6529	6545	16
81	65*61*	6577	6593	6610	6626	6642	6659	6675	6691	6708	16
82	67*24*	6740	6757	6773	6790	6806	6823	6839	6856	6872	17
83	68*89*	6906	6922	6939	6956	6972	6989	7006	7022	7039	17
84	70*56*	7073	7090	7106	7123	7140	7157	7174	7191	7208	17
85	72*25*	7242	7259	7276	7293	7310	7327	7344	7362	7379	17
86	73*96*	7413	7430	7448	7465	7482	7500	7517	7534	7552	17
87	75*69*	7586	7604	7621	7639	7656	7674	7691	7709	7726	18
88	77*44*	7762	7779	7797	7815	7832	7850	7868	7885	7903	18
89	79*21*	7939	7957	7974	7992	8010	8028	8046	8064	8082	18
90	81*00*	8118	8136	8154	8172	8190	8208	8226	8245	8263	18
91	82*81*	8299	8317	8336	8354	8372	8391	8409	8427	8446	18
92	84*64*	8482	8501	8519	8538	8556	8575	8593	8612	8630	19
93	86*49*	8668	8686	8705	8724	8742	8761	8780	8798	8817	19
94	88*36*	8855	8874	8892	8911	8930	8949	8968	8987	9006	19
95	90*25*	9044	9063	9082	9101	9120	9139	9158	9178	9197	19
96	92*16*	9235	9254	9274	9293	9312	9332	9351	9370	9390	19
97	94*09*	9428	9448	9467	9487	9506	9526	9545	9565	9584	20
98	96*04*	9624	9643	9663	9683	9702	9722	9742	9761	9781	20
99	98*01*	9821	9841	9860	9880	9900	9920	9940	9960	9980	20

PP						
	35	34	33	32	31	30
1	3.5	3.4	3.3	3.2	3.1	3.0
2	7.0	6.8	6.6	6.4	6.2	6.0
3	10.5	10.2	9.9	9.6	9.3	9.0
4	14.0	13.6	13.2	12.8	12.4	12.0
5	17.5	17.0	16.5	16.0	15.5	15.0
6	21.0	20.4	19.8	19.2	18.6	18.0
7	24.5	23.8	23.1	22.4	21.7	21.0
8	28.0	27.2	26.4	25.6	24.8	24.0
9	31.5	30.6	29.7	28.8	27.9	27.0
	29	28	27	26	25	24
1	2.9	2.8	2.7	2.6	2.5	2.4
2	5.8	5.6	5.4	5.2	5.0	4.8
3	8.7	8.4	8.1	7.8	7.5	7.2
4	11.6	11.2	10.8	10.4	10.0	9.6
5	14.5	14.0	13.5	13.0	12.5	12.0
6	17.4	16.8	16.2	15.6	15.0	14.4
7	20.3	19.6	18.9	18.2	17.5	16.8
8	23.2	22.4	21.6	20.8	20.0	19.2
9	26.1	25.2	24.3	23.4	22.5	21.6
	23	22	21	20	19	18
1	2.3	2.2	2.1	2.0	1.9	1.8
2	4.6	4.4	4.2	4.0	3.8	3.6
3	6.9	6.6	6.3	6.0	5.7	5.4
4	9.2	8.8	8.4	8.0	7.6	7.2
5	11.5	11.0	10.5	10.0	9.5	9.0
6	13.8	13.2	12.6	12.0	11.4	10.8
7	16.1	15.4	14.7	14.0	13.3	12.6
8	18.4	17.6	16.8	16.0	15.2	14.4
9	20.7	19.8	18.9	18.0	17.1	16.2
	17	16	15	14	13	12
1	1.7	1.6	1.5	1.4	1.3	1.2
2	3.4	3.2	3.0	2.8	2.6	2.4
3	5.1	4.8	4.5	4.2	3.9	3.6
4	6.8	6.4	6.0	5.6	5.2	4.8
5	8.5	8.0	7.5	7.0	6.5	6.0
6	10.2	9.6	9.0	8.4	7.8	7.2
7	11.9	11.2	10.5	9.8	9.1	8.4
8	13.6	12.8	12.0	11.2	10.4	9.6
9	15.3	14.4	13.5	12.6	11.7	10.8
	11	10	9	8	7	6
1	1.1	1.0	0.9	0.8	0.7	0.6
2	2.2	2.0	1.8	1.6	1.4	1.2
3	3.3	3.0	2.7	2.4	2.1	1.8
4	4.4	4.0	3.6	3.2	2.8	2.4
5	5.5	5.0	4.5	4.0	3.5	3.0
6	6.6	6.0	5.4	4.8	4.2	3.6
7	7.7	7.0	6.3	5.6	4.9	4.2
8	8.8	8.0	7.2	6.4	5.6	4.8
9	9.9	9.0	8.1	7.2	6.3	5.4

number, for example, opposite 19 the italic figures are *61;* use these to correct the round number: if the last two figures of 619² are 61, and its value as a round number is 383,200, then its exact value must be 383,161 (not 383261, for the latter would round off to 383300).

FOUR-PLACE LOGARITHMS

N	0	1	2	3	4	5	6	7	8	9	*D*
10	0000*	0043*	0086*	0128*	0170*	0212*	0253*	0294*	0334*	0374*	40
11	0414	0453	0492	0531	0569	0607	0645	0682	0719	0755	37
12	0792	0828	0864	0899	0934	0969	1004	1038	1072	1106	33
13	1139	1173	1206	1239	1271	1303	1335	1367	1399	1430	31
14	1461	1492	1523	1553	1584	1614	1644	1673	1703	1732	29
15	1761	1790	1818	1847	1875	1903	1931	1959	1987	2014	27
16	2041	2068	2095	2122	2148	2175	2201	2227	2253	2279	25
17	2304	2330	2355	2380	2405	2430	2455	2480	2504	2529	24
18	2553	2577	2601	2625	2648	2672	2695	2718	2742	2765	23
19	2788	2810	2833	2856	2878	2900	2923	2945	2967	2989	21
20	3010	3032	3054	3075	3096	3118	3139	3160	3181	3201	21
21	3222	3243	3263	3284	3304	3324	3345	3365	3385	3404	20
22	3424	3444	3464	3483	3502	3522	3541	3560	3579	3598	19
23	3617	3636	3655	3674	3692	3711	3729	3747	3766	3784	18
24	3802	3820	3838	3856	3874	3892	3909	3927	3945	3962	17
25	3979	3997	4014	4031	4048	4065	4082	4099	4116	4133	17
26	4150	4166	4183	4200	4216	4232	4249	4265	4281	4298	16
27	4314	4330	4346	4362	4378	4393	4409	4425	4440	4456	16
28	4472	4487	4502	4518	4533	4548	4564	4579	4594	4609	15
29	4624	4639	4654	4669	4683	4698	4713	4728	4742	4757	14
30	4771	4786	4800	4814	4829	4843	4857	4871	4886	4900	14
31	4914	4928	4942	4955	4969	4983	4997	5011	5024	5038	13
32	5051	5065	5079	5092	5105	5119	5132	5145	5159	5172	13
33	5185	5198	5211	5224	5237	5250	5263	5276	5289	5302	13
34	5315	5328	5340	5353	5366	5378	5391	5403	5416	5428	13
35	5441	5453	5465	5478	5490	5502	5514	5527	5539	5551	12
36	5563	5575	5587	5599	5611	5623	5635	5647	5658	5670	12
37	5682	5694	5705	5717	5729	5740	5752	5763	5775	5786	12
38	5798	5809	5821	5832	5843	5855	5866	5877	5888	5899	12
39	5911	5922	5933	5944	5955	5966	5977	5988	5999	6010	11
40	6021	6031	6042	6053	6064	6075	6085	6096	6107	6117	11
41	6128	6138	6149	6160	6170	6180	6191	6201	6212	6222	10
42	6232	6243	6253	6263	6274	6284	6294	6304	6314	6325	10
43	6335	6345	6355	6365	6375	6385	6395	6405	6415	6425	10
44	6435	6444	6454	6464	6474	6484	6493	6503	6513	6522	10
45	6532	6542	6551	6561	6571	6580	6590	6599	6609	6618	10
46	6628	6637	6646	6656	6665	6675	6684	6693	6702	6712	9
47	6721	6730	6739	6749	6758	6767	6776	6785	6794	6803	9
48	6812	6821	6830	6839	6848	6857	6866	6875	6884	6893	9
49	6902	6911	6920	6928	6937	6946	6955	6964	6972	6981	9
50	6990	6998	7007	7016	7024	7033	7042	7050	7059	7067	9
51	7076	7084	7093	7101	7110	7118	7126	7135	7143	7152	8
52	7160	7168	7177	7185	7193	7202	7210	7218	7226	7235	8
53	7243	7251	7259	7267	7275	7284	7292	7300	7308	7316	8
54	7324	7332	7340	7348	7356	7364	7372	7380	7388	7396	8
55	7404	7412	7419	7427	7435	7443	7451	7459	7466	7474	8

PP

	43	42	41	40
1	4.3	4.2	4.1	4.0
2	8.6	8.4	8.2	8.0
3	12.9	12.6	12.3	12.0
4	17.2	16.8	16.4	16.0
5	21.5	21.0	20.5	20.0
6	25.8	25.2	24.6	24.0
7	30.1	29.4	28.7	28.0
8	34.4	33.6	32.8	32.0
9	38.7	37.8	36.9	36.0

	39	38	37	36
1	3.9	3.8	3.7	3.6
2	7.8	7.6	7.4	7.2
3	11.7	11.4	11.1	10.8
4	15.6	15.2	14.8	14.4
5	19.5	19.0	18.5	18.0
6	23.4	22.8	22.2	21.6
7	27.3	26.6	25.9	25.2
8	31.2	30.4	29.6	28.8
9	35.1	34.2	33.3	32.4

	35	34	33	32
1	3.5	3.4	3.3	3.2
2	7.0	6.8	6.6	6.4
3	10.5	10.2	9.9	9.6
4	14.0	13.6	13.2	12.8
5	17.5	17.0	16.5	16.0
6	21.0	20.4	19.8	19.2
7	24.5	23.8	23.1	22.4
8	28.0	27.2	26.4	25.6
9	31.5	30.6	29.7	28.8

	31	30	29	28
1	3.1	3.0	2.9	2.8
2	6.2	6.0	5.8	5.6
3	9.3	9.0	8.7	8.4
4	12.4	12.0	11.6	11.2
5	15.5	15.0	14.5	14.0
6	18.6	18.0	17.4	16.8
7	21.7	21.0	20.3	19.6
8	24.8	24.0	23.2	22.4
9	27.9	27.0	26.1	25.2

	27	26	25	24
1	2.7	2.6	2.5	2.4
2	5.4	5.2	5.0	4.8
3	8.1	7.8	7.5	7.2
4	10.8	10.4	10.0	9.6
5	13.5	13.0	12.5	12.0
6	16.2	15.6	15.0	14.4
7	18.9	18.2	17.5	16.8
8	21.6	20.8	20.0	19.2
9	24.3	23.4	22.5	21.6

* Interpolated values are given on another page. In the table of five-figure values (p. 323) notice the minus sign after certain final 5's; *e.g.*, log 1.055 = .0233, but log 1.065 = .0273.

FOUR-PLACE LOGARITHMS

N	0	1	2	3	4	5	6	7	8	9	D
55	7404	7412	7419	7427	7435	7443	7451	7459	7466	7474	8
56	7482	7490	7497	7505	7513	7520	7528	7536	7543	7551	
57	7559	7566	7574	7582	7589	7597	7604	7612	7619	7627	
58	7634	7642	7649	7657	7664	7672	7679	7686	7694	7701	
59	7709	7716	7723	7731	7738	7745	7752	7760	7767	7774	
60	7782	7789	7796	7803	7810	7818	7825	7832	7839	7846	7
61	7853	7860	7868	7875	7882	7889	7896	7903	7910	7917	
62	7924	7931	7938	7945	7952	7959	7966	7973	7980	7987	
63	7993	8000	8007	8014	8021	8028	8035	8041	8048	8055	
64	8062	8069	8075	8082	8089	8096	8102	8109	8116	8122	
65	8129	8136	8142	8149	8156	8162	8169	8176	8182	8189	6
66	8195	8202	8209	8215	8222	8228	8235	8241	8248	8254	
67	8261	8267	8274	8280	8287	8293	8299	8306	8312	8319	
68	8325	8331	8338	8344	8351	8357	8363	8370	8376	8382	
69	8388	8395	8401	8407	8414	8420	8426	8432	8439	8445	
70	8451	8457	8463	8470	8476	8482	8488	8494	8500	8506	7
71	8513	8519	8525	8531	8537	8543	8549	8555	8561	8567	
72	8573	8579	8585	8591	8597	8603	8609	8615	8621	8627	
73	8633	8639	8645	8651	8657	8663	8669	8675	8681	8686	
74	8692	8698	8704	8710	8716	8722	8727	8733	8739	8745	
75	8751	8756	8762	8768	8774	8779	8785	8791	8797	8802	6
76	8808	8814	8820	8825	8831	8837	8842	8848	8854	8859	
77	8865	8871	8876	8882	8887	8893	8899	8904	8910	8915	
78	8921	8927	8932	8938	8943	8949	8954	8960	8965	8971	
79	8976	8982	8987	8993	8998	9004	9009	9015	9020	9025	
80	9031	9036	9042	9047	9053	9058	9063	9069	9074	9079	6
81	9085	9090	9096	9101	9106	9112	9117	9122	9128	9133	
82	9138	9143	9149	9154	9159	9165	9170	9175	9180	9186	
83	9191	9196	9201	9206	9212	9217	9222	9227	9232	9238	
84	9243	9248	9253	9258	9263	9269	9274	9279	9284	9289	
85	9294	9299	9304	9309	9315	9320	9325	9330	9335	9340	5
86	9345	9350	9355	9360	9365	9370	9375	9380	9385	9390	
87	9395	9400	9405	9410	9415	9420	9425	9430	9435	9440	
88	9445	9450	9455	9460	9465	9469	9474	9479	9484	9489	
89	9494	9499	9504	9509	9513	9518	9523	9528	9533	9538	
90	9542	9547	9552	9557	9562	9566	9571	9576	9581	9586	4
91	9590	9595	9600	9605	9609	9614	9619	9624	9628	9633	
92	9638	9643	9647	9652	9657	9661	9666	9671	9675	9680	
93	9685	9689	9694	9699	9703	9708	9713	9717	9722	9727	
94	9731	9736	9741	9745	9750	9754	9759	9763	9768	9773	
95	9777	9782	9786	9791	9795	9800	9805	9809	9814	9818	5
96	9823	9827	9832	9836	9841	9845	9850	9854	9859	9863	
97	9868	9872	9877	9881	9886	9890	9894	9899	9903	9908	
98	9912	9917	9921	9926	9930	9934	9939	9943	9948	9952	
99	9956	9961	9965	9969	9974	9978	9983	9987	9991	9096	
100	0000	0004	0009	0013	0017	0022	0026	0030	0035	0039	4

PP

	23	22	21	20
1	2.3	2.2	2.1	2.0
2	4.6	4.4	4.2	4.0
3	6.9	6.6	6.3	6.0
4	9.2	8.8	8.4	8.0
5	11.5	11.0	10.5	10.0
6	13.8	13.2	12.6	12.0
7	16.1	15.4	14.7	14.0
8	18.4	17.6	16.8	16.0
9	20.7	19.8	18.9	18.0

	19	18	17	16
1	1.9	1.8	1.7	1.6
2	3.8	3.6	3.4	3.2
3	5.7	5.4	5.1	4.8
4	7.6	7.2	6.8	6.4
5	9.5	9.0	8.5	8.0
6	11.4	10.8	10.2	9.6
7	13.3	12.6	11.9	11.2
8	15.2	14.4	13.6	12.8
9	17.1	16.2	15.3	14.4

	15	14	13	12
1	1.5	1.4	1.3	1.2
2	3.0	2.8	2.6	2.4
3	4.5	4.2	3.9	3.6
4	6.0	5.6	5.2	4.8
5	7.5	7.0	6.5	6.0
6	9.0	8.4	7.8	7.2
7	10.5	9.8	9.1	8.4
8	12.0	11.2	10.4	9.6
9	13.5	12.6	11.7	10.8

	11	10	9	8
1	1.1	1.0	0.9	0.8
2	2.2	2.0	1.8	1.6
3	3.3	3.0	2.7	2.4
4	4.4	4.0	3.6	3.2
5	5.5	5.0	4.5	4.0
6	6.6	6.0	5.4	4.8
7	7.7	7.0	6.3	5.6
8	8.8	8.0	7.2	6.4
9	9.9	9.0	8.1	7.2

	7	6	5	4
1	0.7	0.6	0.5	0.4
2	1.4	1.2	1.0	0.8
3	2.1	1.8	1.5	1.2
4	2.8	2.4	2.0	1.6
5	3.5	3 0	2.5	2.0
6	4.2	3.6	3.0	2.4
7	4.9	4.2	3.5	2.8
8	5.6	4.8	4.0	3.2
9	6.3	5.4	4.5	3.6

SQUARES

n	n^2	$\frac{(n \pm \frac{1}{2})^2}{.67449^2}$	n	n^2	$\frac{(n \pm \frac{1}{2})^2}{.67449^2}$
10	100	21762	60	3600	77819
11	121	24234	61	3721	80456
12	144	29070	62	3844	83138
13	169	34345	63	3969	85864
14	196	40061	64	4096	88633
15	225	46215	65	4225	91447
16	256	52810	66	4356	94304
17	289	59843	67	4489	97206
18	324	67317	68	4624	10015
19	361	75230	69	4761	10314
		83583			10617
20	400	92375	70	4900	10925
21	441	10161	71	5041	11237
22	484	11128	72	5184	11554
23	529	12139	73	5329	11875
24	576	13194	74	5476	12200
25	625	14293	75	5625	12530
26	676	15436	76	5776	12864
27	729	16623	77	5929	13202
28	784	17854	78	6084	13545
29	841		79	6241	
		19120			13893
30	900	20448	80	6400	14244
31	961	21811	81	6561	14600
32	1024	23218	82	6724	14961
33	1089	24668	83	6889	15326
34	1156	26163	84	7056	15695
35	1225	27702	85	7225	16069
36	1296	29284	86	7396	16447
37	1369	30911	87	7569	16829
38	1444	32581	88	7744	17216
39	1521		89	7921	
		34296			17607
40	1600	36054	90	8100	18003
41	1681	37857	91	8281	18403
42	1764	39703	92	8464	18808
43	1849	41594	93	8649	19217
44	1936	43528	94	8836	19630
45	2025	45506	95	9025	20047
46	2116	47529	96	9216	20469
47	2209	49595	97	9409	20896
48	2304	51705	98	9604	21327
49	2401		99	9801	
		53859			21762
50	2500	56057	100	10000	22201
51	2601	58299	101	10201	22645
52	2704	60585	102	10404	23094
53	2809	62915	103	10609	23547
54	2016	65289	104	10816	24004
55	3025	67707	105	11025	24465
56	3136	70169	106	11236	24932
57	3249	72675	107	11449	25402
58	3364	75225	108	11664	25877
59	3481	77819	109	11881	26356

Locate x between two numbers of the third column, and the first column will give $0.67449\sqrt{x}$. Avoid $0.67449\sqrt{10x}$ by a rough mental calculation.

CIRCULAR FUNCTIONS

RAD	DEG	TAN	SIN	LOG SIN	LOG COS	COS	COT	1′=.0003 (30′s & a 3)	
0000	0	0000	0000	$-\infty$	0	1	∞	90	$\pi/2$
01745*	1	0175	0175	2419	9999	9998	5729	89	1553
03491	2	0349	0349	5428	9997	9994	2864	88	1536
05236	3	0524	0523	7188	9994	9986	1908	87	1518
06981	4	0699	0698	8436	9989	9976	1430	86	1501
08727	5	0875	0872	9403	9983	9962	1143	85	1484
1047	6	1051	1045	0192	9976	9945	9514	84	1466
1222	7	1228	1219	0859	9968	9925	8144	83	1449
1396	8	1405	1392	1436	9958	9903	7115	82	1431
1571	9	1584	1564	1943	9946	9877	6314	81	1414
1745	10	1763	1736	2397	9934	9848	5671	80	1396
1920	11	1944	1908	2806	9919	9816	5145	79	1379
2094	12	2126	2079	3179	9904	9781	4705	78	1361
2269	13	2309	2250	3521	9887	9744	4331	77	1344
2443	14	2493	2419	3837	9869	9703	4011	76	1326
2618	15	2679	2588	4130	9849	9659	3732	75	1309
2793	16	2867	2756	4403	9828	9613	3487	74	1292
2967	17	3057	2924	4659	9806	9563	3271	73	1274
3142	18	3249	3090	4900	9782	9511	3078	72	1257
3316	19	3443	3256	5126	9757	9455	2904	71	1239
3491	20	3640	3420	5341	9730	9397	2747	70	1222
3665	21	3839	3584	5543	9702	9336	2605	69	1204
3840	22	4040	3746	5736	9672	9272	2475	68	1187
4014	23	4245	3907	5919	9640	9205	2356	67	1169
4189	24	4452	4067	6093	9607	9135	2246	66	1152
4363	25	4663	4226	6259	9573	9063	2145	65	1134
4538	26	4877	4384	6418	9537	8988	2050	64	1117
4712	27	5095	4540	6570	9499	8910	1963	63	1100
4887	28	5317	4695	6716	9459	8829	1881	62	1082
5061	20	5543	4848	6856	9418	8746	1804	61	1065
5236	30	5774	5000	6990	9375	8660	1732	60	1047
5411	31	6009	5150	7118	9331	8572	1664	59	1030
5585	32	6249	5299	7242	9284	8480	1600	58	1012
5760	33	6494	5446	7361	9235	8387	1540	57	9948
5934	34	6745	5592	7476	9186	8290	1483	56	9774
6109	35	7002	5736	7586	9134	8192	1428	55	9599
6283	36	7265	5878	7692	9080	8090	1376	54	9425
6458	37	7536	6018	7795	9023	7986	1327	53	9250
6632	38	7813	6157	7893	8965	7880	1280	52	9076
6807	39	8098	6293	7989	8905	7771	1235	51	8901
6981	40	8391	6428	8081	8843	7660	1192	50	8727
7156	41	8693	6561	8169	8778	7547	1150	49	8552
7330	42	9004	6691	8255	8711	7431	1111	48	8378
7505	43	9325	6820	8338	8641	7314	1072	47	8203
7679	44	9657	6947	8418	8569	7193	1036	46	8029
7854	45	1	7071	8495	8495	7071	1000	45	7854
1″=.000005 (5 0′s & a 5)		COT	COS	LOG COS	LOG SIN	SIN	TAN	DEG	RAD

Tangents can be safely interpolated up to tan 70°; and all sines and cosines. Subtract log sin and log cos from each other and from zero to obtain log tan, log cot, log sec, and log csc.

CONSTANTS

SYMBOL	CONSTANT	LOGARITHM
π	3.141593	0.4971499
∠1	180°/π = 57°17′45″	
	= 57°.29578	1.7581226
	= 3437′.747	3.5362739
	= 206264″.80625	5.3144251
e	2.718282	0.4342945
M	0.4342945	9.6377843

CIRCULAR MEASURE

1″=.00000 48481	1′=.00029 08882	1°=.01745 32925
2″=.00000 96963	2′=.00058 17764	2°=.03490 65850
3″=.00001 45444	3′=.00087 26646	3°=.05235 98776
4″=.00001 93925	4′=.00116 35528	4°=.06981 31701
5″=.00002 42407	5′=.00145 44410	5°=.08726 64626
10″=.00004 84814	10′=.00290 88821	10°=.17453 29252
20″=.00009 69627	20′=.00581 77642	20°=.34906 58504
30″=.00014 54441	30′=.00872 66463	30°=.52359 87756

*log π/180 = log .01745329 = 8.2418774.

FIVE-PLACE LOGARITHMS

N	0	1	2	3	4	5	6	7	8	9	*D*
100	00000	00043	00087	00130	00173	00217	00260	00303	00346	00389	43
101	00432	00475	00518	00561	00604	00647	00689	00732	00775	00817	43
102	00860	00903	00945	00988	01030	01072	01115	01157	01199	01242	42
103	01284	01326	01368	01410	01452	01494	01536	01578	01620	01662	41
104	01703	01745	01787	01828	01870	01912	01953	01995	02036	02078	41
105	02119	02160	02202	02243	02284	02325	02366	02407	02449	02490	41
106	02531	02572	02612	02653	02694	02735	02776	02816	02857	02898	40
107	02938	02979	03019	03060	03100	03141	03181	03222	03262	03302	40
108	03342	03383	03423	03463	03503	03543	03583	03623	03663	03703	40
109	03743	03782	03822	03862	03902	03941	03981	04021	04060	04100	39

EXPONENTIALS

x	e^x	$\log_{10} e^x$	e^{-x}	e^{-x^2}
.0001	1.0001	$0_4 4343$	.9999	1.0000
.0002	1.0002	$0_4 8686$	.9998	1.0000
.0003	1.0003	$0_3 1303$	.9997	1.0000
.0004	1.0004	$0_3 1737$	.9996	1.0000
.0005	1.0005	$0_3 2171$	.9995	1.0000
.0006	1.0006	$0_3 2606$	.9994	1.0000
.0007	1.0007	$0_3 3040$	.9993	1.0000
.0008	1.0008	$0_3 3474$	.9992	1.0000
.0009	1.0009	$0_3 3909$	.9991	1.0000
.001	1.0010	$0_3 4343$	.9990	1.0000
.002	1.0020	$0_3 8686$	.9980	1.0000
.003	1.0030	$0_2 1303$	.9970	1.0000
.004	1.0040	$0_2 1737$	.9960	1.0000
.005	1.0050	$0_2 2171$	.9950	1.0000
.006	1.0060	$0_2 2606$	.9940	1.0000
.007	1.0070	$0_2 3040$	.9930	1.0000
.008	1.0080	$0_2 3474$	.9920	1.0000
.000	1.0090	$0_2 3909$	.9910	1.0000
.01	1.0100	$0_2 4343$	.9900	.9999
.02	1.0202	$0_2 8686$	.9802	.9996
.03	1.0305	$0_1 1303$	.9704	.9991
.04	1.0408	$0_1 1737$	.9608	.9984
.05	1.0513	$0_1 2171$	.9512	.9975
.06	1.0618	$0_1 2606$	.9418	.9964
.07	1.0725	$0_1 3040$	.9324	.9951
.08	1.0833	$0_1 3474$	.9231	.9936
.09	1.0942	$0_1 3909$	.9139	.9919
.1	1.1052	$0_1 4343$	.9048	.9900
.2	1.2214	$0_1 8686$	.8187	.9608
.3	1.3499	0.1303	.7408	.9139
.4	1.4918	0.1737	.6703	.8521
.5	1.6487	0.2171	.6065	.7788
.6	1.8221	0.2606	.5488	.6977
.7	2.0138	0.3040	.4966	.6126
.8	2.2255	0.3474	.4493	.5273
.9	2.4596	0.3909	.4066	.4449
1.	2.7183	0.4343	.3679	.3679
2.	7.3891	0.8686	.1353	$_1 1832$
3.	20.086	1.3029	.0498	$_3 1234$
4.	54.598	1.7372	.0183	$_6 1125$
5.	148.41	2.1715	.0067	$_{10} 1389$
6.	403.43	2.6058	.0025	$_{15} 2320$
7.	1096.6	3.0401	.0009	$_{21} 5196$
8.	2981.0	3.4744	.0003	$_{27} 1604$
9.	8103.1	3.9087	.0001	$_{35} 8359$
10.	22029.	4.3429	.0000	$_{43} 7203$

A subscript 4 means .0000; etc.

NAPIERIAN OR NATURAL LOGARITHMS

	1–2	2–3	3–4	4–5	5–6	6–7	7–8	8–9	9–10	No.	Log.
0·00	0·000	0·693	1·099	1·386	1·609	1·792	1·946	2·079	2·197	15	2·708
0·05	0·049	0·718	1·115	1·399	1·619	1·800	1·953	2·086	2·203	20	2·996
0·10	0·095	0·742	1·131	1·411	1·629	1·808	1·960	2·092	2·208	25	3·219
0·15	0·140	0·765	1·147	1·423	1·639	1·816	1·967	2·098	2·214	30	3·401
0·20	0·182	0·788	1·163	1·435	1·649	1·824	1·974	2·104	2·219	35	3·555
0·25	0·223	0·811	1·179	1·447	1·658	1·833	1·981	2·110	2·225	40	3·689
0·30	0·262	0·833	1·194	1·459	1·668	1·841	1·988	2·116	2·230	45	3·807
0·35	0·300	0·854	1·209	1·470	1·677	1·848	1·995	2·122	2·235	50	3·912
0·40	0·336	0·875	1·224	1·482	1·686	1·856	2·001	2·128	2·241	55	4·007
0·45	0·372	0·896	1·238	1·493	1·696	1·864	2·008	2·134	2·246	60	4·094
0·50	0·405	0·916	1·253	1·504	1·705	1·872	2·015	2·140	2·251	65	4·174
0·55	0·438	0·936	1·267	1·515	1·714	1·879	2·022	2·146	2·257	70	4·248
0·60	0·470	0·956	1·281	1·526	1·723	1·887	2·028	2·152	2·262	75	4·317
0·65	0·500	0·975	1·295	1·537	1·732	1·895	2·035	2·158	2·267	80	4·382
0·70	0·531	0·993	1·308	1·548	1·740	1·902	2·041	2·163	2·272	85	4·443
0·75	0·560	1·012	1·322	1·558	1·749	1·910	2·048	2·169	2·277	90	4·500
0·80	0·588	1·030	1·335	1·569	1·758	1·917	2·054	2·175	2·282	95	4·554
0·85	0·615	1·047	1·348	1·579	1·766	1·924	2·061	2·180	2·287	100	4·605
0·90	0·642	1·065	1·361	1·589	1·775	1·931	2·067	2·186	2·293	150	5·010
0·95	0·668	1·082	1·374	1·599	1·783	1·939	2·073	2·192	2·298	200	5·298
1·00	0·693	1·099	1·386	1·609	1·792	1·946	2·079	2·197	2·303	1000	6·908

THE PROBABILITY INTEGRAL $y = \frac{2}{\pi}\int_0^x \epsilon^{-x^2}\,dx$

The tabular value with a prefixed decimal point gives the area under the curve $y=\epsilon^{-x^2}$ between the ordinates 0 and x in terms of the total area between $x=0$ and $x=\infty$. For Chauvenet's criterion see § 202.

x	0	1	2	3	4	5	6	7	8	9
0·0	0000	0113	0226	0338	0451	0564	0676	0789	0901	1013
0·1	1125	1236	1348	1459	1570	1680	1790	1900	2009	2118
0·2	2227	2335	2443	2550	2657	2763	2869	2974	3079	3183
0·3	3286	3389	3491	3593	3694	3794	3893	3992	4090	4187
0·4	4284	4380	4475	4569	4662	4755	4847	4937	5027	5117
0·5	5205	5292	5379	5465	5549	5633	5716	5798	5879	5959
0·6	6039	6117	6194	6270	6346	6420	6494	6566	6638	6708
0·7	6778	6847	6914	6981	7047	7112	7175	7238	7300	7361
0·8	7421	7480	7538	7595	7651	7707	7761	7814	7867	7918
0·9	7969	8019	8068	8116	8163	8209	8254	8299	8342	8385
1·0	8427	8468	8508	8548	8587	8624	8661	8698	8733	8768
1·1	8802	8835	8868	8900	8931	8961	8991	9020	9048	9076
1·2	9103	9130	9155	9181	9205	9229	9252	9275	9297	9319
1·3	9340	9361	9381	9400	9419	9438	9456	9473	9490	9507
1·4	9523	9539	9554	9569	9583	9597	9611	9624	9637	9649
1·5	9661	9673	9684	9695	9706	9716	9726	9736	9746	9755
1·6	9763	9772	9780	9788	9796	9804	9811	9818	9825	9832
1·7	9838	9844	9850	9856	9861	9867	9872	9877	9882	9886
1·8	9891	9895	9899	9903	9907	9911	9915	9918	9922	9925
1·9	9928	9931	9934	9937	9939	9942	9944	9947	9949	9951
2·0	9953	9955	9957	9959	9961	9963	9964	9966	9967	9969
2·1	9970	9972	9973	9974	9975	9976	9977	9979	9980	9980
2·2	9981	9982	9983	9984	9985	9985	9986	9987	9987	9988
2·3	9989	9989	9990	9990	9991	9991	9992	9992	9992	9993
2·4	9993	9993	9994	9994	9994	9995	9995	9995	9995	9996
2·5	9996	9996	9996	9997	9997	9997	9997	9997	9998	9998
2·6	9998	9998	9998	9998	9998	9998	9998	9998	9998	9999
∞	1·0000									

$\frac{x}{0·4769}$	y
0·0	0000
0·5	2641
1·0	5000
1·1	5419
1·2	5817
1·3	6194
1·4	6550
1·5	6883
1·6	7195
1·7	7485
1·8	7753
1·9	8000
2·0	8227
2·1	8434
2·2	8622
2·3	8792
2·4	8945
2·5	9083
2·6	9205
2·7	9314
2·8	9411
2·9	9495
3·0	9570
3·1	96[illegible]5
3·2	9691
3·3	9740
3·4	9782
3·5	9818
3·6	9848
3·7	9874
3·8	9896
3·9	9915
4·0	9930
4·5	9976
5·0	9993

When x=0·4769 area =0·5000.

On Nomograms.—Nomograms are ready-reckoner diagrams. They are of great use if many calculations of the same type have to be made. The scales of the factors entering into the calculation are drawn on paper in the correct positions, and all that has to be done is to lay a straight-edge across these scales and read off the result on a final scale. As an example * we show an interesting nomogram giving the duration of sunlight for every day of the year for places between latitudes 42° N. and 60° N. There are three scales—a scale of latitude, a scale of days, and the scale of hours of sunlight. The graduations and positions of these scales have to be worked out mathematically beforehand. The dotted line in the figure shows that at a place whose latitude is 49° 50′ the duration of sunlight on April 21 (also Aug. 23) is 14·1 hours.

The subject of Nomography has been well dealt with in the text-books of Brodetsky and Lipka (see the Bibliography), and to them the student is referred for further information.

* This diagram was made by W. H. Herbert, B.Sc., and published by the Minister of the Interior, Canada, to whom we are indebted for permission to use the block.

HON.

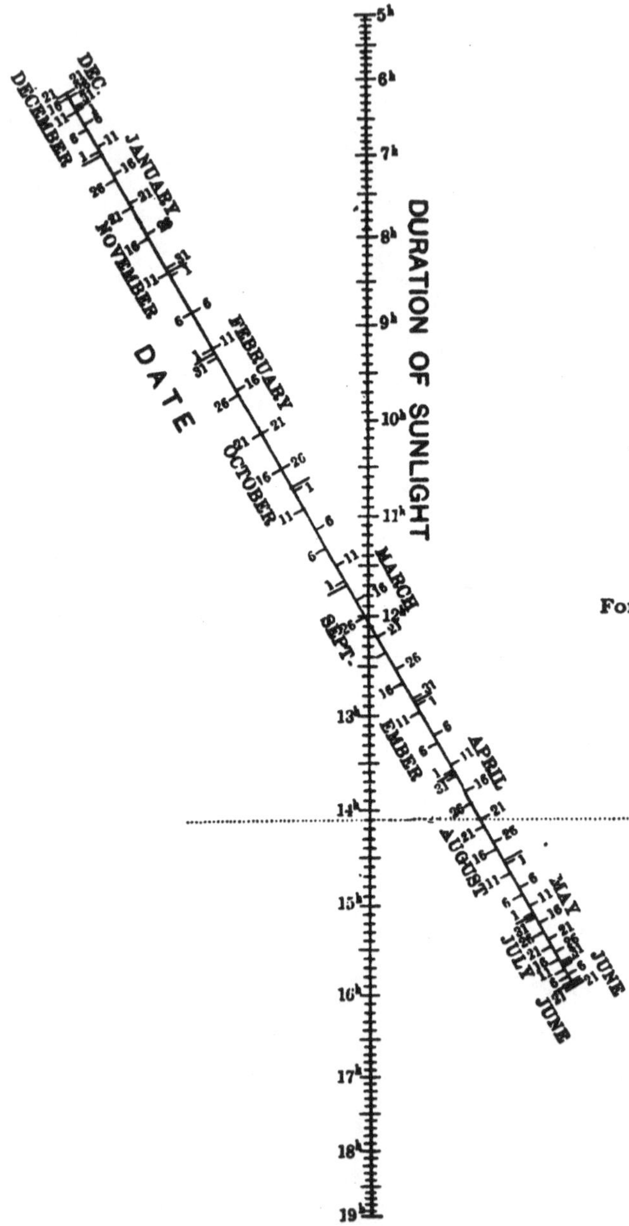

DU

For all p

artment of the Interior
EWART, MINISTER; W. W. CORY, DEPUTY MINISTER
APHICAL SURVEY OF CANADA
F. H. PETERS
DIRECTOR
NOMOGRAM
Showing
ON OF SUNLIGHT
For every day in the year
nada between latitudes 42° N. and 60° N.
LATITUDE
60°
59°
58°
57°
56°
55°
54°
53°
52°
51°
50°
49°
48°
47°
46°
45°
44°
43°
42°

[To face p. 326.

BIBLIOGRAPHY

1. Mainly Mathematical

M. Merriman, *Method of Least Squares* (Wiley).

W. W. Johnson, *The Theory of Errors and Method of Least Squares* (Wiley).

J. S. Stevens, *Theory of Measurements* (Van Nostrand).

L. W. Weld, *Theory of Errors and Least Squares* (Macmillan).

D. Brunt, *The Combination of Observations* (Cambridge).

D. C. Jones, *A First Course in Statistics* (Bell).

E. T. Whittaker and G. Robinson, *Calculus of Observations* (Blackie).

E. T. Whittaker and G. Robinson, *A Short Course of Interpolation* (Blackie).

J. Lipka, *Graphical and Mechanical Computation* (Wiley).

G. A. Carse and G. Shearer, *Fourier's Analysis and Periodogram Analysis.*

J. W. Mellor, *Higher Mathematics for Students of Chemistry and Physics* (Longmans).

S. Brodetsky, *A First Course in Nomography* (Bell).

H. W. Goodwin, *Precision of Measurements and Graphical Methods* (McGraw-Hill).

J. Perry, *Elementary Practical Mathematics* (Macmillan).

G. A. Gibson, *Treatise on Graphs* (Macmillan).

S. Lupton, *Notes on Observations* (Macmillan).

S. W. Holman, *Discussion of the Precision of Measurements* (Wiley).

J. G. MacGregor, *Physical Laws and Observations* (Edinburgh University).

2. Mainly Biological and Statistical

L. Doncaster, *Heredity in the Light of Recent Research* (Cambridge University Press).

C. B. Davenport, *Statistical Methods with Special Reference to Biological Variation* (Wiley).

R. H. Lock, *Recent Progress in the Study of Variation, Heredity, and Evolution* (Murray).

G. U. Yule, *An Introduction to the Theory of Statistics* (Griffin).

W. P. and E. M. Elderton, *Primer of Statistics* (Black).

W. P. Elderton, *Frequency Curves and Correlation* (Layton).

A. L. Bowley, *Elements of Statistics* (King).

Publications of the Galton Laboratory for National Eugenics (University of London).

K. Pearson, *The Grammar of Science*, 3rd ed., Part II (Black), and numerous papers in " Biometrika."

G. C. Whipple, *Vital Statistics* (Wiley).

E. B. Babcock and R. E. Clausen, *Genetics in Relation to Agriculture* (McGraw-Hill).

D'Arcy Thompson, *Growth and Form* (Cambridge University Press).

W. M. Feldman, *Biomathematics* (Griffin).

R. Pearl, *Medical Biometry and Statistics* (Saunders).

Mathematical and Physical Tables

The best for elementary use is Clark's (Oliver and Boyd). For more advanced students Kaye and Laby's (Longmans) is more suitable.

INDEX

[*The numbers refer to pages.*]

A

ABRIDGED division, 7
  multiplication, 8
Accidental errors, 167
Accuracy, Ch. xii
  of abridged methods, 41
  decimal, 35
  finer degrees of, 144
  ideal, 34
  infinite, 142
  relative, 37
  superfluous, 144
Adjustment of errors (see *least squares*)
Advantage of dispersion, 186
Analysis, graphic, Ch. ix
Angle(s), Ch. iii
  as a coordinate, 86
  small, 54
Approximate calculations, 5
Areas, Ch. xxii
  by approximate rules, 290
  by mensuration, 16, 291
  measurement of, 16
Arithmetical mean, 175
Asymmetrical frequency distributions, 176, 257
Asymptotes, 80
Average, 157, 175, 179
  deviation, 184
  as a least-square value, 226
  by symmetry, 179
  weighted, 200
Averaging experimental data, 157
Axes, change of, 85

B

Balance (see *beam balances*)
Beam balance
  double weighing 53
Bernoulli's formula, 270
Binomial expansion, 51, 173
Biological Applications, Ch. xxi
Black-thread method, 98

C

C.G.S. system, 12
Caliper, vernier, 156
Centiles, 265
Change of axes, 85
  of variables, 102
Characteristic deviations, 184, 192, 316
Chauvenet's criterion, 206
Choice of scales, 71
Circle, equation of, 86
Circular functions, Ch. iii, 317, 322
  with slide rule, 67
  measure, 21
Class interval, 170
Classification of errors, 166
Coefficient of variability, 256
  correlation, 262
Coincidence, measurement by, Ch. xiii, 154
  principle of, Ch. xiii
Comparison of characteristic deviations, 192
Compound errors, Ch. xviii
Conditioned measurements, 164
  and least squares, 226, 241
Constant error(s), 167, Ch. xx
Contour lines, 137
Coordinates, polar, 91
  rectilinear, 70
  in space, Ch. xi
Correlation, 257
Cosine, 28
Counting squares, 16, 291
Criterion, criteria, Ch. xvii
  Chauvenet's, 206
  Wright's, 211
Cube-root with slide rule, 68
Curve(s) (see also *graphic diagrams*)
  and equation, 74
  of errors, 84, 173
  exponential, 80
  interpolation along, 119
  logarithmic, 82
  probability, 173, 177, 192, 206, 254

D

Deciles, 265
Decimal accuracy, 35
Delta(s)
  properties of, 52, 54
Density, determination of, 19, 223, 250
Density, variation of water, 100, 317
Dependent measurements, 164

Derived units, 12
Deviation and Dispersion, Ch. xv
Deviation(s), 178
 average, 184
 (root-) mean-square, 185
 standard, 185
Diagonal scale, 24
Diagram, frequency, 170
 graphic (see also *graphic*)
Dimodal variation, 256
Direct measurements, 164
Disagreement of measurements, 165
Dispersion(s), 185 (see also *probable error*)
 of average(s), 189
 and deviation, Ch. xv
 of individual measurements, 185
 relative, 191
 and semi-inter-quartile range, 186
 table of, 188
 and weighting, 201
Distribution, frequency, 170, 171
Division, abridged, 7
 with slide rule, 64

E

Element, of a smooth curve, 119
Ellipse, 86
Equal intervals, 157, 160, 232
Equation(s) (see also *graph*)
 graph of, 75
 of a graph, 85, 106
 interpretation of, 97
 list of, 106
 polar, 91
 roots of, 120
Equivalent(s), 19, 315
 for slide rule, 65
Error(s) and measurements, Ch. xiv
 accidental, 167
 classification of, 166
 constant, 167, Ch. xx
 curve of, 47, 84, 171, 192, 206
 periodic, 248
 possible, 145
 probable, (see *probable error*)
 progressive, 250
 relative, 39, 143
 size of, 145
 sources of, 167
 systematic, Ch. xx
Estimation, of tenths, 31
Expectancy, 270
Exponential curve, 81
 variation, 105
Extrapolation, Ch. x, 123
 danger of, 125

F

Fatigue, law of, 272
Figures, significant, Ch. iv
Finite differences, 126
Fletcher's trolley, 159
Form, standard, 41
Frequency curve, 84, 170, 173
 equation of, 256
 distribution, 171
 polygon, 169
Function(s), 27
 circular, 28
$F(x)$, $f(x)$, 28
Fundamental units, 12

G

Geometric mean, 175
Grade, grade angle, 9
Gradient, 9, 25 (see also *slope*)
Graph (see *graphic diagram*)
Graphic analysis, Ch. ix, 271–274
 diagram(s), Ch. viii
  equation of, 85, 106
  of an equation, 75
  of exponential curve, 80
  of a frequency distribution, 169
  of hyperbola, 80
  of logarithmic curve, 82
  on logarithmic paper, 103, 105
  of parabola, 79, 99, 100
  slope of, 87
  smoothing of, 73
  of straight line, 77, 96
 interpolation, 121
Gregory-Newton interpolation formula, 127
Growth and temperature, 273

H

Half, rounding off, 15
Half-table methods, 160
Harmonic mean, 175
 motion, 95, 125
Hatchet planimeter, 292
Heat, specific, 149
Histogram, 170, 264
Hooke's law, 122, 158
Hyperbola, 81

I

Ideal accuracy, 36
Inclination, 25
Independent measurements, 164
Indirect measurements, 164, Ch. xviii
Infinite accuracy, 142
Inflection, point of, 84, 192
Integral, probability, 209, 325
Interpolation, Ch. x
 formulæ, 125
 logarithmic, 130

Interquartile range, 180
Interval(s), class, 170
Irregular areas, 16
  by approximate rules, Ch. xxii

L

Large deviations, 205, 211
Least squares, Ch. xix
  and average, 231
  for consecutive equal intervals, 232
  for a linear law, 227
  for a parabolic law, 228
  for proportionality, 229
Line, straight, law of, 77, 96, 106
Linear equation, 97
    and black thread, 98
    intercept form, 79, 99
  interpolation, 120
Locus, loci, 75 (see also *graphic diagram*)
Logarithm(s), Ch. v
  common, 44, 320
  natural, 47, 324
  properties of, 44
  with slide rule, 67
Logarithmic curve, 82
  paper, 103
  scales, 60

M

Magnitudes small, negligible, 50, Ch. vi
Mass, measurement of, 18
  true, by weighing, 53
Mean(s), 175
  arithmetical, 175
  choice of, 176
  geometrical, 175
  harmonic, 175
  quadratic, 175
Measurements(s) and errors, Ch. xiv
  agreement of, 165
  coarse, crude, 189
  by coincidence, 154
  conditioned, 164
    and least squares, 226, 241
  dependent, 164
  direct, 164
  disagreement of, 165
  errors of, 166
  by estimation, 153
  harmony of, 165
  independent, 164
  indirect, 164, Ch. xviii
  possible error of, 145
  statement of, 190
Median, 176, 180, 254
Mendelism, 271
Method of coincidence, 154
Mid-ordinate rule, 290
Minimum $\Sigma(v^2)$, 226
Modal value, mode, 176, 254
Multiplication, abridged, 8
  with slide rule, 63

N

Natural laws and graphs, 74
  logarithms, 47
Newton interpolation formula, 125
Nomograms, 326
Normal equations, 238
Notebooks, 3

O

Observations, weighting of, Ch. xvi
Observational equations, 237
Ogee (ogive) curve, 266, 269
Operands, 55
Ordinates, summation of, 290
Origin, cha ge of, 85

P

π, angle, 22
  value of, 33
Paper, logarithmic, 103
  semi-logarithmic, 105
Parabola, 79, 99
  finding equation for, 100
Pendulum, period of, 158, 223
Per cent. slope, 10
Periodic errors, 248
Peters' formulæ, 190
Physical arithmetic, 5
Planimeter, 291, 292
Plotting, principles of, 71, 72
Point of inflection, 84
Polar coordinates, 91
Polygon, frequency, 169
Population and extrapolation, 123, 132
Positive and negative asymmetry, 257
Possible error, 145
Power, probable error of, 218
Powers, formula for, 51
Probability function, calculation of, 47
  curve, 84, 173, 256
  integral table of, 209, 325
Probability paper, 263
Probable error(s), 147, 189, 194 (see also *dispersion*)
    and other characteristic deviations, 192
    of indirect measurements, Ch. xviii
      associative law, 217
      of average, 190
      of a difference, 217
      of coefficient of correlation, 262

Probable error(s) of distributive law, 219
of an individual, 186, 194
of median, 256
of a multiple, 217
of a power, 218
of probable error, 194
of a product, 218
of a sum, 216
Product, possible error of, 146
probable error of, 218
Progressive errors, 250
Proportion and least squares, 229
with slide rule, 64
Proportionality, 5
in a graph, 98
Protractor, 23

Q

Quadratic mean, 175
Quartiles, upper and lower, 180, 265
Quotient, possible error of, 146

R

Racing records, 272
Radian measure, 55
Range, semi-interquartile, 180
Rate of growth, 56, 87, 273
Rates of increase, 56, 87
Ratio(s), slide rule, 64
Rectangular hyperbola, 80
Regression, 259
Rejection of observations, Ch. xvii
Relative accuracy, 37
dispersion, 191
errors, 143
Repeated values and weights, 199
Representative magnitude(s), 175
Root(s), by double position, 120
with slide rule, 66
of $y = f(x)$, 120
Rounding off, 15
Rule, slide, Ch. vii

S

Σ(sigma), 189
Scale(s), diagonal, 24
logarithmic, 60
Semi-interquartile range, 180
and dispersion, 186
Semi-logarithmic paper, 105
Sensations, law of, 274
Significant figures, Ch. iv, 36, 142
versus decimal places, 39
Significance of the dispersion, 186
Similitude, laws of, 275, 280
Simpson's rules, 290
Sine(s), 27
Size of errors, 145
Skewness, 177, 257
Slide rule, Ch. vii
ratios, 65, 157
Slope, 9
of a graph, 87, 90
of a straight line, 77
Small magnitudes, Ch. vi
Smooth(ed) curve, 73
Solution by double position, 120
Sources of error, 167
Space-time curve, 89
Specific gravity of water, 100, 317
Specific heat, 149
Spherometer, 224
Spiral spring, 158
Squares, least (see *least squares*)
and roots, slide rule, 66
Standard(s) deviation, 185
form of expression, 41
Statement of a measurement, 190
Statistical methods, Ch. xiv, Ch. xxi
representation of, 278
Straight line, 77, 96
equation, 77, 78, 97, 99
Sum, possible error of, 146
probable error of, 216
Superfluous accuracy, 144
Symmetry of a curve, 176, 257
Systematic errors, Ch. xx

T

Tables, characteristic deviations, 316
Chauvenet's criterion, 208
circular functions, 322
coin tossing, 174
concentration of variates or errors, 194
267
constants, 316, 322
δ-formulæ, 55
density of water, 317
equations of graphs, 106
equivalents, 315
errors, classification of, 166
size of, 145
exponential functions, 323
general sources of error, 167
logarithms, 320, 323
inverse tangents, 317
inverse circular measure, 317
natural logarithms, 324
probability integral, 325
probable errors (indirect), 219
slide-rule equivalents, 65
squares, 318
Tabular difference, 126
Tangent(s), 25

INDEX

Target diagram, 202
Tenths, estimation of, 31
Test for systematic errors, 247
Thread method of straight line, 98
Three-dimensions, Ch. xi
Time intervals, averaging, 158, 232
Transfo mation to linear equation, 106
Triangle, angles of, 241
Typical curves, 78–84
  (see also *line*, *parabola*, etc.)
Types of frequency distribution, 171

V

Variability, 256
Variable(s), 27
  change of, 102, 106
Variates, 168, 254
  and errors, 168
Variation(s), 4, 178
Vector sum, 220
Velocity, angular, 22
  variable, 89
Vernier, 155
Vibration, determination of period of, 158
Volume, irregular, 17, 297
  of cylinder, 222

W

Water, density of, 100, 317
Weber's law, 274
Weighing, double, 53
Weight(s), arbitrary, 200
  and dispersion, 201
Weights and measures, Ch. ii
  for repeated values, 199
Weighted average, 200
Weighting of observations, Ch. xvi
Wounds, equations of, 271
Wright's criterion, 211

X

x-axis, 71
x-intercept, 99

Y

y-axis, 71
y-intercept, 99

Z

Zero as a dispersion, 191
  of a scale, 33
  significant, 36
  the most probable error, 194
Zero-sum method, 161

ERRATA

p. 26, line 3. For "sine" read "tangent"; for "tangent" read "sine."

p. 47, §61. For "Probabilty" read "Probability."

p. 51, 9 lines from bottom. For "$\frac{n(n-1)(n-2)}{1.2.3}x^2$" read "$\frac{n(n-1)(n-2)}{1.2.3.}x^3$."

p. 63, 2 lines from bottom. For "3000" read "30000."

p. 65, small type under table. For "355/133," "3550/1330," "35500/13300," "355000/133000" read "355/113," "3550/1130," "35500/11300," "355000/113000," respectively.

p. 93, Q. 31, line 6. For "height" read "time."

p. 101, heading of middle column of table. For "0.69x" read "$0.69x^2$."

p. 101, 4 lines from bottom. For "$-069x^2$" read "$-0.69x^2$."

p. 120, line 27. For "of 0.4 equals" read "of 0.04 equals."

p. 127, For the heading of first column of table read "x".

p. 169, line 2, small type under Fig. 58. For "Trdinates" read "ordinates."

p. 187, 9 lines from bottom. For "The value of d" read "The value of p."

p. 194, middle table, second column, line 2. For "46:1" read "4.6:1."

p. 201, 8 lines from bottom. For "$\frac{67}{32}$" read "$\frac{67}{22}$."

p. 218, line 9. For "$(a_1a_2ap_3)^2$" read "$(a_1a_2p_3)^2$."

p. 221, 9 lines from bottom. For "P the probable error in M" read "P the probable error in A."

p. 229, foot of second column of table. For "y" read "Σy."

p. 232, line 8. For "(§145)" read "(§146)."

p. 259, line 16. For "were" read "was."

p. 261, 2 lines from bottom. For "= 14" read "= 12."

p. 262, line 4. For "= 12" read "= 14."

p. 271, last line. For "$25\frac{1}{2}\ 10^{-0.033t}$" read "$25\frac{1}{2} \times 10^{-0.033t}$."

p. 290, 15 lines from bottom. For "7 or 10 or 13 or 16, etc." read "4 or 7 or 10, etc."

6 lines from bottom. For "4 or 7 or 10 ordinates" read "7 or 11 ordinates."

5 lines from bottom. For "With 10 ordinates" read "With 11 ordinates."

ERRATA

p. 305, Ex. II, 5(b). For "sp. gr. < density" read "sp. gr. > density."
9. For "70.3 x x" read ".488 x x."

p. 305, Ex. IV, 16. For "1/150000 per cent." read ".0000086 per cent."
For "1 1/3 per cent." read "1 1/4 per cent."
18. For "0.003903" read "0.00387."

Ex. VI, 12. For "17.2705" read "17.2683."

Ex. VII, 7(a). For "50.2" read "50.4."
7(d). For "7" read "6.24"; for "0.7" read "0.62."
10. For "30.6 sq. yds." read "30.7 sq. yds."

p. 305, Ex. VIII, Each number is one too high; the first answer is number 9.
11 (corrected). For "$y = -0.79\ x + 3.4$" read "$y = -0.80\ x + 3.4$."
24 (corrected). For "48.4°F." read "49.4°F."
31 (corrected). For "$s = 1600\ t = 32\ t^2$" read "$s = 1600\ t - 16\ t^2$."

p. 306, Ex. IX, 57. For "$\log (A - x) = \log A - 0.00217\ t$" read "$\log (A - x) = \log A - 0.00213\ t$."
58. For "$C_1/C_2^2 = 0.50$" read "$C_1/C_2^2 = 5.0$."

Ex. X, 14. For "2.36 and 4.43 inches" read "3.37 and 4.45 inches."

p. 307, Ex. XII, 8. For "168.7 ± 1 inches^2" read "168.7 ± 0.1 inches^2."

Ex. XIV, 3. line 5 of column 2. For "0.14" read "0.13."

Ex. XV, 7. Add "p.e. = 2.0."

p. 308, Ex. XVII, 4. line 1. For "0.075" read "0.11."

Ex. XVIII, 6. For "($2.08 \times 10^{12} \pm 1.8$ per cent)" read "($2.11 \times 10^{12} \pm 3.2$ per cent)."

p. 309, Ex. XIX, 18. For "$-0.07653\ x^2$" read "$-0.7653\ x^2$."

Ex. XXI, 16. For "0.16 per cent." read "0.14 per cent."

p. 310, Ex. XXII, 4. For "5139" read "5.39."
20. For "143,800 foot-pounds" read "1,438,000 foot-pounds."
23. For "261" read "265."

www.ingramcontent.com/pod-product-compliance
Lightning Source LLC
LaVergne TN
LVHW100252210926
848480LV00003B/904